Computational Biology

Volume 23

The *Computational Biology* series publishes the very latest, high-quality research devoted to specific issues in computer-assisted analysis of biological data. The main emphasis is on current scientific developments and innovative techniques in computational biology (bioinformatics), bringing to light methods from mathematics, statistics and computer science that directly address biological problems currently under investigation.

The series offers publications that present the state-of-the-art regarding the problems in question; show computational biology/bioinformatics methods at work; and finally discuss anticipated demands regarding developments in future methodology. Titles can range from focused monographs, to undergraduate and graduate textbooks, and professional text/reference works.

More information about this series at http://www.springer.com/series/5769

K. Erciyes

Distributed and Sequential Algorithms for Bioinformatics

 Springer

K. Erciyes
Computer Engineering Department
Izmir University
Uckuyular, Izmir
Turkey

ISSN 1568-2684
Computational Biology
ISBN 978-3-319-38748-2   ISBN 978-3-319-24966-7 (eBook)
DOI 10.1007/978-3-319-24966-7

Springer Cham Heidelberg New York Dordrecht London
© Springer International Publishing Switzerland 2015
Softcover re-print of the Hardcover 1st edition 2015

Printed on acid-free paper

Springer International Publishing AG Switzerland is part of Springer Science+Business Media
(www.springer.com)

To the memory of Atilla Ozerdim and all disabled people in his name.
Atilla was a former student and then a colleague. He was highly intelligent however was severely disabled lower than his neck. He used a computer by typing with a stick in his mouth and wrote thousands of lines of code and researched this way.

Preface

Recent technological advancements in the last few decades provided vast and unprecedented amounts of biological data including data of DNA and cell, and biological networks. This data comes in two basic formats as DNA nucleotide and protein amino acid sequences, and more recently, topological data of biological networks. Analysis of this huge data is a task on its own and the problems encountered are NP-hard most of the time, defying solutions in polynomial time. Such analysis is required as it provides a fundamental understanding of the functioning of a cell which can help understand human health and disease states and the diagnosis of diseases, which can further aid development of biotechnological processes to be used for medical purposes such as treatment of diseases.

Instead of searching for optimal solutions to these difficult problems, approximation algorithms that provide sub-optimal solutions are usually preferred. An approximation algorithm should guarantee a solution within an approximation factor for all input combinations. In many cases, even approximation algorithms are not known to date and using heuristics that are shown to work for most of the input cases experimentally are considered as solutions.

Under these circumstances, there is an increasing demand and interest in research community for parallel/distributed algorithms to solve these problems efficiently using a number of processing elements. This book is about both sequential and distributed algorithms for the analysis and modeling of biological data and as such, it is one of the first ones in this topic to the best of our knowledge. In the context of this book, we will assume a distributed algorithm as a parallel algorithm executed on a distributed memory processing system using *message-passing* rather than special purpose parallel architectures. For the cases of shared memory parallel computing, we will use the term *parallel algorithm* explicitly. We also cover algorithms for biological sequences (DNA and protein) and biological network (protein interaction networks, gene regulation, etc.) data in the same volume. Although algorithms for DNA sequences have a longer history of study, even the sequential algorithms for biological networks such as the protein interaction networks are rare and are at an early stage of development in research studies. We aim to give a unified view of algorithms for sequences and networks of biological systems where possible. These two views are not totally unrelated; for example, the function of a protein is influenced by both its position in a network and its amino acid sequence, and also by its 3-D structure. It can also be seen that the problems in the sequence and network domains are analogous to some extent; for example, sequence alignment and network alignment, sequence motifs and network motifs, sequence clustering and network clustering are analogous problems in these two related worlds. It is not difficult to predict that the analysis of biological networks will have a profound effect on our understanding the origins of life, health and disease states, as analysis of DNA/RNA and protein sequences have provided.

The parallel and distributed algorithms are needed to solve bioinformatics problems simply for the speedup they provide. Even the linear time algorithms may be difficult to realize in bioinformatics due to the size of the data involved. For example, suffix trees are fundamental data structures in bioinformatics, and constructing them takes $O(n)$ time by relatively new algorithms such as Ukkonen's or Farach's. Considering human DNA which consists of over 3 billion base pairs, even these linear time algorithms are time-consuming. However, by using distributed suffix trees, the time can be reduced to $O(n/k)$ where k is the number of processors.

One wonders then about the scarcity of the research efforts in the design and implementation of distributed algorithms for these time-consuming difficult tasks. A possible reason would be that a number of problems have been introduced recently and the general approach in the research community has been to search for sequential algorithmic solutions first and then investigate ways of parallelizing these algorithms or design totally new parallel/distributed algorithms. Moreover, the parallel and distributed computing is a principle on its own where researchers in this field may not be familiar with bioinformatics problems in general, and the multidisciplinary efforts in this discipline and bioinformatics seem to be just starting. This book is an effort to contribute to the filling of the aforementioned gap between the distributed computing and bioinformatics. Our main goal is to first introduce the fundamental sequential algorithms to understand the problem and then describe distributed algorithms that can be used for fundamental bioinformatics problems such as sequence and network alignment, and finding sequence and network motifs, and clustering. We review the most fundamental sequential algorithms which aid the understanding of the problem better and yield parallel/distributed versions. In other words, we try to be as comprehensive as possible in the coverage of parallel/distributed algorithms for the fundamental bioinformatics problems with an in-depth analysis of sequential algorithms.

Writing a book on bioinformatics is a challenging task for a number of reasons. First of all, there are so many diverse topics to consider, from clustering to genome rearrangement, from network motif search to phylogeny, and one has to be selective not to divert greatly from the initially set objectives. We had to carefully choose a subset of topics to be included in this book in line with the book title and aim; tasks that require substantial computation power due to their data sizes and therefore are good candidates for parallelization. Second, bioinformatics is a very dynamic area of study with frequent new technological advances and results which requires a thorough survey of contemporary literature on presented topics. The two worlds of bioinformatics, namely biological sequences and biological networks, both have similar challenging problems. A closer look reveals these two worlds in fact have analogous problems as mentioned; sequence alignment and network alignment; sequence motif search and network motif search; sequence clustering and network clustering which may be comforting first. However, these problems are not closely related in general, other than the clustering problem which can be converted from the sequence domain to the network domain with some effort.

We have a uniform chapter layout by first starting with an informal description of the problem at hand. We then define it formally and review the significant sequential algorithms in this topic briefly. We describe parallel/distributed algorithms for the same problem, which is the main theme of the book, and briefly discuss software packages if there is any available. When distributed algorithms for the topic are scarce, we provide clues and possible approaches for distributed algorithms as possible extensions to sequential ones or as totally new algorithms, to aid starting researchers in the topic. There are several coarsely designed and unpublished distributed algorithms, approximately 2-3 algorithms in some chapter, for the stated problems in biological sequences and networks which can be taken as starting points for potential researchers in this field. In the final part of each chapter, we emphasize main points, compare the described algorithms, give a contemporary review of the related literature, and show possible open research areas in the chapter notes section.

The intended audience for this book is the graduate students and researchers of computer science, biology, genetics, mathematics, and engineering, or any person with basic background in discrete mathematics and algorithms. The Web page for the book is at: http://eng.izmir.edu.tr/~kerciyes/DSAB.

I would like to thank graduate students at Izmir University who have taken complex networks and distributed algorithms courses for their valuable feedback when parts of the material covered in the book were presented during lectures. I would like to thank Esra Rüzgar for her review and comments on parallel network motif search algorithms of Chap. 14. I would also like to thank Springer senior editor Wayne Wheeler and associate editor Simon Rees for their continuous support and patience during the course of this project.

Uckuyular, Izmir, Turkey   K. Erciyes

Contents

Part II Biological Sequences

Introduction

1.1 Introduction

Biology is the science of life and living organisms. An organism is a living entity that may consist of organs that are made of tissues. Cells are the building blocks of organisms and form tissues of organs. Cells consist of molecules and molecular biology is the science of studying the cell at molecular level. The nucleus of a cell contains deoxyribonucleic acid (DNA) which stores all of the genetic information. DNA is a double helix structure consisting of four types of molecules called *nucleotides*. It consists of a long sequence of nucleotides of about 3 billion pairs. From the viewpoint of computing, DNA is simply a string that has a four-letter alphabet. The ribonucleic acid (RNA) has one strand and also consists of four types of nucleotides like DNA, with one different nucleotide. Proteins are large molecules outside the nucleus of the cell and perform vital life functions. A protein is basically a linear chain of molecules called amino acids. Molecules of the cell interact to perform all necessary functions for living.

Recent technological advancements have provided vast amounts of biological data at molecular level. Analysis and extracting meaningful information from this data requires new methods and approaches. This data comes in two basic forms as sequence and network data. On one hand, we are provided with sequence data of DNA/RNA and proteins, and on the other hand, we have topological information about the connectivity of various networks within the cell. Analysis of this data is a task on its own due to its huge size.

We first describe the problems encountered in the analysis of biological sequences and networks and we then describe why distributed algorithms are imperative as computational methods in this chapter. It seems design and implementation of distributed algorithms are inevitable for these time-consuming difficult problems and their scarcity can be attributed to the relatively recent provision of the biological data and the field being multidisciplinary in nature. We conclude by providing the outline of the book.

© Springer International Publishing Switzerland 2015
K. Erciyes, *Distributed and Sequential Algorithms for Bioinformatics*,
Computational Biology 23, DOI 10.1007/978-3-319-24966-7_1

1.2 Biological Sequences

The biological sequences in the cell we are referring are the nucleotide sequences in DNA, RNA, and amino acid sequences in proteins. DNA contains four nucleotides in a double-helix-shaped two-strand structure: Adenine (A), Cytosine (C), Guanine (G), and Thymine (T). Adenine always pairs with thymine, and guanine with cytosine. The primary structure of a protein consists of a linear chain of amino acids and the order of these affects its 3D shape. Figure 1.1 shows a simplified diagram of DNA and a protein.

Analysis of these biological sequences involves the following tasks:

- *Comparison of sequences*: A basic requirement to analyze a newly discovered sequence is to compare it with the known sequences. The basic assumption here is that the similar structures may indicate similar functions. In the very basic form, we attempt to find the approximately nucleotides in two or more sequences. This process is commonly called sequence alignment and can be solved in polynomial time by dynamic algorithms. A sequence alignment algorithm provides the similarities and distances between a number of input sequences which can be used for further processing.
- *Clustering*: The grouping of similar sequences is called clustering and this process is at a higher level than sequence alignment as it needs the distances between the sequences as computed by the alignment. Clustering of sequences aims to find their affinities and infer ancestral relationships based on the groupings. The functions of sequences in clusters can be analyzed more easily and also this information can be used for disease analysis of organisms.
- *Sequence patterns*: DNA and proteins contain repeating subsequences which are called *sequence repeats*. These repeats may be consecutive or dispersed, and the former is commonly referred to as *tandem repeats* and the latter as *sequence motifs*. In various diseases, the number of the repeats is more than expected, and hence discovering them helps to understand the disease mechanism in an organism. They also reside near genes and may be used to find the location of genes. The number

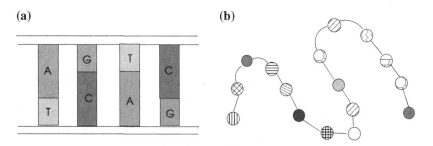

Fig. 1.1 **a** DNA double helix structure. **b** A protein structure having a linear sequence of amino acids

and locations of repeats are unique for individuals and can be used in forensics to identify individuals.

- *Gene finding*: A gene codes for a polypeptide which can form or be a part of a protein. There are over 20,000 genes in human DNA which occupy only about 3 % of human genome. Finding genes is a fundamental step in their analysis. A mutated gene may cause the formation of a wrong protein, disturbing the healthy state of an organism, but mutations are harmless in many cases.
- *Genome rearrangements*: Mutations of DNA at a coarser level than point mutations of nucleotides involve certain alterations of segments or genes in DNA. These changes include reversals, duplications, and transfer to different locations in DNA. Genome rearrangements may result in the production of new species but in many cases, they are considered as the causes of complex diseases.
- *Haplotype inference*: The DNA sequencing methods of humans provide the order of DNA nucleotides from two chromosomes as this approach is more cost-effective and practical. However, the sequence information from a single chromosome called a *haplotype* is needed for disease analysis and also to discover ancestral relationships. Discovering single chromosome data from the two chromosomes is called haplotype inference and is needed for the data to be meaningful

All of the above-described tasks are fundamental areas of research in the analysis of biological sequences. Comparison of sequences, clustering, and finding repeats apply both to DNA/RNA and protein sequences. Protein sequences may also be employed to discover genes as they are the product of genes; however, genome rearrangements and haplotype inference problems are commonly associated with the DNA/RNA sequences. Except for the sequence alignment problem, there are hardly any polynomial algorithms for these problems. Even when there is a solution in polynomial time, the size of data necessitates the use of approximation algorithm if they are available. As we will see, the heuristic algorithms that can be shown to work for a wide range of input combinations experimentally are the only choice in many cases.

1.3 Biological Networks

Biological networks consist of biological entities which interact in some form. The modeling and analysis of biological networks are fundamental areas of research in bioinformatics. The number of nodes in a biological network is large and these nodes have complex relations among them. We can represent a biological network by a graph where an edge between two entities indicates an interaction between them. This way, many results in graph theory and also various graph algorithms become available for immediate use to help solve a number of problems in biological networks.

We can make coarse distinction between the networks in the cell and other biological networks. The cell contains DNA, RNA, proteins, and metabolites at the molecular level. Networks at biological level are gene regulation networks, signal transduction networks, protein–protein interaction (PPI) networks, and metabolic networks. DNA is static containing the genetic code and proteins take part in various vital functions in the cell. Genes in DNA code for proteins in a process called *gene expression*.

DNA, RNA, proteins, and metabolites all have their own networks. Intracellular biological networks have molecules within the cell as their nodes and their interactions as links. PPI networks have proteins which interact with each other to cooperate to accomplish various vital functions for life. PPI networks are not constructed randomly and they show specific structures. They have a few number of highly connected nodes called *hubs* and the rest of the nodes in the network have very few connections to other proteins. The distance between two farthest proteins in such a network is very small compared to the size of the network. This type of networks is called *small-world* network as it is relatively easy to reach a node from any other node in the network. Hubs in PPI networks form dense regions of the network and these clusters of nodes called *protein complexes* usually have some fundamental functions attributed to them. It is therefore desirable to detect these clusters in PPI networks and we will investigate algorithms for this purpose in Chap. 12. A PPI network has very few hubs and many low-degree nodes. Such networks are commonly called scale-free networks and a hub in a PPI network is presumed to act as a gateway between a large number of nodes. Figure 1.2 shows a subnetwork of a PPI network of the immune system where small-world and scale-free properties can be visually detected along with a number of hubs.

The main areas of investigation in biological networks are the following:

- *Analysis of networks*: Given a graph $G(V, E)$ representing a biological network, our interest is to analyze the topological properties of this graph. In most of the biological networks, we find them as small-world and scale-free networks. Another point of interest is to find the importance of nodes and links in these networks. The *centrality* parameters provide evaluation of the importance of vertices and edges in the network graph.
- *Clustering*: Clustering in this case involves finding dense regions of the network graph. The nodes in these regions have more connections between them than the rest of the network presumably due to a significant function performed in these subgraphs. A disease state of an organism may also cause highly populated subgraphs of a PPI network for example. Detecting these clusters is an important research area in biological networks.
- *Network motifs*: Network motifs are the frequent occurring subgraph patterns in biological networks. They are the analogous structures to DNA and protein sequence motifs. A network motif is commonly perceived as the building block of a biological network and therefore it is assumed to have some basic functional significance. Discovering these motifs is a computationally difficult task since the number of possible motifs grows exponentially with increased size. For example,

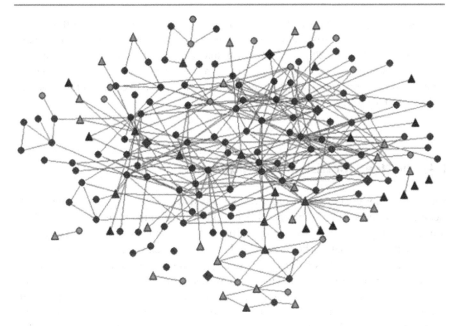

Fig. 1.2 A subnetwork of the SpA PPI network involved in innate immune response (taken from [1]). *Triangles* are the genes

there are only 13 possible motifs with 4 nodes, while the number of all subgraphs of size 5 is 51. Comparison of networks and hence deducing ancestral relationships are also possible by searching for common motifs in them.

- *Network alignment*: In many cases, two or more biological networks may be similar and finding the similar subgraphs may show the conserved and therefore important topological structures in them. A fundamental problem in biological networks is the assessment of this similarity between two or more biological networks. Analogous to sequence alignment of DNA/RNA and proteins, the network alignment process aims to align two or more biological networks. This time, we basically search for topological similarities between the networks. Node similarity is also considered together with topological similarity to have a better view of the affinity of two or more networks. Our aim in alignment is very similar to sequence alignment in a different domain, and we try to compare two or more networks, find their conserved modules, and infer ancestral relationships between them.
- *Phyloegeny*: Phylogeny is the study and analysis of evolutionary relationships among organisms. Phylogenetics aims to discover these relationships using molecular sequence and morphological data to reconstruct evolutionary dependencies among the organisms. Phylogenetics is one of the most studied and researched topics in bioinformatics and its implications are numerous. It can help disease analysis and design of drug therapies by analyzing phylogenetic dependencies of pathogens. Transmission characters of diseases can also be predicted using phylogenetics.

Most of the problems outlined above do not have solutions in polynomial time and we are left with approximation algorithms in rare cases but mostly with heuristic algorithms. Since the size of the networks under consideration is huge, sampling the network and implementing the algorithm in this sample is typically followed in many methods. This approach provides a solution in reasonable time at the expense of decreased accuracy.

1.4   The Need for Distributed Algorithms

There are many challenges in bioinformatics and we can first state that obtaining meaningful data is difficult in general as it is noisy most of the time. The size of data is large making it difficult to find time-efficient algorithms. Complexity class P is related to problems that can be solved in polynomial time and some problems we need to solve in bioinformatics belong to this class. Many problems, however, can only be solved in exponential time and in some cases, solutions to the problem at hand do not even exist. These problems belong to the class nondeterministic polynomial hard (NP-hard) and we need to rely on approximation algorithms which find suboptimal solutions with bounded approximation ratios for such hard problems. Most of the time, there are no known approximation algorithms and we need to rely on heuristic methods which show experimentally that the algorithm works for most of input combinations.

Parallel algorithms and distributed algorithms provide faster execution as a number of processing elements are used in parallel. Here, we make a distinction between shared memory parallel processing where processing elements communicate using a shared and protected memory and distributed processing where computational nodes of a communication network communicate by the transfer of messages without any shared memory. Our main focus in this book is about distributed, that is, distributed memory, message-passing algorithms although we will also cover a number of shared memory parallel algorithms. The efficiency of a parallel/distributed algorithm is basically measured by the *speedup* it provides which is expressed as the ratio of the number of time steps of the sequential algorithm $T(s)$ to the number of time steps of the parallel algorithm $T(p)$. If k processing elements are used, the speedup should approach k ideally. However, there are the synchronization delays in shared memory parallel processing and the message delays of the interconnection network in distributed processing which means the speedup value could be far from ideal.

Whether the algorithm works in polynomial time to find an exact solution or an approximate solution again in polynomial time; bioinformatics problems have an additional property; the size of the input data is huge. We may therefore attempt to provide a parallel or distributed version of an algorithm even if it does solve the problem in polynomial time. For example, suffix trees are versatile data structures that find various implementations in algorithmic solutions to bioinformatics problems. There exist algorithms that construct suffix trees in a maximum of n steps where n is the length of the sequence from which a suffix tree is built. Although this time complexity is acceptable for a majority of computational tasks, we may need to search for faster solutions when a suffix tree of a DNA sequence of about 3 billion pairs is to be constructed. Theoretically, it is difficult to find an algorithm that has better time than linear-time execution; however, we can distribute this task to a number of processing elements and achieve significant gains in time. There is also the case of double gain in time by first designing an approximate algorithm, if this is possible, with better execution time and then providing a distributed version of this algorithm to provide further speedup.

1.5   Outline of the Book

Our description of algorithms in each chapter follows the same model where the problem is first introduced informally and the related concepts are described. We then present the problem formally with the notation used and review the fundamental sequential algorithms with emphasis on the ones that have potential to be modified and executed in parallel/distributed environments. After this step, we usually provide a template of a generic algorithm that summarizes the general approaches to have a distributed algorithm for the problem at hand. This task can be accomplished basically by partitioning data, partitioning code or both among the processing elements. We also propose new algorithms for specific problems under consideration that can be implemented conveniently in this step. We then review published work about parallel and distributed algorithms for the problem described, and describe fundamental research studies that have received attention in more detail at algorithmic level. Finally, the main points of the chapter are emphasized with pointers for future work in the Chapter Notes section.

We have three parts, and start with Part I which is about the basic background about bioinformatics with three chapters on molecular biology; graphs and algorithms; and parallel and distributed processing. This part is not a comprehensive treatment of three main areas of research, but rather a dense coverage of these three major topics with pointers for further study. It is self-contained, however, covering most of the fundamental concepts needed for parallel/distributed processing in bioinformatics.

Part II describes the problems in the sequence analysis of biological data. We start this part with string algorithms and describe algorithms for sequence alignment which is one of the most researched topics in the sequence world. We then investigate sequence motif search algorithm and describe genome analysis problems such as gene finding, genome rearrangement, and haplotype inference at a coarser level.

In Part III, our main focus is on parallel and distributed algorithms for biological networks, the PPI networks being the center of focus. We will see that the main problems in biological networks can be classified into cluster detection, network motif discovery, network alignment, and phylogenetics. Cluster detection algorithms aim to find clusters of high inter-node connections and these dense regions may implicate a specific function or health and disease states of an organism. Network motif algorithms attempt to discover repeating patterns of subgraphs and they may indicate again some specific function attributed to these patterns. Network alignment algorithms investigate the similarities between two or more graphs and show whether these organisms have common ancestors and how well they are related. Analysis of phylogenetic trees and networks which construct ancestor–descendant relationships among existing individuals and organisms is also reviewed in this part which is concluded by discussing the main challenges and future prospects of bioinformatics in the Epilogue chapter.

Reference

1. Zhao J, Chen J, Yang T-H, Holme P (2012) Insights into the pathogenesis of axial spondyloarthropathy from network and pathway analysis. BMC Syst Biol. 6(Suppl 1):S4. doi:10.1186/1752-0509-6-S1-S4

Part I

Background

Introduction to Molecular Biology

2

2.1 Introduction

Modern biology has its roots at the work of Gregor Mendel who identified the fundamental rules of hereditary in 1865. The discovery of chromosomes and genes followed later and in 1952 Watson and Crick disclosed the double helix structure of DNA. All living organisms have common characteristics such as replication, nutrition, growing and interaction with their environment. An *organism* is composed of *organs* which perform specific functions. Organs are made of *tissues* which are composed of aggregation of cells that have similar functions. The *cell* is the basic unit of life in all living organisms and it has molecules that have fundamental functions for life. Molecular biology is the study of these molecules in the cell. Two of these molecules called *proteins* and *nucleotides* have fundamental roles to sustain life. Proteins are the key components in everything related to life. DNA is made of nucleotides and parts of DNA called *genes* code for proteins which perform all the fundamental processes for living using biochemical reactions.

Cells synthesize new molecules and break large molecules into smaller ones using complex networks of chemical reactions called *pathways*. Genome is the complete set of DNA of an organism and human genome consists of chromosomes which contain many genes. A gene is the basic physical and functional unit of hereditary that codes for a protein which is a large molecule made from a sequence of amino acids. Three critical molecules of life are deoxyribonucleic acid (DNA), ribonucleic acid (RNA), and proteins. A central paradigm in molecular biology states that biological function is heavily dependent on the biological structure.

In this chapter, we first review the functions performed by the cell and its ingredients. The DNA contained in the nucleus, the proteins, and various other molecules all have important functionalities and we describe these in detail. The central dogma of life is the process of building up a protein from the code in the genes as we will outline. We will also briefly describe biotechnological methods and introduce some of the commonly used databases that store information about DNA, proteins, and other molecules in the cell.

© Springer International Publishing Switzerland 2015
K. Erciyes, *Distributed and Sequential Algorithms for Bioinformatics*,
Computational Biology 23, DOI 10.1007/978-3-319-24966-7_2

2.2 The Cell

Cells are the fundamental building blocks of all living things. The cell serves as a structural building block to form tissues and organs. Each cell is independent and can live on its own. All cells have a metabolism to take in nutrients and convert them into molecules and energy to be used. Another important property of cells is *replication* in which a cell produces another cell that has the same properties as itself. Cells are composed of approximately 70 % water; 7 % small molecules like amino acids, nucleotides, salts, and lipids, and 23 % macromolecules such as proteins and polysaccharids. A cell consists of molecules in a dense liquid surrounded by a membrane as shown in Fig. 2.1.

The *eukaryotic cells* have nuclei containing the genetic material which is separated from the rest of the cell by a membrane and the *prokaryotic cells* do not have nuclei. Prokaryotes include bacteria and archaea; and plants, animals, and fungi are examples of eukaryotes. The tasks performed by the cells include taking nutrients from food, converting these to energy, and performing various special missions. A cell is composed of many parts each with a different purpose. The following are the important parts of an eukaryotic cell with their functions:

- **Nucleus**: Storage of DNA molecules, and RNA and ribosome synthesis.
- **Endoplasmic reticulum**: Synthesis of lipids and proteins
- **Golgi apparatus**: Distribution of proteins and lipids and posttranslational processing of proteins.
- **Mitochondria**: Generation of energy by oxidizing nutrients.
- **Vesicles**: *Transport vesicles* move molecules such as proteins from endoplasmic reticulum to Golgi apparatus, *Secretory vesicles* have material to be excreted from the cell and *lysosomes* provide cellular digestion.

The nucleus is at the center of the cell and is responsible for vital functions such as cell growth, maturation, division, or death. *Cytoplasm* consists of jellylike fluid which surrounds the nucleus and it contains various other structures. *Endoplasmic*

Fig. 2.1 Parts of a cell

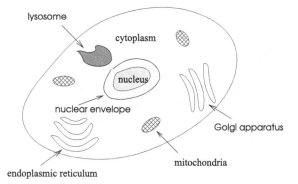

reticulum enwraps the nucleus, and processes molecules made by the cell and transports them to their destinations. Conversion of energy from food to a form that can be used by the cell is performed by *mitochondria* which have their own genetic material. These components of the cell are shown in Fig. 2.1. The cell contains various other structures than the ones we have outlined here.

Chemically, cell is composed of few elements only. Carbon (C), hydrogen (H), oxygen (O), and nitrogen (N) are the dominant ones with phosphorus (P) and sulfur (S) appearing in less proportions. These elements combine to form molecules in the cell, using covalent bonds in which electrons in their outer orbits are shared between the atoms. A *nucleotide* is one such molecule in the cell which is a chain of three components: a base B, a sugar S, and a phosphoric acid P. The three basic macromolecules in the cell that are essential for life are the DNA, RNA, and proteins.

2.2.1 DNA

James Watson and Francis Crick discovered the Deoxyribonucleic Acid (DNA) structure in the cell in 1953 using X-ray diffraction patterns which showed that the DNA molecule is long, thin, and has a spiral-like shape [5]. The DNA is contained in the nuclei of eukaryotic cells and is composed of small molecules called *nucleotides*. Each nucleotide consists of a five-carbon sugar, a phosphate group, and a base. The carbon atoms in a sugar molecule are labeled $1'$ to $5'$ and using this notation, DNA molecules start at $5'$ end and finish at $3'$ end as shown in Fig. 2.2. There are four nucleotides in the DNA which are distinguished by the bases they have: Adenine (A), Cytosine (C), Guanine (G), and Thymine (T). We can therefore think of DNA as a string with a four letter alphabet $\sum = \{A,C,G,T\}$. Human DNA consists approximately of three billion bases. Nucleotide A pairs only with T, and C pairs only with G, we can say A and T are complementary and so are G and C as shown in Fig. 2.2.

Given the sequence S of a DNA strand, we can construct the other strand S' by taking the complement of bases in this strand. If we take the complement of the

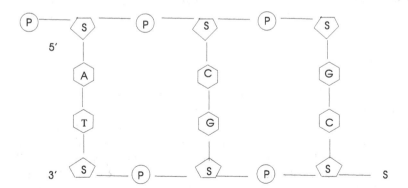

Fig. 2.2 DNA structure

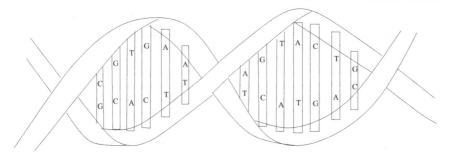

Fig. 2.3 DNA double helix structure

resulting strand we will obtain the original strand. This process is used and essential for protein production. Physically, DNA consists of two strands held together by hydrogen bonds, arranged in a double helix as shown in Fig. 2.3. The *complement* of a DNA sequence consists of complements of its bases. The DNA therefore consists of two complementary strands which bind to each other tightly providing a stable structure. This structure also provides the means to replicate in which the double DNA helix structure is separated into two strands and each of these strands are then used as templates to synthesize their complements.

The DNA molecule is wrapped around proteins called *histones* into complex-walled structures called *chromosomes* in the nucleus of each cell. The number of chromosomes depends on the type of eukaryote species. Each chromosome consists of two *chromatides* which are coil-shaped structures connected near the middle forming an x-like structure. Chromosomes are kept in the nucleus of a cell in a highly packed and hierarchically organized form. A single set of chromosomes in an organism is called *haploid*, two sets of chromosomes is called *diploid*, and more than two sets is called *polyploid*. Humans are diploid where each chromosome is inherited from a parent to have two chromosomes for each of the 23 chromosome set. The sex chromosome is chromosome number 23 which either has two chromosomes shaped X resulting in a female, or has X and Y resulting in a male. The type of chromosome inherited from father determines the sex of the child in this case.

2.2.2 RNA

The ribonucleic acid (RNA) is an important molecule that is used to transfer genetic information. It has a similar structure to DNA but consists of only one strand and does not form a helix structure like DNA. It also has nucleotides which consist of a sugar, phosphate, and a base. The sugar however is a *ribose* instead of deoxyribose and hence the name RNA. Also, DNA base thymine (T) is replaced with uracil (U) in RNA. The fundamental kinds of RNA are the messenger RNA (mRNA), transfer RNA (tRNA), and ribosomal RNA (rRNA) which perform different functions in the cell. RNA provides a flow of information in the cell. First, DNA is copied to mRNA in the nucleus and the mRNA is then translated to protein in the cytoplasm. During translation, tRNA and rRNA have important functions. The tRNA is responsible for forming the amino acids which make up the protein, as prescribed in the mRNA; and the rRNA molecules are the fundamental building blocks of the ribosomes which carry out translation of mRNA to protein.

2.2.3   Genes

A *gene* is the basic unit of hereditary in a living organism determining its character as a whole. A gene physically is a sequence of DNA that codes for an RNA (mRNA, tRNA, or rRNA) and the mRNA codes for a protein. The study of genes is called *genetics*. Gregor Mendel in the 1860s was first to experiment and set principles of passing hereditary information to offsprings.

There are 23 pairs of chromosomes in humans and between 20000–25000 genes are located in these chromosomes. The starting and stopping locations of a gene are identified by specific sequences. The protein coding parts of a gene are called *exons* and the regions between exons with no specific function are called *introns*. Genes have varying lengths and also, exons and introns within a gene have varying lengths. A gene can combine with other genes or can be nested within another gene to yield some functionality, and can be mutated which may change its functionality at varying degrees in some cases leading to diseases. The complete set of genes of an organism is called its *genotype*. Each gene has a specific function in the physiology and morphology of an organism. The physical manifestation or expression of the genotype is the *phenotype* which is the physiology and morphology of an organism cite. A gene may have different varieties called *alleles* resulting in different phenotyping characteristics. Humans are diploid meaning we inherit a chromosome from each parent, therefore we have two alleles of each gene. The genes that code for proteins constitute about 1.5 % of total DNA and the rest contains RNA encoding genes and sequences that are not known to have any function. This part of DNA is called *junk DNA*. There is no relatedness between the size of genome, number of genes, and organism complexity. In fact, some single cell organisms have a larger genome than humans.

2.2.4   Proteins

Proteins are large molecules of the cell and they carry out many important functions. For example, they form the antibodies which bind to foreign particles such as viruses and bacteria. As enzymes, they work as catalysts for various chemical reactions; the messenger proteins transmit signals to coordinate biological processes between different cells, tissues, and organs, also they transport small molecules within the cell and the body. Proteins are made from the information contained in genes. A protein consists of a chain of amino acids connected by *peptide bonds*. Since such a bond releases a water molecule, what we have inside a protein is a chain of amino acid

Table 2.1 Amino acids

Name	Abbrv.	Code	Pol.	Name	Abbrv.	Code	Pol.
Alanine	Ala	A	H	Methionine	Met	M	H
Cysteine	Cys	C	P	Asparagine	Asn	N	P
Aspartic acid	Asp	D	P	Proline	Pro	P	H
Glutamic acid	Glu	E	P	Glutamine	Gln	Q	P
Phenylalanine	Phe	F	H	Arginine	Arg	R	P
Glycine	Gly	G	P	Serine	Ser	S	P
Histidine	His	H	P	Threonine	Thr	T	P
Isoleucine	Ile	I	H	Valine	Val	V	H
Lysine	Lys	K	P	Tryptophan	Trp	W	H
Leucine	Leu	L	H	Tyrosine	Tyr	Y	P

residues. Typically, a protein has about 300 amino acid residues which can reach 5000 in large proteins. The essential 20 amino acids that make up the proteins is shown in Table 2.1 with their abbreviations, codes, and polarities.

Proteins have highly complex structures and can be analyzed at four hierarchical structures. The *primary structure* of a protein is specified by a sequence of amino acids that are linked in a chain and the *secondary structure* is formed by linear regions of amino acids. A *protein domain* is a segment of amino acid sequences in a protein which has independent functions than the rest of the protein. The protein also has a 3D structure called *tertiary structure* which affects its functionality and several protein molecules are arranged in *quaternary structure*. The function of a protein is determined by its four layer structure. A protein has the ability to fold in 3D and its shape formed as such affects its function. Using its 3D shape, it can bind to certain molecules and interact. For example, mad cow disease is believed to be caused by the wrong folding of a protein. For this reason, predicting the folding structure of a protein from its primary sequence and finding the relationship between its 3D structure and its functionality has become one of the main research areas in bioinformatics.

2.3 Central Dogma of Life

The central dogma of molecular biology and hence life was formulated by F. Crick in 1958 and it describes the flow of information between DNA, RNA, and proteins. This flow can be specified as DNA → mRNA → protein as shown in Fig. 2.4. The forming of mRNA from a DNA strand is called *transcription* and the production of a protein based on the nucleotide sequence of the mRNA is called *translation* as described next.

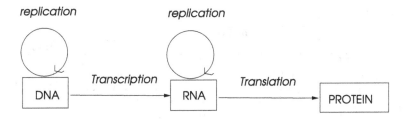

Fig. 2.4 Central dogma of life

2.3.1 Transcription

In the transcription phase of protein coding, a single stranded RNA molecule called mRNA is produced which is complementary to the DNA strand it is transcribed. The transcription process in eukaryotes takes place in the nucleus. The enzyme called *RNA polymerase* starts transcription by first detecting and binding a *promoter* region of a gene. This special pattern of DNA shown in Fig. 2.5 is used by RNA polymerase to find where to begin transcription. The reverse copy of the gene is then synthesized by this enzyme and a terminating signal sequence in DNA results in the ending of this process after which pre-mRNA which contains exons and introns is released. A post-processing called *splicing* involves removing the introns received from the gene and reconnecting the exons to form the mature and much shorter mRNA which is transferred to cytoplasm for the second phase called *translation*. The complete gene contained in the chromosome is called *genomic* DNA and the sequence with exons only is called *complementary* DNA or cDNA [25].

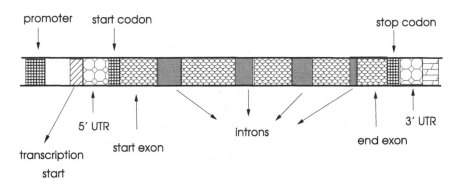

Fig. 2.5 Structure of a gene

Table 2.2 The genetic code

1st L.	2nd Letter				3rd L.
	U	C	A	G	
U	UUU } Phe UUC UUA } Leu UUG	UCU UCC } Ser UCA UCG	UAU } Tyr UAC UAA **Stop** UAG **Stop**	UGU } Cys UGC UGA **Stop** UGG Trp	U C A G
C	CUU CUC } Leu CUA CUG	CCU CCC } Pro CCA CCG	CAU } His CAC CAA } Gln CAG	CGU CGC } Arg CGA CGG	U C A G
A	AUU } Ile AUC AUA AUG **Met**	ACU ACC } Thr ACA ACG	AAU } Asn AAC AAA } Lys AAG	AGU } Ser AGC AGA } Arg AGG	U C A G
G	GUU GUC } Val GUA GUG	GCU GCC } Ala GCA GCG	GAU } Asp GAC GAA } Glu GAG	GGU GGC } Gly GGA GGG	U C A G

2.3.2 The Genetic Code

The genetic code provides the mapping between the sequence of nucleotides and the type of amino acids in proteins. This code is in triplets of nucleotide bases called *codons* where each codon encodes one amino acid. Since there are four nucleotide bases, possible total number of codons is $4^3 = 64$. However, proteins are made of 20 amino acids only which means many amino acids are specified by more than one codon. Table 2.2 displays the genetic code.

Such redundancy provides fault tolerance in case of mutations in the nucleotide sequences in DNA or mRNA. For example, a change in the codon UUA may result in UUG in mRNA but the amino acid *leucine* corresponding to each of these sequences is formed in both cases. Similarly, all of the three codons UAA, UAG, and UGA cause termination of the polypeptide sequence and hence a single mutation from A to G or from G to A still causes termination preventing unwanted growth due to mutations. Watson et al. showed that the sequence order of codons in DNA correspond directly to the sequence order of amino acids in proteins [28]. The codon AUG specifies the beginning of a protein amino acid sequence, therefore, the amino acid *methionine* is found as the first amino acid in all proteins.

2.3.3 Translation

The translation phase is the process where a mature mRNA is used as a template to form proteins. It is carried out by the large molecules called *ribosomes* which consist of proteins and the ribosomal RNA (rRNA) [5]. A ribosome uses tRNA to

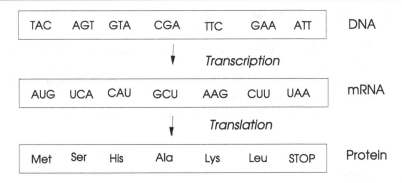

Fig. 2.6 Construction of a protein

first detect the start codon in the mRNA which is the nucleotide base sequence AUG. The tRNA has three bases called *anticodons* which are complementary to the codons it reads. The amino acids as prescribed by the mRNA are then formed and added to the linear protein structure according to the genetic code. Translation to the protein is concluded by detecting one of the three stop codons. Once a protein is formed, a protein may be transferred to the needed location by the signals in the amino acid sequence. The new protein must fold into a 3D structure before it can function [27]. Figure 2.6 displays the transcription and translation phases of a superficial protein made of six amino acids as prescribed by the mRNA.

2.3.4 Mutations

Mutations are changes in genetic code due to a variety of reasons. As an example, a stop codon UAA may be formed instead of an amino acid coding codon UCA (Serine) by a single point mutation of A to C, which will result in a shorter protein that will likely be nonfunctional. An amino acid may be replaced by another one, again by a point mutation such as the forming of CCU (Proline) instead of CAU (Histidine) by the mutation of C to A. These types of mutations may have important or trivial effects depending on the functioning of the mutated amino acid in the protein [5]. Furthermore, a nucleotide may be added or deleted to the sequence, resulting in the shifting of all codons by one nucleotide which will result in a very different sequence.

Mutations can be caused by radiations from various sources such as solar, radioactive materials, X-rays, and UV light. Chemical pollution is also responsible for many cases of mutations and viruses which insert themselves into DNA cause mutations. The inherited mutations are responsible for genetic diseases such as multiple sclerosis and Alzheimer disease. In many cases though, mutations result in better and improved characteristics in an organism such as better eyesight.

2.4 Biotechnological Methods

Biotechnology is defined as any technological application that uses biological systems, living organisms, or derivatives thereof, to make or modify products or processes for specific use, as defined by the Convention on Biological Diversity (CBD) [7]. The main biological methods are the *cloning* and *polymerase chain reaction* to amplify it, and *sequencing* to determine the nucleotide sequence in a DNA segment.

2.4.1 Cloning

DNA needs to be in sufficient quantities to experiment. DNA *cloning* is a method to amplify the DNA segment which could be very small. In this method, the DNA to be amplified called *insert* is inserted into the genome of an organism which is called the *host* or the *vector*. The host is then allowed to multiply during which the DNA inserted to it also multiplies. The host can then be disposed of, leaving only the amplified DNA segment. The commonly used organisms for cloning DNA are *plasmids*, *cosmids*, *phages*, and *yeast artificial chromosomes* (YACs) [25]. A plasmid is a circular DNA in bacteria and is used for cloning DNA of sizes up to 15 kbp. Phages are viruses and DNA segment inserted in them gets replicated when the virus infects an organism and multiplies itself. In YAC-based cloning, an artificial chromosome of yeast is constructed by the DNA insert sequence and the yeast chromosome control sections. The yeast multiplies its chromosomes including the YAC and hence multiplying the insert. YAC-based cloning can be used for very large segments of a million base pairs [25].

2.4.2 Polymerase Chain Reaction

The polymerase chain reaction (PCR) developed by Kary Mullis [3] in 1983, is a biomedical technology used to amplify selected DNA segment over several orders of magnitude. The amplification of DNA is needed for a number of applications including analysis of genes, discovery of DNA motifs, and diagnosis of hereditary diseases. PCR uses *thermal cycling* in which two phases are employed. In the first phase, the DNA is separated into two strands by heat and then, a single strand is enlarged to a double strand by the inclusion of a primer and polymerase processing. DNA polymerase is a type of enzyme that synthesizes new strands of DNA complementary to the target sequence. These two steps are repeated many times resulting in an exponential growth of the initial DNA segment. There are some limitations of PCR processing such as the accumulation of pyrophosphate molecules and the existence of inhibitors of the polymerase in the DNA sample which results in the stopping of the amplification.

2.4.3 DNA Sequencing

The sequence order of bases in DNA is needed to find the genetic information. *DNA sequencing* is the process of obtaining the order of nucleotides in DNA. The obtained sequence data can then be used to analyze DNA for various tasks such as finding evolutionary relationships between organisms and treatment of diseases. The exons are the parts of DNA that contain genes to code for proteins and all exons in a genome is called *exome*. Sequencing exomes is known as *whole exome sequencing*. However, research reveals DNA sequences external to the exons also affect protein coding and health state of an individual. In *whole genome sequencing*, the whole genome of an individual is sequenced. The new generation technologies are developed for both of these processes. A number of methods exist for DNA sequencing and we will briefly describe only the few fundamental ones.

The sequencing technology called *Sanger sequencing* named after Frederick Sanger who developed it [23,24], used deoxynucleotide triphosphates (dNTPs) and di-deoxynucleotide triphosphates (ddNTPs) which are essentially nucleotides with minor modifications. The DNA strand is copied using these altered bases and when these are entered into a sequence, they stop the copying process which results in different lengths of short DNA segments. These segments are ordered by size and the nucleotides are read from the shortest to the longest segment. Sanger method is slow and new technologies are developed. The *shotgun* method of sequencing was used to sequence larger DNA segments. The DNA segment is broken into many overlapping short segments and these segments are then cloned. These short segments are selected at random and sequenced in the next step. The final step of this method involves assembling the short segments in the most likely order to determine the sequence of the long segment, using the overlapping data of the short segments.

Next generation DNA sequencing methods employ massively parallel processing to overcome the problems of the previous sequencing methods. Three platforms are widely used for this purpose: the Roche/454 FLX [21], the Illumina/Solexa Genome Analyzer [4], and the Ion Torrent: Proton/PGM Sequencing [12]. The Roche/454 FLX uses the *pyrosequencing* method in which the input DNA strand is divided into shorter segments which are amplified by the PCR method. Afterward, multiple reads are sequenced in parallel by detecting optical signals as bases are added. The Illumina sequencing uses a similar method, the input sample fragment is cleaved into short segments and each short segment is amplified by PCR. The fragments are located in a slide which is flooded with nucleotides that are labeled with colors and DNA polymerase. By taking images of the slide and adding bases, and repeating this process, bases at each site can be detected to construct the sequence. The Ion proton sequencing makes use of the fact that addition of a dNTP to a DNA polymer releases an H^+ ion. The preparation of the slide is similar to other two methods and the slide is flooded with dNTPs. Since each H^+ decreases pH, the changes in pH level is used to detect nucleotides [8].

2.5 Databases

A database is a collection of structured data and there are hundreds of databases in bioinformatics. Many of these databases are generated by filtering and transforming data from other databases which contain raw data. Some of these databases are privately owned by companies and access is provided with a charge. In most cases however, bioinformatics databases are publicly accessible by anyone. We can classify bioinformatics databases broadly as nucleotide databases which contain DNA/RNA sequences; protein sequence databases with amino acid sequences of proteins, microarray databases storing gene expression data, and pathway databases which provide access to metabolic pathway data.

2.5.1 Nucleotide Databases

The major databases for nucleotides are the GenBank [10], the European Molecular Biology Laboratory-European Bioinformatics Institute (EMBL-EBI) database [19], and the DNA Databank of Japan (DDJB) [26]. GenBank is maintained by the National Center for Biotechnology Information (NCBI), U.S. and contains sequences for various organisms including primates, plants, mammals, and bacteria. It is a fundamental nucleic acid database and genomic data is submitted to GenBank from research projects and laboratories. Searches in this database can be performed by keywords or by sequences. The EMBL-EBI database is based on EMBL Nucleotide Sequence Data Library which was the first nucleotide database in the world and receives contributions from projects and authors. EMBL supports text-based retrieval tools including SRS and BLAST and FASTA for sequence-based retrieval [9].

2.5.2 Protein Sequence Databases

Protein sequence databases provide storage of protein amino acid sequence information. Two commonly used protein databases are the Protein Identification Resource (PIR) [16,31] and the UniProt [15] containing SwissProt [2]. The PIR contains protein amino acid sequences and structures of proteins to support genomic and proteomic research. It was founded by the National Biomedical Research Foundation (NBRF) for the identification and interpretation of protein sequence information, and the Munich Information Center for Protein Sequences (MIPS) [22] in Germany, and the Japan International Protein Information Database later joined this database. SwissProt protein sequence database was established in 1986 and provided protein functions, their hierarchical structures, and diseases related to proteins. The Universal Protein Resource (UniProt) is formed by the collaboration of EMBL-EBI, Swiss Institute of Bioinformatics (SIB) and PIR in 2003 and SwissProt was incorporated into UniProt. PDBj (Protein Data Bank Japan) is a protein database in Japan providing an archive of macromolecular structures and integrated tools [17].

2.6   Human Genome Project

The human genome project (HGP) is an international scientific research project to produce a complete human DNA sequence and identifying genes of human genome as well as other organisms such as mice, bacteria, and flies. This project was planned in 1984, started in 1990 and was finished in 2003. About 20 universities and research centers in United States, Japan, China, France, Germany, and the United Kingdom participated in this project. It aimed to sequence the three billion base pairs in human genome to analyze and search for the genetic causes of diseases to find cure for them, along with analysis of various other problems in molecular biology.

The results of this project are that there are between 20,000–25,000 genes in humans, and the human genome has more repeated DNA segments than other mammalian genomes. The work on results are ongoing but the results started to appear even before the completion of the project. Many companies are offering genetic tests which can show the tendencies of an individual to various illnesses. Comparing human genome with the genomes of other organisms will help our understanding of evolution better. Some ethical, legal, and social issues are questioned as a result of this project. Possible discrimination based on the genetic structure of an individual by the employers is one such concern and may result in unbalance in societies. However, this project has provided data that can be used to find molecular roots of diseases and search for cures.

2.7   Chapter Notes

We have reviewed the basic concepts of molecular biology at introductory level. The processes are evidently much more complex than outlined here. More detailed treatment of this topic can be found [1,20,29,30]. The cell is the basic unit of life in all organisms. The two types of cell are the eurokaryotes which are cells with nuclei and prokaryotes which do not have nuclei. Both of these life forms have the genetic material embedded in their DNA. Human DNA consists of a sequence of smaller molecules called nucleotides which are placed in 23 pairs of structures called chromosomes. A sequence in a chromosome that codes for a protein is called a gene. Genes identify amino acid sequences which form the proteins. The central dogma of life consists of two fundamental steps called transcription and translation. During transcription, a complementary copy of a DNA strand is formed and then the introns are extracted to form mRNA which is carried out of nucleus and the ribosomes form the amino acids prescribed using cRNA and tRNA. The three nucleotides that prescribe an amino acid is called a codon and the genetic code provides the mapping from a codon to an amino acid. Proteins also interact with other proteins forming protein–protein interaction (PPI) networks and their function is very much related to their hierarchical structure and also their position in the PPI networks.

We also briefly reviewed the biotechnologies for DNA multiplying, namely cloning and PCR technologies. These techniques are needed to provide sufficient amount of DNA to experiment in the laboratories. DNA sequencing is the process of obtaining nucleotide sequence of DNA. The databases for DNA and protein sequences contain data obtained by various bioinformatics projects and are presented for public use. DNA microarrays provide snapshots of DNA expression levels of vast number of genes simultaneously and gene expression omnibus (GEO) [11] from NCBI and ArrayExpress [14] from EBI are the two databases for microarray-based gene expression data. There are also pathway databases which provide data for biochemical pathways, reactions, and enzymes. Kyoto Encyclopedia of Genes and Genomes (KEGG) [13, 18] and BioCyc [6] are two such databases.

The computer science point of view can be confined to analysis of two levels of data in bioinformatics: the DNA/RNA and protein sequence data and the data of biological networks such as the PPI networks. Our main focus in this book will be the sequential and distributed algorithms for the analysis of these sequence and network data.

Exercises

1. For the DNA base sequence S = AACGTAGGCTAAT, work out the complementary sequence S' and then the complementary of the sequence S'.
2. A superficial gene has the sequence CCGTATCAATTGGCATC. Assuming this gene has exons only, work out the amino acid of the protein to be formed.
3. Discuss the functions of three RNA molecules named tRNA, cRNA, and mRNA.
4. A protein consists of the amino acid sequence A-B-N-V. Find three gene sequences that could have resulted in this protein.
5. Why is DNA multiplying needed? Compare the cloning and PCR methods of multiplying DNA in terms of technology used and their performances.

References

1. Alberts B, Bray D, Lewis J, Raff M, Roberts K (1994) Molecular biology of the cell. Garland Publishing, New York
2. Bairoch A, Apweiler R (1999) The SWISS-PROT protein sequence data bank and its supplement TrEMBL in 1999. Nucleic Acids Res 27(1):49–54
3. Bartlett JMS, Stirling D (2003) A short history of the polymerase chain reaction. PCR Protoc 226:36
4. Bentley DR (2006) Whole-genome resequencing. Curr Opin Genet Dev 16:545–552
5. Brandenberg O, Dhlamini Z, Sensi A, Ghosh K, Sonnino A (2011) Introduction to molecular biology and genetic engineering. Biosafety Resource Book, Food and Agriculture Organization of the United Nations
6. Caspi R, Altman T, Dreher K, Fulcher CA, Subhraveti P, Keseler IM, Kothari A, Krummenacker M, Latendresse M, Mueller LA, Ong Q, Paley S, Pujar A, Shearer AG, Travers M, Weerasinghe D, Zhang P, Karp PD (2011) The MetaCyc database of metabolic pathways and enzymes and the BioCyc collection of pathway/genome databases. Nucleic Acids Res 40(Database issue):D742D753

7. CBD (Convention on Biological Diversity). 5 June 1992. Rio de Janeiro. United Nations
8. http://www.ebi.ac.uk/training/online/course/ebi-next-generation-sequencing-practical-course
9. http://www.ebi.ac.uk/embl/
10. http://www.ncbi.nlm.nih.gov/genbank
11. http://www.ncbi.nlm.nih.gov/geo/
12. http://www.iontorrent.com/. Ion Torrent official page
13. http://www.genome.jp/kegg/pathway.html
14. http://www.ebi.ac.uk/arrayexpress/. Website of ArrayExpress
15. http://www.uniprot.org/. Official website of PIR at Georgetown University
16. http://pir.georgetown.edu/. Official website of PIR at Georgetown University
17. http://www.pdbj.org/. Official website of Protein Databank Japan
18. Kanehisa M, Goto S (2000) KEGG: Kyoto encyclopedia of genes and genomes. Nucleic Acids Res 28(1):2730
19. Kneale G, Kennard O (1984) The EMBL nucleotide sequence data library. Biochem Soc Trans 12(6):1011–1014
20. Lewin B (1994) Genes V. Oxford University Press, Oxford
21. Margulies M, Egholm M, Altman WE, Attiya S, Bader JS et al (2005) Genome sequencing in microfabricated high-density picolitre reactors. Nature 437:376–380
22. Mewes HW, Andreas R, Fabian T, Thomas R, Mathias W, Dmitrij F, Karsten S, Manuel S, Mayer KFX, Stmpflen V, Antonov A (2011) MIPS: curated databases and comprehensive secondary data resources in 2010. Nucleic Acids Res (England) 39
23. Sanger F, Coulson AR (1975) A rapid method for determining sequences in DNA by primed synthesis with DNA polymerase. J Mol Biol 94(3):441–448
24. Sanger F, Nicklen S, Coulson AR (1977) DNA sequencing with chain-terminating inhibitors. Proc Natl Acad Sci USA 74(12):5463–5467
25. Setubal JC, Meidanis J (1997) Introduction to computational molecular biology. PWS Publishing Company, Boston
26. Tateno Y, Imanishi T, Miyazaki S, Fukami-Kobayashi K, Saitou N, Sugawara H et al (2002) DNA data bank of Japan (DDBJ) for genome scale research in life science. Nucleic Acids Res 30(1):27–30
27. Voet D, Voet JG (2004) Biochemistry, 3rd edn. Wiley
28. Watson J, Baker T, Bell S, Gann A, Levine M, Losick R (2008) Molecular biology of the gene. Addison-Wesley Longman, Amsterdam
29. Watson JD (1986) Molecular biology of the gene, vol 1. Benjamin/Cummings, Redwood City
30. Watson JD, Hopkins NH, Roberts JW, Steitz JA, Weiner AM (1987) Molecular biology of the gene, vol 2. Benjamin/Cummings, Redwood City
31. Wu C, Nebert DW (2004) Update on genome completion and annotations: protein information resource. Hum Genomics 1(3):229–233

Graphs, Algorithms, and Complexity

3

3.1 Introduction

Graphs are discrete structures which are frequently used to model biological networks. We start this chapter with a brief review of basics of graph theory concepts and then describe concepts such as connectivity and distance that are used for the analysis of biological networks. Trees are graphs with no cycles and are used for various representations of biological networks. Spectral properties of graphs provide means to explore structures such as clusters of biological networks and are briefly described.

Algorithms have key functionality to solve bioinformatics problems and we provide a short review of fundamental algorithmic methods starting by describing the complexity of the algorithms. Our emphasis in this section is again on methods such as dynamic programming and graph algorithms that find many applications in bioinformatics. Basic graph traversal procedures such as the breadth-first search and depth-first search are frequently employed for various problems in biological networks and are described in detail. We then introduce complexity classes NP, NP-hard, and NP-complete for problems that cannot be solved in polynomial time. Using approximation algorithms with known approximation ratios provides suboptimal solutions to difficult problems and sometimes our only choice in tackling a bioinformatics problem is the use of heuristics which are commonsense rules that work for most cases of the input. We present a rather dense review of all of these key concepts in relation to bioinformatics problems in this chapter.

3.2 Graphs

Graphs are frequently used to model networks of any kind. A graph $G(V, E)$ has a nonempty vertex set V and a possibly nonempty edge set E. The number of vertices of a graph is called its *order* and the number of its edges is called its *size*. An edge

© Springer International Publishing Switzerland 2015
K. Erciyes, *Distributed and Sequential Algorithms for Bioinformatics*,
Computational Biology 23, DOI 10.1007/978-3-319-24966-7_3

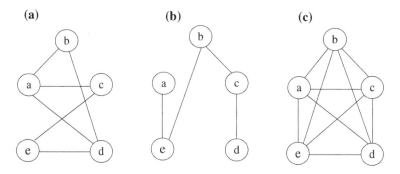

Fig. 3.1 **a** A graph G. **b** Complement of G. **c** Clique of order 5

e of a graph is incident to two vertices called its *endpoints*. A *loop* is an edge that starts and ends at the same vertex. *Multiple edges* e_1 and e_2 are incident to the same endpoint vertices. A *simple graph* does not have any loops or multiple edges. In the context of this book, we will consider only simple graphs. The complement of a graph $G(V, E)$ is the graph $\overline{G}(V, E')$ which has the same vertex set as G and an edge $(u, v) \in E'$ if and only if $(u, v) \notin E$. A *clique* is a graph with the maximum number of connections between its vertices. Each vertex in a clique has $n - 1$ connections to all other vertices where n is the order of the clique. A maximal clique is a clique which is not a proper subset of another clique. A maximum clique of a graph G is a clique of maximum size in G. Figure 3.1 displays example graphs.

The edges of a *directed graph* have orientation between their endpoints. An oriented edge (u, v) of a directed graph shown by an arrow from u to v, starts from vertex u and ends at vertex v. The degree of a vertex v of a graph G is the number of edges that are incident to v. The maximum degree of a graph is shown by $\Delta(G)$. The *in-degree* of a vertex v in a directed graph is the number of edges that end at v and the *out-degree* of a vertex v in such a graph is the number of edges that start at v. A directed graph is depicted in Fig. 3.2a. The in-degrees of vertices a, b, c, d, e in

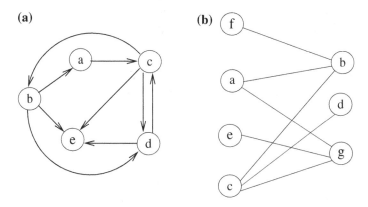

Fig. 3.2 **a** A directed graph. **b** A bipartite graph

this graph are $1, 1, 2, 2, 3$ and the out-degrees are $1, 3, 3, 2, 0$, respectively. Given a graph $G(V, E)$; if edge $(u, v) \in E$, then vertices u and v are called *neighbors*. An *open neighborhood* $N(v)$ (or $\Gamma(v)$) of a vertex v includes all vertices of v, whereas its *closed neighborhood* $N[v] = N(v) \cup \{v\}$ is the union of its open neighborhood and itself.

Given two graphs $G_1(V_1, E_1)$ and $G_2(V_2, E_2)$; G_1 is called a subgraph of G_2 if $V_1 \subseteq V_2$ and $E_1 \subseteq E_2$. An induced subgraph $G_1(V_1, E_1)$ of a graph $G_2(V_2, E_2)$ preserves in G_1 all of the incident edges between the vertices of V_1 that exist in G_2.

3.2.1 Types of Graphs

Certain classes of graphs have important applications in biological networks. We will now describe weighted graphs and bipartite graphs which may be used to represent biological networks. An *edge-weighted* graph $G(V, E, w)$ has weights associated with its edges such that $w : E \to \mathbb{R}$. Similarly, a vertex-weighted graph has weights associated with its vertices such that $w : V \to \mathbb{R}$.

A *bipartite graph* $G(V_1, V_2, E)$ has V_1 and V_2 as subsets of its vertices such that no edges exist between any two vertices of V_1, and similarly there are no edges between any two vertices of V_2. In other words, for any edge $(u, v) \in E$ of a bipartite graph, if $u \in V_1$, then $v \in V_2$ and $(u, v) \notin E$. A bipartite graph $G(V_1, V_2, E)$ is shown in Fig. 3.2b with $V_1 = \{f, a, e, c\}$ and $V_2 = \{b, d, g\}$.

3.2.2 Graph Representations

A graph $G(V, E)$ is commonly represented using an adjacency matrix A. An element a_{ij} of A is equal to 1 if there is an edge joining vertices v_i and v_j, otherwise a_{ij} is 0. For a directed graph, a_{ij} is 1 if there is an edge that starts at vertex i and ends at vertex j. Another way to represent a graph G is using *linked lists*. Each vertex is connected to its neighbors in a linked list and the graph is represented by n such lists. The *incidence matrix* I can also be used to display the connection of vertices of a graph G. In this case, an element i_{ij} of this matrix is equal to unity if vertex v_i is incident to edge (v_i, v_j). Figure 3.3 displays a graph and its adjacency list.

The adjacency matrix A and the incidence matrix I of this graph are shown below.

$$
A = \begin{bmatrix} 0 & 1 & 0 & 1 & 1 & 1 \\ 1 & 0 & 1 & 1 & 0 & 0 \\ 0 & 1 & 0 & 1 & 0 & 0 \\ 1 & 1 & 1 & 0 & 1 & 0 \\ 1 & 0 & 0 & 1 & 0 & 1 \\ 1 & 0 & 0 & 0 & 1 & 0 \end{bmatrix} \qquad I = \begin{bmatrix} 1 & 1 & 0 & 0 & 0 & 0 & 1 & 1 & 0 \\ 0 & 1 & 1 & 0 & 0 & 1 & 0 & 0 & 1 \\ 0 & 0 & 1 & 1 & 0 & 0 & 0 & 0 & 0 \\ 0 & 0 & 0 & 1 & 1 & 0 & 0 & 1 & 1 \\ 0 & 0 & 0 & 0 & 1 & 1 & 1 & 0 & 0 \\ 1 & 0 & 0 & 0 & 1 & 0 & 0 & 0 & 0 \end{bmatrix}
$$

Using an adjacency matrix, we can check in constant time whether an edge (i, j) exists between vertices i and j in a graph. However, this structure occupies a maximum of n^2 memory locations for the case of directed graphs, and searching for a

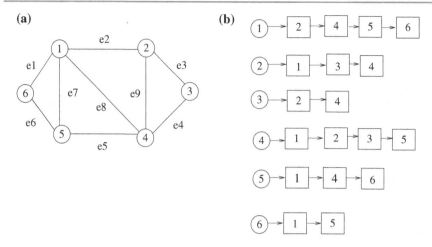

Fig. 3.3 a A graph G. **b** Its adjacency list

neighbor requires n steps. For sparse graphs where the density of the graph is low with relatively much less number of edges than dense graphs ($m \ll n$), this structure has the disadvantage of occupying unnecessary memory space. The adjacency list takes $m + n$ memory locations and examining all neighbors of a given node requires m/n time on average. However, checking the existence of an edge requires m/n steps whereas we can achieve this in one step using an adjacency matrix. In general, adjacency lists should be preferred for sparse graphs and adjacency matrices for dense graphs.

3.2.3 Paths, Cycles, and Connectivity

A *walk* of a graph $G(V, E)$ is a finite sequence of edges $(v_0, v_1), (v_1, v_2), ...,$ (v_{m-1}, v_m) where any two consecutive edges are adjacent or identical. The vertex v_0 is the initial vertex and v_m is the final vertex of the walk. The length of the walk is the number of edges in the walk. A *trail* is a walk in which all edges are distinct, and a *path* is a trail with all distinct vertices. A *cycle* is a trail or a path with the same initial and final vertices containing at least one edge. The *girth* of a graph is the length of the shortest cycle it has. In a connected graph G, the *distance* $d(u, v)$ between vertices u and v is the length of the shortest path from u to v. The *diameter* of a graph is the greatest distance between any two of its vertices.

A graph is *connected* if and only if there is a path between each vertex pair. A *cutset* of a graph is the set of edges of minimum size, removal of which results in a disconnected graph. In other words, a cutset of a graph G does not contain another cutset of G. The *edge connectivity* $\lambda(G)$ is the size of the smallest cutset of the graph

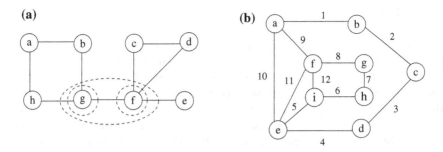

Fig. 3.4 a The bridge (g, f) of a graph. The endpoints of this bridge are also cut vertices of the graph. **b** The Hamiltonian cycle is $\{a, b, c, d, e, i, h, g, f, a\}$ and the sequence of Eularian trail starting from vertex a is shown by the numbers on edges

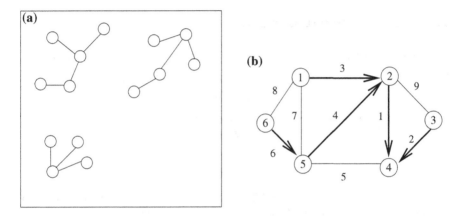

Fig. 3.5 a A forest. **b** An MST of a graph rooted at vertex 4

G. A graph G is *k-edge connected* if $\lambda(G) \geq k$. If the cutset of a graph has only one edge, this edge is called a *bridge*. The *vertex connectivity* $\kappa(G)$ of G on the other hand, is the minimum number of vertices needed to delete from G to leave it disconnected. All incident edges to a vertex should be removed when removing it from the graph. The graph is called *k-connected* if $\kappa(G) \geq k$. A *cut vertex* of a graph G is a vertex removal of which disconnects G. Figure 3.5a displays a bridge and cut vertices of a graph.

A trail containing every edge of a graph G exactly once is called an *Eularian trail* and G is called *Eularian*. It can be shown that a connected graph G is Eularian if and only if each vertex of G has even degree (see Exercise 3.2). An *Eularian cycle* is an Eularian trail that starts and ends at the same vertex. A cycle of a graph G that passes through each vertex of G exactly once is called a *Hamiltonian cycle*. An Eularian trail and a Hamiltonian cycle are shown in Fig. 3.4b.

3.2.4 Trees

A *tree* is a graph with no cycles and a *forest* is a graph that has two or more disconnected trees. A spanning tree $T(V, E')$ of a graph $G(V, E)$ is a tree that covers all vertices of G where $|E'| \leq |E|$. A tree where a special vertex is designated as root with all vertices having an orientation toward it is called a *rooted tree*, otherwise the tree is *unrooted*. Any vertex u except the root in a rooted tree is connected to a vertex v on its path to the root called its *parent*, and the vertex u is called the *child* of vertex v. For an edge-weighted graph $G(V, E, w)$, the *minimum spanning tree* (MST) of G has the minimum total sum of weights among all possible spanning trees of G. The MST of a graph is unique if all of its edges have distinct weights. Figure 3.5 shows a forest consisting of unrooted trees and a rooted MST of a graph.

3.2.5 Spectral Properties of Graphs

Given a square matrix $A[n \times n]$, an *eigenvalue* λ and the corresponding *eigenvector* x of A satisfy the following equation:

$$Ax = \lambda x \tag{3.1}$$

which can be written as

$$Ax - \lambda x = 0 \tag{3.2}$$

$$(A - \lambda I)x = 0 \tag{3.3}$$

The necessary condition for Eq. 3.2 to hold is $det(A - \lambda I)$ to be 0. In order to find an eigenvalue of a matrix A, we can do the following [3]:

1. Form matrix $(A - \lambda I)$.
2. Solve the equation for λ values.
3. Substitute each eigenvalue in Eq. 3.1 to find x vectors corresponding to λ values.

Eigenvalues are useful in solving many problems in engineering and basic sciences. Our main interest in eigenvalues and eigenvectors will be their efficient use for clustering in biological networks. The *Laplacian matrix* of a graph is obtained by subtracting its adjacency matrix A from its degree matrix D which has $d_{ii} = deg(i)$ at diagonal elements and $d_{ij} = 0$ otherwise as $L = D - A$. An element of the Laplacian matrix therefore is given by

$$L_{ij} = \begin{cases} 1 & \text{if } i = j \text{ and } deg(j) \neq 0 \\ -1 & \text{if } i \text{ and } j \text{ are adjacent} \\ 0 & \text{otherwise} \end{cases} \tag{3.4}$$

The Laplacian matrix L provides useful information about the connectivity of a graph [4]. It is positive semidefinite with all eigenvalues except the smallest one being positive and the smallest eigenvalue is 0. The number of components of a graph

G is given by the number of eigenvalues of its Laplacian which are 0. The second smallest eigenvalue of $L(G)$ shows whether graph G is connected or not, a positive value showing connectedness. A greater second eigenvalue shows a better connected graph.

3.3 Algorithms

An algorithm consists of a number of instructions to solve a given problem. It has inputs and execute the instructions on the input to provide an output. An algorithm has *assignment*, *decision*, and *repetition* as three main subtasks. Assignment is attributing values to variables, decision is the act of changing the flow of the algorithm based on some test condition and repetition is performed by executing the same code on possibly different data. An obvious requirement from any algorithm is that it should find the results correctly. Also, given two algorithms that provide the same result to a problem, we would prefer the one that solves the problem in shorter time. Algorithm 3.1 finds the number of occurrences of a given key value in an integer array and we can observe all of the described main subtasks in this algorithm.

Algorithm 3.1 *Alg_Example*

1: **Input** : Array $A[1..n]=\{...\}$
2: key = 5
3: **Output** : *count*
4: *count* $\leftarrow 0$ ▷ assignment
5: **for** $i = 1$ to n **do**   ▷ repetition
6:     **if** $A[i] = key$ **then**   ▷ decision
7:         *count* $\leftarrow count + 1$
8:     **end if**
9: **end for**

3.3.1 Time and Space Complexities

The time complexity of an algorithm is the number of steps needed to obtain the output and is commonly expressed in terms of the size of the input. In Algorithm 3.1, we would need at least n steps since the *for* loop is executed n times. Also, the assignment outside the loop needs 1 step. However, we would be more interested in the part of the algorithm that has a running time dependent on the input size, as the execution times of the other parts of an algorithm would diminish when the input size gets larger. In the above example, the algorithm has to be executed exactly n times. However, if we change the requirement to find the first occurrence of the input *key*, then the execution time would be at most n steps.

We can state the time complexity of an algorithm formally as follows. If f and g are two functions from $\mathbb{N}$ to $\mathbb{R}^+$, then

- $f(n) = O(g(n))$, if there is a constant $c > 0$ such that $f(n) \leq cg(n) \; \forall n \geq n_0$;
- $f(n) = \Omega(g(n))$, if there is a constant $c > 0$ such that $f(n) \geq cg(n) \; \forall n \geq n_0$;
- $f(n) = \Theta(g(n))$, if $f(n) = O(g(n))$ and $f(n) = \Omega(g(n))$.

The $O(g(n))$ states that when the input size is equal to or greater than a threshold n_0, the running time is always less than $cg(n)$. In other words, the $g(n)$ function provides an upper bound on the running time. The $\Omega(g(n))$ however, provides a lower bound on the running time, ensuring that the algorithm requires at least $cg(n)$ steps after a threshold n_0. If there are two constants $c_1, c_2 > 0$, such that the running time of the algorithm is greater than or equal to $c_1.g(n)$ and less than or equal to $c_2.g(n)$ after a threshold n_0; the running time of the algorithms is $\Theta(g(n))$. In most cases, we are interested in the worst case running time of an algorithm. The space complexity of an algorithm specifies the size of the memory required as a function of the input size.

3.3.2 Recurrences

A *recursive algorithm* calls itself with smaller values until a base case is encountered. A check for the base case is typically made at the start of the algorithm and if this condition is not met, the algorithm calls itself with possibly a smaller value of the input. Let us consider finding the nth power of an integer x using a recursive algorithm called *Power* as shown in Algorithm 3.2. The base case which is the start of the returning point from the recursive calls is when n equals 0, and each returned value is multiplied by the value of x when called which results in the value x^n.

Algorithm 3.2 *Recursive_Ex*

1: **procedure** $Power(x, n)$
2:     **if** $n = 0$ **then**
3:         **return** 1
4:     **else**
5:         **return** $x \times power(x, n - 1)$
6:     **end if**
7: **end procedure**

Recursive algorithms result in simpler code in general but their analysis may not be simple. They are typically analyzed using *recurrence relations* where time to solve the algorithm for an input size is expressed in terms of the time for smaller input sizes. The following recurrence equation for the *Power* algorithm defines the relation between the time taken for the algorithm for various input sizes:

$$T(n) = T(n-1) + 1 \qquad (3.5)$$
$$= T(n-1) + 1 = T(n-2) + 2$$
$$= T(n-2) + 2 = T(n-3) + 3$$

Proceeding in this manner, we observe that $T(n) = T(n-k) + k$ and when $n = k$, $T(n) = T(0) + n = n$ steps are required, assuming that the base case does not take any time. We could have noticed that there are n recursive calls until the base case of $n = 0$ and each return results in one multiplication resulting in a total of n steps. The analysis of recurrences may be more complicated than our simple example and the use of more advanced techniques such as the Master theorem may be needed [7].

3.3.3 Fundamental Approaches

The fundamental classes of algorithms are the *greedy algorithms*, *divide and conquer algorithms*, *dynamic programming*, and the *graph algorithms*. A greedy algorithm makes the locally optimal choice at each step and these algorithms do not provide an optimal solution in general but may find suboptimal solutions in reasonable times. We will see a greedy algorithm that finds MST of a graph in reasonable time in graph algorithms.

Divide and conquer algorithms on the other hand, divide the problem into a number of subproblems and these subproblems are solved recursively, and the solutions of the smaller subproblems may be combined to find the solution to the original problem. As an example of this method, we will describe the *mergesort* algorithm which sorts elements of an array recursively. The array is divided into two parts, each part is sorted recursively and the sorted smaller arrays are merged into a larger sorted array. Merging is the key operation in this algorithm and when merging two smaller sorted arrays, the larger array is formed by finding the smallest value found in both smaller arrays iteratively. For example, merging the two sorted arrays $\{2, 8, 10, 12\}$ and $\{1, 4, 6, 9\}$ results in $\{1, 2, 4, 6, 8, 9, 10, 12\}$. The dynamic programming method and the graph algorithms have important applications in bioinformatics and they are described next.

3.3.4 Dynamic Programming

Dynamic programming is a powerful algorithmic method that divides the problem into smaller subproblems as in the divide and conquer strategy. However, the solutions to the subproblems during this process are stored to be used in the latter stages of the algorithm. We will show an example of this approach using Fibonacci sequence which starts with 0 and 1 and each number in this sequence is the sum of the two previous numbers. This sequence can be stated as the reccurrence:

$$F(i) = F(i-1) + F(i-2) \qquad for \quad i \geq 2 \qquad (3.6)$$

Algorithm 3.3 shows how Fibonacci sequence can be computed using dynamic programming. The array F is filled with the n members of this sequence at the end of the algorithm which requires $\Theta(n)$ steps. Dynamic programming is frequently used in bioinformatics problems such as sequence alignment and DNA motif search as we will see in Chaps. 6 and 8.

Algorithm 3.3 $Fibo\_dyn$

1: **Input** : $n \geq 2$
2: **int** $F[n]$
3: $F[0] \leftarrow 0; F[1] \leftarrow 1$
4: **for** $i \leftarrow 2$ to n **do**
5:     $F[i] \leftarrow F[i-1] + F[i-2]$
6: **end for**
7: **return** $F(n)$

3.3.5 Graph Algorithms

Graph algorithms are implemented on graph structures which may be represented by adjacency matrices or lists. Two commonly used algorithms executed on graphs are the *breadth-first search* and the *depth-first search* algorithms which both build spanning trees that can be used for a variety of applications in bioinformatics.

3.3.5.1 Breadth-First Search

The breadth-first search (BFS) algorithm calculates distances from a source vertex s to all other vertices in a graph. It has many applications including finding distances and testing bipartiteness of a graph. Intuitively, any vertex u that is a neighbor of s will have a distance of 1 to s and the neighbors of u that are two hops away will have a distance of 2 to s. Algorithm 3.4 shows one way of implementing the BFS algorithm [3]. The source vertex s is initialized to zero and all other vertices are labeled with ∞ distances. The source vertex is inserted in the queue que and during each iteration of the algorithm, a vertex v is dequeued from Q, its neighbors are labeled with a distance of distance v plus 1 if they are not already labeled. The algorithm also labels parents of vertices on their BFS path to the source vertex s. Different order of queueing may result in different parents for vertices but their distance to s will not change.

A BFS tree in a sample graph is depicted in Fig. 3.6a.

Theorem 3.1 *The time complexity of BFS algorithm is $\Theta(n + m)$ for a graph of order n and size m.*

Algorithm 3.4 *BFS*

1: **Input** : $G(V, E)$ ▷ undirected and connected graph
2: **Output** : d_v and $prev[v]$, $\forall v \in E$ ▷ distances and predecessors of vertices in BFS tree
3: **for all** $v \in V \setminus \{s\}$ **do**   ▷ initialize all vertices except source s
4:      $d_v \leftarrow \infty$
5:      $prev[v] \leftarrow \perp$
6: **end for**
7: $d_s \leftarrow 0$ ▷ initialize source s
8: $prev[s] \leftarrow s$
9: $que \leftarrow s$
10: **while** $que \neq \emptyset$ **do**   ▷ do until que is empty
11:      $v \leftarrow deque(que)$ ▷ deque the first element u
12:      **for all** $(u, v) \in E$ **do**   ▷ process all neighbors of u
13:          **if** $d_u = \infty$ **then**
14:              $d_u \leftarrow d_v + 1$
15:              $prev[u] \leftarrow v$
16:              $enque(que, u)$
17:          **end if**
18:      **end for**
19: **end while**

Proof The initialization between lines 3 and 6 takes $\Theta(n)$ time. The *while* loop is executed at most n times and the *for* loop between the lines 12 and 18 is run at most $deg(u)+1$ times considering the vertices with no neighbors. Total running time in this case is

$$n + \sum_{u \in V}(deg(u) + 1) = n + \sum_{u \in V} deg(u) + n = 2n + 2m \in \Theta(n + m) \quad (3.7)$$

For general dense graphs excluding the sparse graphs, $m \gg n$, and the time complexity can be considered to be $\Theta(m)$.

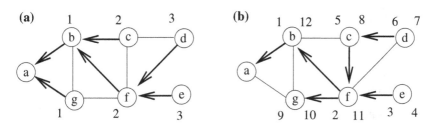

Fig. 3.6 **a** A BFS tree from vertex a. **b** A DFS tree from vertex a in the same graph. The first and last visit times of a vertex are shown in teh *left* and *right* of a vertex consecutively

3.3.5.2 Depth-First Search

The depth-first search (DFS) algorithm visits vertices of a graph recursively by visiting any unvisited neighbor of the vertex it currently is visiting and it goes as deep as it can rather than widely as in the BFS algorithm. When there are no unvisited neighbors of a vertex, the algorithm returns to the vertex where it came from and the algorithm stops when it returns to the starting vertex. Algorithm 3.5 shows one way of implementing DFS algorithm for a forest as in [1]. The first loop is executed once for each connected component of the graph and the second loop forms DFS trees recursively for each of these components. There is a color associated with a vertex; $white$ is for unvisited vertices, $gray$ is for a visited but not finished vertices, and $black$ means a vertex is visited and return from that vertex is done. There are also two times associated with each vertex, $d[u]$ records the global time when the vertex u is first visited and $f[u]$ shows when the visit is finished. As the output is a tree rooted at starting vertex, the array $pred$ stores the predecessors of each vertex in this tree. A DFS tree in an example graph is shown in Fig. 3.6b.

Algorithm 3.5 DFS_forest

1: **Input** : $G(V, E)$, directed or undirected graph
2: **Output** : $pred[n]; d[n], f[n]$ ▷ place of a vertex in DFS tree and its visit times
3: **int** $time \leftarrow 0; color[n]$
4: **for all** $u \in V$ **do**   ▷ initialize distances
5:      $color[u] \leftarrow false$
6:      $pred[u] \leftarrow \perp$
7: **end for**
8: **for all** $u \in V$ **do**
9:      **if** $color[u] = white$ **then**
10:        $DFS(u)$ ▷ call for each connected component
11:      **end if**
12: **end for**
13:
14: **procedure** $DFS(u)$
15:      $color[u] \leftarrow gray$
16:      $time \leftarrow time + 1; d[u] \leftarrow time$ ▷ first visit
17:      **for all** $(u, v) \in E$ **do**   ▷ initialize distances
18:          **if** $color[v] = white$ **then**
19:              $pred[v] \leftarrow u$
20:              $DFS(v)$
21:          **end if**
22:      **end for**
23:      $color[u] \leftarrow black$
24:      $time \leftarrow time + 1 ; f[u] \leftarrow time$ ▷ return visit
25: **end procedure**

There are two loops in the algorithm; the first loop considers each vertex at a maximum of $O(n)$ time and the second loop considers all neighbors of a vertex in a total of $\sum_u N(u) = 2m$ time. Time complexity of DFS algorithm is, therefore, $\Theta(n + m)$.

3.3.5.3 Prim's Minimum Spanning Tree Algorithm

This algorithm was initially proposed by Jarnik and then by Prim to find MST of a connected, weighted, directed, or undirected graph $G(V, E, w)$. It starts from an arbitrary vertex s and includes it in the MST. Then, at each step of the algorithm, the lowest weight outgoing edge (u, v) from the current tree fragment T_f such that $u \in T_f$ and $v \in G \setminus T_f$ is found and added to the tree T_f. If T_f is an MST fragment of G, it can be shown that $T \cup (u, v)$ is also an MST fragment of G. Proceeding in this manner, the algorithm finishes when all vertices are included in the final MST as shown in Algorithm 3.6.

Algorithm 3.6 *Prim_MST*

1: **Input** : $G(V, E, w)$
2: **Output** : MST T of G
3: $V' \leftarrow s$
4: $T \leftarrow \emptyset$
5: **while** $V' \neq V$ **do**   ▷ continue until all vertices are visited
6:     select the edge (u, v) with minimal weight such that $u \in T$ and $v \in G \setminus T$
7:     $V' \leftarrow V' \cup \{v\}$
8:     $T \leftarrow T \cup \{(u, v)\}$
9: **end while**

Figure 3.7 shows the iterations of Prim's algorithm in a sample graph. The complexity of this algorithm is $O(n^2)$ as the *while* loop is executed for all vertices and the search for the minimum weight edge for each vertex will take $O(n)$ time. This complexity can be reduced to $O(m \log n)$ time using binary heaps and adjacency lists.

3.3.5.4 Dijkstra's Shortest Path Algorithm

BFS algorithm finds shortest paths from a source vertex s to all other vertices in an unweighted graph. In a weighed graph, Dijkstra's shortest path algorithm uses the greedy approach similar to Prim's algorithm. It checks all of the outgoing edges from the current tree fragment as in Prim's MST algorithm. However, when selecting an edge (u, v) to be included in the shortest path tree T, total distance of its endpoint vertex v from the source vertex s is considered. When the minimum distance vertex is found, it is included in the partial tree and the distance of all affected vertices are updated. Algorithm 3.7 shows the pseudocode of Dijkstra's shortest path algorithm.

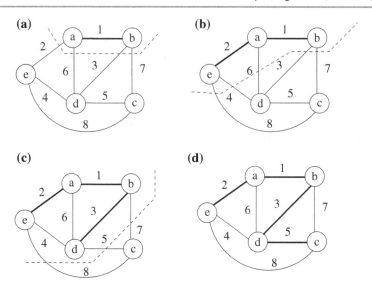

Fig. 3.7 Running of Prim's MST algorithm for vertex a

Algorithm 3.7 *Dijkstra's*

1: **Input** : $G(V, E, w)$
2: **Output** : T ▷ the shortest path tree rooted at s
3: $D[1 : n] \leftarrow \infty$ ▷ array showing distance to s
4: $T \leftarrow \{s\}$ ▷ T starts with the source vertex
5: $D[s] \leftarrow 0$
6: **while** $cut(T) \neq \emptyset$ **do**
7:     **select** $(u, v) \in cut(T)$ such that $D[u] + w_{uv}$ is minimum
8:         $D[v] \leftarrow D[u] + w_{uv}$
9:         $T \leftarrow T \cup (u, v)$
10: **end while**

The execution of this algorithm is shown in Fig. 3.8 where shortest path tree T is formed after four steps. Since there are n vertices, $n - 1$ edges are added to the initial tree T consisting of the source vertex s only. The time complexity of the algorithm depends on the data structures used. As with the Prim's algorithm, there are two nested loops and implementation using adjacency list requires $O(n^2)$ time. Using adjacency lists and Fibonacci heaps, this complexity may be reduced to $O(m + n \log n)$.

This algorithm only finds the shortest path tree from a source vertex and can be executed for all vertices to find all shortest paths. It can be modified so that the predecessors of vertices are stored (See Exercise 3.6).

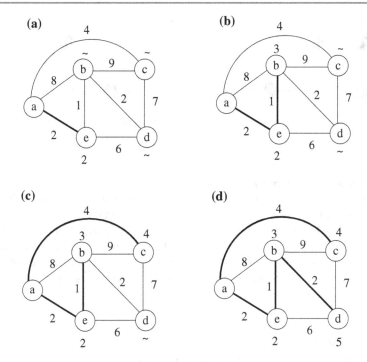

Fig. 3.8 Running of Dijkstra's shortest path algorithm

3.3.6 Special Subgraphs

Given a simple and undirected graph $G(V, E)$, a special subgraph $G' \subseteq G$ has some property which may be implemented to discover a structure in biological networks. The special subgraphs we will consider are the independent sets, dominating sets, matching and the vertex cover.

3.3.6.1 Independent Sets

An *independent set* I of a simple and undirected graph $G(V, E)$ is a subset of vertices G such that there is not an edge joining any two vertices in I. Formally, $\forall u$ and $v \in I$, $(u, v) \notin E$. The independent set problem (IND) asks to find I with the maximum number of vertices as an optimization problem where we try to find the best solution from a number of solutions. The decision version of this problem requires an answer as *yes* or *no* to the question: Is there an independent size of at least k in a graph G where $k < n$ with $n = |V|$? Figure 3.9a shows an independent set which is maximum.

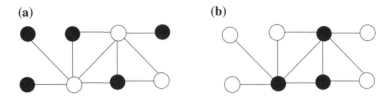

Fig. 3.9 a A maximum independent set of a graph. **b** The minimum dominating set of the same graph

3.3.6.2 Dominating Sets

A *dominating set* of a graph $G(V, E)$ is a subset D of its vertices such that any vertex of G is either in this set or a neighbor of a vertex in this set. The dominating set problem (DOM) asks to find D with the minimum number of vertices as an optimization problem. The decision version of this problem requires an answer as *yes* or *no* to the question: Is there a dominating set with a maximum size of k in a graph G where $k < n$? Dominating sets can be used to find clusters in networks. A minimum dominating set is shown in Fig. 3.9b.

3.3.6.3 Matching

A *matching* in a graph $G(V, E)$ is a subset E' of its edges such that the edges in E' do not share any endpoints. Matching finds various applications including routing in computer networks. A maximum matching of a graph G has the maximum size among all matchings of G. A maximal matching of G is a matching in G which cannot be enlarged. Finding maximum matching in a graph can be performed in polynomial time [2] and hence is in P. A maximum matching of size 4 is depicted in Fig. 3.10a.

3.3.6.4 Vertex Cover

A vertex cover of a graph $G(V, E)$ is a subset V' of vertices of G such that every edge in G is incident to at least one vertex in V'. Formally, for $V' \subseteq V$ to be a vertex cover; $\forall(u, v) \in E$, either u or v or both should be an element of V'. The optimization version of this problem asks to find a vertex cover of minimum size

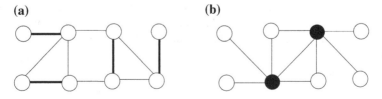

Fig. 3.10 a A maximum matching of a graph. **b** A minimum vertex cover of a graph

in a graph G. The decision version of this problem requires an answer as *yes* or *no* to the question: Is there a vertex cover of size of at least k in a graph G where $k < n$? A minimal vertex cover V' of a graph $G(V, E)$ is a vertex cover such that removal of a vertex from V' leaves at least one edge of G uncovered. Vertex cover has many applications including facility location in networks. Figure 3.10b displays a minimum vertex cover in a graph.

3.4 NP-Completeness

All of the algorithms we have considered up to this point have polynomial execution times which can be expressed as $O(n^k)$ where n is the input size and $k \geq 0$. These algorithms are in complexity class P which contains algorithms with polynomial execution times. Many algorithms, however, do not belong to P as either they have exponential running times or sometimes the problem at hand does not even have a known solution with either polynomial time or exponential time. Problems with solutions in P are called *tractable* and any other problem outside P is called *intractable*.

We need to define some concepts before investigating complexity classes other than P. Problems we want to solve are either *optimization problems* where we try to obtain the best solution to the problem at hand, or *decision problems* where we try to find an answer in the form of a *yes* or *no* to a given instance of the problem as described before. A *certificate* is an input combination for an algorithm and a *verifier* (or a *certifier*) is an algorithm that checks a certifier for a problem and provides an answer in the form of a *yes* or *no*. We will exemplify these concepts by the use of *subset sum problem*. Given a set of n numbers $S = \{i_1, \ldots, i_n\}$, this problem asks to find a subset of S sum of which is equal to a given value. For example, given $S = \{3, 2, -5, -2, 9, 6, 1, 12\}$ and the value 4, the subset $\{3, 2, -1\}$ provides a *yes* answer. The input $\{3, 2, -1\}$ is the certificate and the verifier algorithm simply adds the values of integers in the certifier in $k - 1$ steps where k is the size of the certifier.

Clearly, we need to check each subset of S to find a solution if it exits and this can be accomplished in exponential time of 2^n which is the number of subsets of a set having n elements. Now, if we are given an input value m and a specific subset R of S as a certifier, we can easily check in polynomial time using $\Theta(k)$ additions where k is the size of R, whether R is a solution. The certifier is R and the verifier is the algorithm that sums the elements of R and checks whether this sum equals m. Such problems that can be verified with a nondeterministic random input in polynomial time are said to be in complexity class *Nondeterministic Polynomial* (NP). Clearly, all of the problems in P have verifiers that have polynomial running time and hence $P \subseteq NP$. Whether P = NP is not known but the opposite is widely believed by computer scientists.

NP-Complete problems constitute a subset of problems in NP and they are as hard as any problem in NP. Formally, a decision problem C is NP-Complete if it is in NP and every problem in NP can be reduced to C in polynomial time.

Fig. 3.11 Relations between
the complexity classes

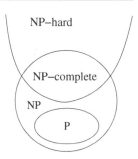

NP-hard problems are the problems which do not have any known polynomial time
algorithms and solving one NP-hard problem in polynomial time implies all of the
NP-hard problems that can be solved in polynomial time. In other words, NP-hard
is a class of problems that are as hard as any problem in NP. For example, finding the
least cost cycle in a weighted graph is an optimization problem which is NP-hard.
Figure 3.11 displays the relations between these complexity classes.

3.4.1 Reductions

If a problem A can be reduced to another problem B in polynomial time, we can
assume they are at the same level of difficulty. Formally, a problem A is polynomially
reducible to another problem B if any input I_A of A can be transformed to an input
I_B of B in polynomial time, and the answer to input I_B for problem B is true if and
only if the answer is true for problem A with the input I_A [3].

We will illustrate this concept by reduction of the independent set problem to the
clique problem. We have seen that cliques are complete graphs with edges between
all pairs of vertices. The clique optimization problem (CLIQUE) asks to find a clique
with the maximum number of vertices in a given simple and undirected graph. The
decision version of this problem searches an answer to the question: Is there a clique
of size at least k in a graph G where $k < n$?

We will now prove that IND problem can be reduced to CLIQUE problem in
polynomial time and hence these problems are equivalent. Given a graph $G(V, E)$
with an independent set $I \subset V$, we form $\overline{G}(V, E')$ which is the complement graph of
G with the same vertex set but a complementary edge set E'. Our claim is, if I is an
independent set in G, then I is a clique in $\overline{G}$. From the definition of the independent
set, we know that for any two vertices u and $v \in I$, the edge $(u, v) \notin E$ which means
$(u, v) \in E'$. This would mean checking whether a certificate C is an independent
set in G which is equivalent to checking whether C is a clique in $\overline{G}$. Clearly, the
transformation from G to $\overline{G}$ can be performed in polynomial time and therefore IND
is reducable to CLIQUE which is stated as *IND $\leq$ CLIQUE*. Figure 3.12 illustrates
reduction from an independent set to a clique.

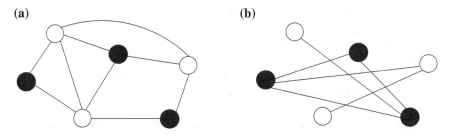

Fig. 3.12 Reduction from independent set to clique. **a** An independent set of a graph G. **b** The clique of G using the same vertices

3.4.2 Coping with NP-Completeness

Determining that a problem is NP-Complete has certain implications. First, there is no need to search for an algorithm in P to solve the problem, however, we can employ one of the following methods to find a solution:

- We can still use an exponential algorithm, if it exists, for a small input size. For bioinformatics problems, the size of the input in general is huge and this option would not be viable.
- We can use clever algorithms that prune the search space. *Backtracking* and *branch and bound* algorithms are two such types of algorithms. The backtracking algorithms construct a space tree of possible solutions and evaluate cases along the branches of this tree. When a case that does not meet the requirements is encountered at a vertex of this tree, all of its subtree is discarded from the search space resulting in a smaller size of search space. A branch and bound algorithm works similar to the backtracking algorithm, but additionally, it records the best solution obtained and whenever the branch of a tree does not provide a better solution, the subtree is discarded.
- *Approximation algorithms* which find suboptimal solutions with proven approximation ratios to the optimal value can be used.
- *Heuristics* which provide suboptimal solutions in reasonable time may be used. Heuristics do not guarantee to find a solution to the problem and typically are tested against a variety of inputs to show they work most of the time.

3.4.2.1 Approximation Algorithms

An approximation algorithm finds a suboptimal solution to a given difficult problem. The *approximation ratio* of an approximation algorithm shows how close the algorithm achieves to the optimal case and this ratio should be proven. For example, if an approximation algorithm to find the minimum dominating set in a graph G is 1.6, we know that the output from this algorithm will be at most 1.6 times the size of the minimum dominating set of G. We will exemplify the use of approximation algorithms by the *vertex cover* problem. As this is a known NP-complete problem

[5], our aim is to search for approximation algorithms for vertex cover. Algorithm 3.8 displays our approximation algorithm for the vertex cover problem. At each iteration, a random edge (u, v) is picked, its endpoint vertices are included in the cover and all edges incident to these edges are deleted from G. This process continues until there are no more edges left and since all edges are covered, the output is a vertex cover. We need to know how close the size of the output is to the optimum VC.

Algorithm 3.8 *Approx_MVC*

1: Input $G(V, E)$
2: $\leftarrow E, MVC \leftarrow \emptyset$
3: **while** $S \neq \emptyset$ **do**
4:     **pick** any $(u, v) \in S$
5:     $MVC \leftarrow MVC \cup \{u, v\}$
6:     **delete** all edges incident to either u or v from S
7: **end while**

This algorithm in fact finds a maximal matching of a graph G and includes the endpoints of edges found in the matching to the vertex cover set. Let M be a maximal matching of G. The minimum vertex cover must include at least one endpoint of each edge $(u, v) \in M$. Hence, the size of minimum vertex cover is at least as the size of M. The size of the cover returned by the algorithm is $2|M|$ as both ends of the matching are included in the cover. Therefore,

$$|VC| = 2|M| \leq 2 \times |MinVC| \qquad (3.8)$$

where VC is the vertex cover returned by the algorithm and MinVC is the minimum vertex cover. Therefore, the approximation ratio of this algorithm is 2. The running of this algorithm in a sample graph as shown in Fig. 3.13 results in a vertex cover that has a size of 6 which is 1.5 times the optimum size of 4 shown in Fig. 3.13b.

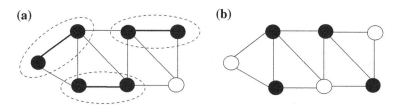

Fig. 3.13 Vertex Cover Examples. **a** The output of the approximation algorithm. **b** The optimal vertex cover

3.4.2.2 Using Heuristics

A *heuristic* is a commonsense observation that may be used as a rule in an algorithm. It is not proven mathematically and therefore extensive tests are required to claim that a certain heuristic works well for most of the input combinations. As an example, we may choose to select the vertex with the highest degree in the current working graph to find the vertex cover. This seems as a good choice as we would be covering the highest number of edges at each iteration and this should result in less number of vertices in the cover. Unfortunately, this heuristic is not better than the random choice and it can be shown its approximation ratio is not constant and dependent on the number of vertices with an approximation ratio of $\Theta(\log n)$.

Approximation algorithms often employ heuristics also, but they have a proven approximation ratio. In many types of bioinformatics problems, the use of heuristics is sometimes the only choice as the problems are NP-hard most of the time and approximation algorithms are not known to date. Extensive testing and validation are needed in this case to show the proposed heuristics perform well for most of the input combinations.

3.5 Chapter Notes

We have reviewed the basic background to analyze biological sequences and biological networks in this chapter. Graph theory with its rich background provides a convenient model for efficient analysis of biological networks. Our review of graphs emphasized concepts that are relevant to biological networks rather than being comprehensive. We then provided a survey of algorithmic methods again with emphasis on the ones used in bioinformatics. The dynamic programming and the graph algorithms are frequently used in bioinformatics as we will see. Finally, we reviewed the complexity classes P, NP, NP-hard, and NP-complete. Most of the problems we encounter in bioinformatics are NP-hard which often require the use of approximation algorithms or heuristics. Since the size of data is huge in these applications, we may use approximation algorithms or heuristics even if an algorithm in P exists for the problem at hand. For example, if we have an exact algorithm that solves a problem in $O(n^2)$ time, we may be interested in an approximation algorithm with a reasonable approximation ratio that finds the solution in $O(n)$ time for $n \gg 1$. Similarly, the size of data being large necessitates the use of parallel and distributed algorithms which provide faster solutions than the sequential ones even if the problem is in P, as we will see in the next chapter.

A comprehensive review of graph theory is provided by West [9] and Harary [6]. Cormen et al. [1], Skiena [8] and Levitin [7] all provide detailed descriptions of key algorithm design concepts.

Exercises

1. Find the adjacency matrix and the adjacency list representation of the graph shown in Fig. 3.14.
2. Prove that each vertex of as graph G has an even degree if G has an Eularian trail.
3. The exchange sort algorithm sorts the numbers in an array of size n by first finding the maximum value of the array, swapping the maximum value with the value in the first place, and then finding the maximum value of the remaining $(n-1)$ elements and continue until there are two elements. Write the pseudocode of this algorithm; show its iteration steps in the array $A = \{3, 2, 5, 4, 6\}$ and work out its time complexity.
4. Modify the algorithm that finds the Fibonacci numbers (Algorithm 3.3) such that only two memory locations which show the last two values of the sequence are used.
5. Work out the BFS and DFS trees rooted at vertex b in the graph of Fig. 3.15 by showing the parents of each vertex except the root.
6. Sort the weights of the graph of Fig. 3.16 in ascending order. Then, starting from the lightest weight edge, include edges in the MST as long as they do not form cycles with the existing edges in the MST fragment obtained so far. This procedure is known as Kruskal's algorithm. Write the pseudocode for this algorithm and work out its time complexity.

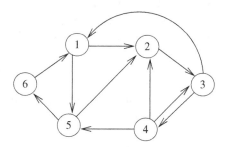

Fig. 3.14 Example graph for Exercise 1

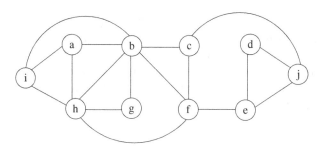

Fig. 3.15 Example graph for Exercise 5

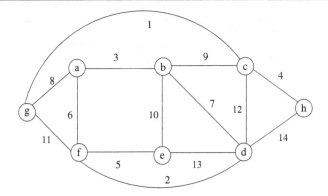

Fig. 3.16 Example graph for Exercise 6

Fig. 3.17 Example graph for
Exercise 7

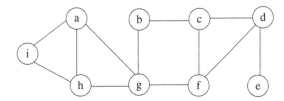

7. Modify Dijkstra's shortest path algorithm (Algorithm 3.7) so that the shortest path tree information in terms of predecessors of vertices are stored.
8. Find the minimum dominating set and the minimum vertex cover of the graph in Fig. 3.17.
9. Show that the independent set problem can be reduced to vertex cover problem in polynomial time. (Hint: If V' is an independent set of $G(V, E)$, then $V \setminus V'$ is a vertex cover).

References

1. Cormen TH, Leiserson CE, Rivest RL, Stein C (2009) Introduction to algorithms, 3rd edn. The MIT Press, Cambridge
2. Edmonds J (1965) Paths, trees, and flowers. Can J Math 17:449–467
3. Erciyes K (2014) Complex networks: an algorithmic perspective. CRC Press, Taylor and Francis
4. Fiedler M (1989) Laplacian of graphs and algebraic connectivity. Comb Graph Theory 25:57–70
5. Garey MR, Johnson DS (1979) Computers and intractability: a guide to the theory of NP-completeness. W. H. Freeman, New York
6. Harary F (1979) Graph theory. Addison-Wesley, Reading

7. Levitin A (2011) Introduction to the design and analysis of algorithms, 3rd edn. Pearson International Edn. ISBN: 0-321-36413-9
8. Skiena S (2008) The algorithm design manual. Springer, ISBN-10: 1849967202
9. West DB (2001) Introduction to graph theory, 2nd edn. Prentice-Hall, ISBN 0-13-014400-2

Parallel and Distributed Computing

4

4.1 Introduction

The terms computing, algorithm and programming are related to each other but
have conceptually different meanings. An algorithm in general is a set of instructions described frequently using *pseudocode* independent of the hardware, the operating system and the programming language used. Programming involves use of
implementation details such as operating system constructs and the programming
language. Computing is more general and includes methodologies, algorithms, programming and architecture.

A sequential algorithm consists of a number of instructions executed consecutively. This algorithm is executed on a central processing unit (CPU) of a computer
which also has memory and input/output units. A program is compiled and stored in
memory and each instruction of the program is fetched from the memory to the CPU
which decodes and executes it. In this so called *Von Neumann model*, both program
code and data are stored in external memory and need to be fetched to the CPU.

Parallel computing is the use of parallel computers to solve computational problems faster than a single computer. It is widely used for computationally difficult
and time consuming problems such as climate estimation, navigation and scientific
computing. A parallel algorithm executes simultaneously on a number of processing
elements with the processing elements being configured into various architectures.
A fundamental issue in parallel computing architecture is the organization of the
interconnection network which provides communication among tasks running on
different processing elements. The tasks of the parallel algorithm may or may not
share a commonly accessible global memory. The term parallel computing in general,
implies tightly coupling between the tasks as we will describe.

Distributed computing on the other hand, assumes the communication between
the tasks running on different nodes of the network communicate using messages
only. Distributed algorithms are sometimes called *message-passing algorithms* for
this reason. A *network algorithm* is a type of distributed algorithm where each node
is typically aware of its position in the network topology; cooperates with neighbor

© Springer International Publishing Switzerland 2015
K. Erciyes, *Distributed and Sequential Algorithms for Bioinformatics*,
Computational Biology 23, DOI 10.1007/978-3-319-24966-7_4

nodes to solve a network problem. For example, a node would search for shortest paths from itself to all other nodes by cooperating with its neighbors in a network routing algorithm. In a distributed memory routing algorithm on the other hand, we would attempt to solve shortest paths between all nodes of a possibly large network graph using few processing elements in parallel. Both can be called distributed algorithms in the general sense.

We start this chapter by the review of fundamental parallel and distributed computing architectures. We then describe the needed system support for shared-memory parallel computing and review multi-threaded programming by providing examples on POSIX threads. The parallel algorithm design techniques are also outlined and the distributed computing section is about message passing paradigm and examples of commonly used message passing software are described. We conclude the analysis by the UNIX operating system and its network support for distributed processing over a computer network.

4.2 Architectures for Parallel and Distributed Computing

Parallel computers may be constructed using special custom developed processors or off-the-shelf general purpose processors used in personal computers and workstations. The interconnection network connecting the processors may be custom built or a general one. Contemporary view of parallel processing hardware uses general purpose processors and general purpose interconnection networks to provide scalability and access to a wider population of users.

4.2.1 Interconnection Networks

An interconnection network connects the processing elements of the parallel or the distributed computing system. The communication medium may be shared among processors where only one message at a time can be transferred. In switched communication, each processor has its own connection to the switch. Switches provide parallel transmission of messages at the cost of more sophisticated and expensive hardware than the shared medium. Figure 4.1 displays a shared medium of computers connected and a switched medium of 2D mesh.

A *hypercube* is another example of a switched network consisting of processors placed at the vertices of a cubic structure. A d-dimension hypercube has 2^d processors labeled $0, \ldots, 2^d - 1$ as shown in Fig. 4.2a. Each node of a hypercube differs by one bit in its binary label representation from each of its neighbors. The farthest two nodes in a hypercube of d dimension has $\log n$ hops between them. A binary tree network has processors as the leaves, and switches as the other nodes of a tree where each switching node has exactly two children as shown in Fig. 4.2b.

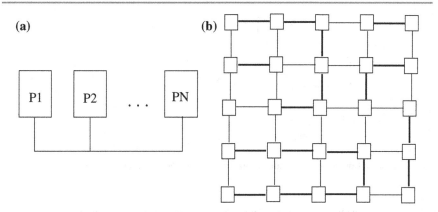

Fig. 4.1 Sample interconnection networks with each processor having switching capabilities. **a** A shared medium. **b** A 2-D mesh. Sample concurrent paths are shown in *bold* in (**b**)

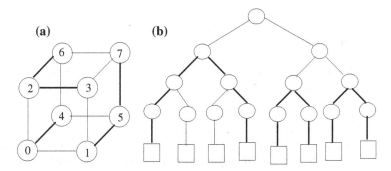

Fig. 4.2 Sample interconnection networks. **a** A 3-dimension hypercube. **b** A binary tree network. *Circles* are the switching elements, *squares* are the processors. Three sample concurrent paths are shown in *bold* in both figures

4.2.2 Multiprocessors and Multicomputers

A *multiprocessor* consists of a number of processors that communicate using a shared memory. The memory can be a single unit or it can be distributed over the network. A single shared memory multiprocessor is called *uniform memory access* (UMA) or a *symmetric multiprocessor* (SMP). An UMA architecture is shown Fig. 4.3a where each processor has its own cache and communicate using the global memory. Commonly, a processor has its own local memory and uses the global shared memory for inter-processor communications in a multiprocessor.

A *multicomputer* however, may and commonly does not have a shared memory and each processor typically communicates by *message passing* where messages are the basic units of communication. Each processing element in a multicomputer has its own local memory and input/output unit and hence is a general purpose computing element. A multicomputer architecture is displayed in Fig. 4.3b. A multicomputer

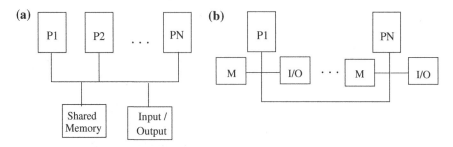

Fig. 4.3 **a** A multiprocessor. **b** A multicomputer both with N processors

is generally more tightly coupled and more homogeneous than a distributed system although technically, a distributed system is a multicomputer system as it does not have shared memory.

4.2.3 Flynn's Taxonomy

Flynn provided a taxonomy of parallel computers based on their ability to execute instructions on data as follows [4]:

- *Single-instruction Single-data* (SISD) computers: The single processing unit computers are in this category.
- *Single-instruction Multiple-data* (SIMD) computers: These are the computers that operate single instructions on multiple data. They have a single control unit and a number of processors. A *processor array* is an SIMD computer commonly used for matrix operations.
- *Multiple-instruction Single-data* (MISD) computers: These computers perform a number of instructions on a single data stream. They have a number of pipelining stages that process a data stream independently.
- *Multiple-instruction Multiple-data* (MIMD) computers: This category is the most versatile category of parallel computing where computers execute independently on different data streams. Multiprocessors and multicomputers are in this class. We will be considering MIMD computers which do not require special hardware for most of the problems considered in this book.

4.3 Parallel Computing

Parallel computing employs a number of processing elements which cooperate to solve a problem faster than a sequential computer. It involves decomposition of the computation and/or data into parts, assigning these to processors and providing the necessary interprocess communication mechanisms. Parallel processing is also employed within the CPU using instruction level parallelism (ILP) by executing independent instructions at the same time, however, the physical limits of this method have been reached recently and parallel processing at a coarser level has re-gained its popularity among researchers.

4.3.1 Complexity of Parallel Algorithms

A parallel algorithm is analyzed in terms of its *time, processor* and *work complexities*. The time complexity $T(n)$ of a parallel algorithm is the number of time steps required to finish it. The processor complexity $P(n)$ specifies the number of processors needed and the work complexity $W(n)$ is the total work done by all processors where $W = P \times T$. A parallel algorithm for a problem A is more efficient than another parallel algorithm B for the same problem if $W_A < W_B$.

Let us consider a sequential algorithm that solves a problem A with a worst case running time of $T_s(n)$ which is also an upper bound for A. Furthermore, let us assume a parallel algorithm does $W(n)$ work to solve the same problem in $T_p(n)$ time. The parallel algorithm is work-optimal if $W(n) = O(T(n))$. The *speedup* S_p obtained by a parallel algorithm is the ratio of the sequential time to parallel time as follows:

$$S_p = \frac{T_s(n)}{T_p(n)} \tag{4.1}$$

and the efficiency E_p is,

$$E_p = \frac{S_p}{p} \tag{4.2}$$

which is a value between zero and one, with p being the number of processors. Ideally S_p should be equal to the number of processors but this will not be possible due to the overheads involved in communication among parallel processes. Therefore, we need to minimize the interprocess communication costs to improve speedup. Placing many processes on the same processor decreases interprocess communication costs as local communication costs are negligible, however, the parallelism achieved will be greatly reduced. *Load balancing* of parallel processes involves distributing the processes to the processors such that the computational load is balanced across the processors and the inter-processor communications are minimized.

4.3.2 Parallel Random Access Memory Model

Parallel Random Access Memory Model (PRAM) is an idealized shared memory model of parallel computation. This model consists of p identical processors each with a local memory and they share a single memory of size m. Each processor works synchronously at each step by reading one local or global memory location, execute a single global RAM operation and write to a local or global memory location.

This model is commonly used for parallel algorithm design as it discards various implementation details such as communication and synchronization. It can further be classified into the following subgroups:

- Exclusive read exclusive write (EREW): Every memory cell can be read or written by only one processor at a time.
- Concurrent read exclusive write (CREW): Multiple processors can read a memory cell but only one can write at a time.
- Concurrent read concurrent write (CRCW): Multiple processors can read and write to the same memory location concurrently. A CRCW PRAM is sometimes called a concurrent random-access machine.

Exclusive read concurrent write (ERCW) is not considered as it does not make sense. Most algorithms for PRAM model use SIMD model of parallel computation and the complexity of a PRAM algorithm is the number of synchronous steps of the algorithm.

4.3.2.1 Summing an Array

We will show an example of an EREW PRAM algorithm that adds the elements of an array A of size $2p$ where p is the number of parallel processors. The idea of this algorithm is that each processor adds two consecutive array locations starting with the even numbered location and stores the result in the even location as shown in Fig. 4.4 for $p = 8$ and array size 16. We then need only half of the processors to add the four resulting numbers in the second step. Proceeding in this manner, the final sum is stored in the first location $A[0]$ of array A.

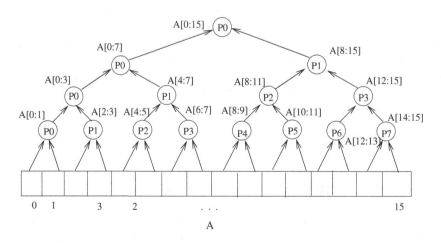

Fig. 4.4 Parallel summing of an array A of size 16 by 8 processors

Algorithm 4.1 shows one way of implementing this algorithm using EREW notation where the active processors are halved in each step for a total of $\log n$ steps. We could modify this algorithm so that contents of the array A are not altered by starting with $2p$ processor which copy A to another array B (see Exercise 4.2).

Algorithm 4.1 *EREW Sum of an array*

1: **Input** : $A[0, ..., 2p - 1], p = 2^m$
2: **Output** : sum $S = \sum_{j=0}^{n-1} A[j]$
3: **for** $k = 1$ to $\log n$ **do**
4:     **for** $i = 0$ to $n/2^k - 1$ **in parallel do**
5:         $A[i] \leftarrow A[2i] + A[2i + 1]$
6:     **end for**
7: **end for**
8: **if** $i = 0$ **then**
9:     $S \leftarrow A[i]$
10: **end if**

The time complexity of this algorithm is $\log n$ and in the first step, each processor p_i performs one addition for a total of $n/2$ additions. In the second step, we have $n/4$ processors each performing 2 additions and it can readily be seen that we have a total of $n/4$ summations at each step. The total work done in this case is $(n \log n)/4$. We could have a sequential algorithm that finds the result in only $n - 1$ steps. In this case, we find the work done by this algorithm is not optimal.

4.3.3 Parallel Algorithm Design Methods

We will assume that a parallel program consists of tasks that can run in parallel. A task in this sense has code, local memory and a number of input/output ports. Tasks communicate by sending data to their output ports and receive data from their input ports [12]. The message queue that connects an output port of a task to an input port of another task is called a *channel*. A task that wants to receive data from one of its input channels is blocked if there is no data available in that port. However, a task sending data to an output channel does not usually need to wait for the reception of this data by the receiving task. This model is commonly adopted in parallel algorithm design where reception of data is synchronous and the sending of it is asynchronous. Full synchronous communication between two tasks requires the sending task to be blocked also until the receiving task has received data. A four-step design process for parallel computing was proposed by Foster consisting of *partitioning*, *communication*, *agglomeration* and *mapping* phases [5].

Partitioning step can be performed on data, computation or both. We may divide data into a number of partitions that may be processed by a number of processing elements which is called *domain decomposition* [12]. For example, if the sum of an array of size n using p processors is needed, we can allocate each processor n/p data items which can perform addition of these elements and we can then transfer all of the partial sums to a single processor which computes the total sum for output. In partitioning computation, we divide the computation into a number of modules and then associate data with each module to be executed on each processor. This strategy is known as *functional decomposition* and sometimes we may employ both domain and functional decompositions for the same problem that needs to be parallelized.

The communication step involves the deciding process of the communication among the tasks such as which task sends or receives data from which other tasks. The communication costs among tasks that reside on the same processor can be ignored, however, the interprocess communication costs using the interconnection network is not trivial and needs to be considered.

The agglomeration step deals with grouping tasks that were designed in the first two steps so that the interprocess communication is reduced. The final step is the allocation of the task groups to processors which is commonly known as the *mapping* problem. There is a trade off however, as grouping tasks onto the same processor decreases communication among them but results in less parallelism. One way of dealing with this problem is to use a *task dependency graph* which is a directed graph representing the precedences and communications of the tasks. The problem is then reduced to the graph partitioning problem in which a vast amount of literature exists [1]. Our aim in general is to find a minimum weight cut set of the task graph such that communication between tasks on different processors is minimized and the load among processors is balanced.

Figure 4.5 shows such a task dependency graph where each task has a known computation time and the costs of communications are also shown. We assumed the

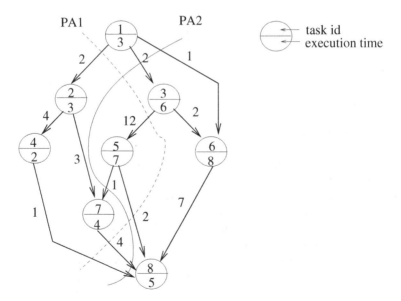

Fig. 4.5 A task dependency graph with 8 tasks

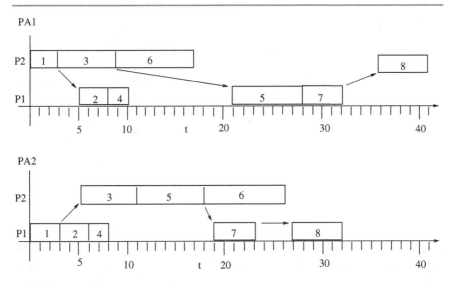

Fig. 4.6 Mapping of tasks of Fig. 4.5

computation times of tasks are known beforehand which is not realistic in a general computing system but is valid for many real-time computing systems. As a general rule, a task T_i that receives some data from its predecessors cannot finish before its predecessors finish, simply because we need to make sure that all of its data has arrived before it can start its execution. Otherwise, we cannot claim that T_i will finish in its declared execution time.

We can partition these tasks to two processors $P1$ and $P2$ with partition PA_1 where we only consider the number of tasks in each partition resulting in a total IPC cost of 21; or $PA2$ in which we consider IPC costs with a total cost of 8 for IPC. These schedules are displayed by the *Gantt charts* [6] which show the start and finishing times of tasks in Fig. 4.6. The resulting total execution times are 41 and 32 for partitions PA_1 and PA_2 from which we can deduce $PA2$ which has a lower IPC cost and shorter overall execution time is relatively more favorable.

4.3.4 Shared Memory Programming

Modern operating systems are based on the concept of a *process*. A process is an instance of a program in execution, and it has various dynamic data structures such as program counter, stack, open file pointers etc. A fundamental task performed by an operating system is the management of processes. Processes can be in one of the basic states as *new*, *ready*, *blocked*, *running* and *terminated* as shown in Fig. 4.7. When a new process enters the system, it is assigned to *ready* state so that it can run when CPU is allocated to it. A process waiting on an event such as a disk operation is blocked and enters *blocked* state, to enable another process to have the CPU while

Fig. 4.7 Basic process states

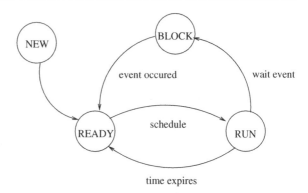

it is waiting. When that event occurs, it is made *ready* again to be scheduled by the operating system. A running process may have its time slice expired in which case it is *preempted* and put in the *ready* state to be scheduled when its turn comes. The data about a process is kept in the structure called process control block (PCB) managed by the operating system.

Having processes in an operating system provides modularity and a level of parallelism where CPU is not kept idle by scheduling a ready process when the current process is blocked. However, we need to provide some mechanisms for these processes to synchronize and communicate so that they can cooperate. There are two types of synchronization among processes; *mutual exclusion* and *conditional synchronization* as described next.

4.3.4.1 The Critical Section Problem

Processes need to access some parts of their code exclusively if this segment of code manipulates a common data with other processes. This segment of code is called a *critical section* and we need mechanisms to provide exclusive operations in these sections. Let us demonstrate the critical section problem by a solid example. Assume two concurrent processes P_1 and P_2 increment a global integer variable x which has a value of 12. The increment operation for each process will be implemented typically by loading a CPU register with the value of x, incrementing the register value and then storing the value of this register in the location of x as shown below.

```
P1                              P2
1. LOAD R2,m[x]                 1. LOAD R5,m[x]
2. INC  R2                      2. INC  R5
3. STO  m[x],R2                 3. STO  m[x],R5
```

Now, let us assume P_1 executes first, and just before it can execute line 3, its time slice expires and the operating system performs a context switch by storing its variables including R2 which has 13 in its PCB. P_2 is executed afterwards which executes by reading the value of 12 into register R15, incrementing R5 and storing its value 13 in x. The operating system now schedules P_1 by loading CPU registers from its PCB and P_1 writes 13 on x location which already contains 13. We have incremented x once instead of twice as required.

As this example shows, some parts of the code of a process has to be executed exclusively. In the above example, if we had provided some means so that P_1 and P_2 executed lines 1–3 without any interruption, the value of x would be consistent. Such a segment of the code of a process which has to be executed exclusively is called a *critical section*. A simple solution to this problem would be disabling of interrupts before line 1 and then enabling them after line 3 for each process. In user C code for example, we would need to enclose the statement $x = x + 1$ by the assembly language codes *disable interrupt* (DI) and *enable interrupt* (EI). However, allowing machine control at user level certainly has shortcomings, for example, if the user forgets enabling interrupts then the computation would stop completely.

There are various methods to provide mutual exclusion while a critical section is executed by a process and the reader is referred to [16] for a comprehensive review. Basically, we can classify the methods that provide mutual exclusion at hardware, operating system or application (algorithmic) level. At hardware level, special instructions that provide mutual exclusion to the memory locations are used whereas algorithms provide mutual exclusion at application level. However, it is not practical to expect a user to implement algorithms when executing a critical section, instead, operating system primitives for this task can be appropriately used. Modern operating systems provide a mutual exclusion data structure (mutex) for each critical section and two operations called *lock* and *unlock* on the mutex structure. The critical section can then be implemented by each process on a shared variable x as follows:

```
mutex m1;
int x     /* shared integer */

process P1{
    ...      /* non-critical section */

    lock(&m1);
    x++;      /* critical section */
    unlock(&m1);
    ...      /* non-critical section */
}
```

The *lock* and *unlock* operations on the mutex variables are *atomic*, that is, they cannot be interrupted, as provided by the operating system.

4.3.4.2 Process Synchronization

Conditional synchronization allows processes to synchronize on events. For example, let us assume a process P_1 is waiting for a resource r that another process P_2 is using. P_1 needs to be notified when the processing of P_1 is over as it has no other means of knowing the availability of the resource r. Operating systems provide various facilities and system calls to wait on an event and to signal an event for synchronization.

A *semaphore* is a special structure that contains at least an integer. Two atomic operations; *wait* and *signal* can be performed on a semaphore. The *wait* system call is used to wait for the event that the semaphore represents and the *signal* primitive signals the occurrence of that event. In a practical implementation of a semaphore, a process queue may be associated with a semaphore. We can define an array of semaphores with each semaphore having an integer value and a process queue as shown below. The *wait* system call then decrements the value of the semaphore, and queues the calling process in the semaphore process queue by changing its state to BLOCKED if this value becomes negative, meaning the event waited has not occurred yet. The *signal* call does the reverse operation by first incrementing the value of the semaphore and if this value is negative or zero, which means there was at least one process waiting in the semaphore queue for the event, it dequeues the first waiting process, changes its state to READY and calls the scheduler so that the freed process can execute.

```
typedef struct { int value;
                 que pr_que;
               }sem_t,
sem_t sem_tab[N_sems], *sp;
pcbptr currptr, procptr;
```

```
void wait(int s)                    void  signal(int s)
{ semptr sp=&(sem_tab[s]);          { semptr sp=&(sem_tab[s]);
  sp->value--;                        sp->value++;
  if ( sp->value < 0 )                if ( sp->value <= 0 )
  { enque(currptr, sp->pr_que);     { procptr=deque(sp->pr_que);
    currptr->state=BLOCKED;           procptr->state=READY;
    Schedule();                       Schedule();
  }                                 }
}                                   }
```

A semaphore can be used for both mutual exclusion and conditional synchronization. The following example demonstrates these two usages of semaphores where two processes *producer* and *consumer* synchronize using semaphores. The semaphore $s1$ is used for mutual exclusion and the semaphores $s2$ and $s3$ are used for waiting and signalling events.

```
sem_t s1,s2,s3;
int shared;
```

```
void producer()             void  consumer()
{ while(true) {             { while(true) {
      input in_data;              wait(s2);
      wait(s2);                   wait(s1);
```

```
        wait(s1);                    out_data=shared;
        shared = in_data;            signal(s1);
        signal(s1);                  signal(s2);
        signal(s3);                  print out_dat;
    }                            }
}                            }
```

The shared memory location which should be mutually accessed is *shared* and is protected by the semaphore $s1$. The *producer* continuously reads data from the keyboard, and first waits on $s2$ to be signalled by the *consumer* indicating it has read the previous data. It then writes the input data to the *shared* location and signals the *consumer* by the $s3$ semaphore so that it can read shared and display the data. Without this synchronization, the *producer* may overwrite previous data before it is read by the *consumer*; or the consumer may read the same data more than once.

4.3.5   Multi-threaded Programming

A *thread* is a lightweight process with an own program counter, stack and register values. It does not have the memory page references, file pointers and other data that an ordinary process has, and each thread belongs to one process. Two different types of threads are the *user level threads* and the *kernel level threads*. User level threads are managed by the run-time system at application level and the *kernel* which is the core of an operating system that handles basic functions such as interprocess communication and synchronization, is not aware of the existence of these threads. Creation and other management functions of a kernel thread is performed by the kernel.

A user level thread should be non-blocking as a single thread of a process that does a blocking system call will block all of the process as the kernel sees it as one main thread. The kernel level threads however can be blocking and even if one thread of a process is blocked, kernel may schedule another one as it manages each thread separately by using *thread control blocks* which store all thread related data. The kernel threads are a magnitude or more times slower than user level threads due to their management overhead in the kernel. The general rule is to use user level threads if they are known to be non-blocking and use kernel level threads if they are blocking, at the expense of slowed down execution. Figure 4.8 displays the user level and kernel level threads in relation to the kernel.

Using threads provide the means to run them on multiprocessors or multi-core CPUs by suitable scheduling policies. Also, they can share data which means they do not need interprocess communication. However, the shared data must be protected by issuing operating system calls as we have seen.

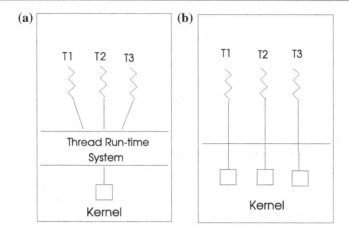

Fig. 4.8 **a** User level threads. **b** Kernel level threads which are attached to thread control blocks in the kernel

4.3.5.1 POSIX Threads

Portable operating system interface (POSIX) is a group of standards specified by IEEE to provide compatibility between different operating systems [11] and POSIX threads is a POSIX standard specifying an application programming interface (API) for thread management. We will briefly review only basic POSIX threads API routines which provide a rich variety of procedures for thread management. One such fundamental routine is *pthread_create* function used to create a thread is as follows:

```
pthread_create(&thread_id,&attributes,start_function,arguments);
```

where *thread_id* is the variable where the created thread identifier will be stored after this system call, *start_function* is the address of the thread code and the *arguments* are the variables passed to the created thread.

The following example illustrates the use of threads for parallel summing of an integer array A with n elements. Each thread sums the portion of the array defined by its identity passed to it during its creation. Note that we are invoking the same thread code with different parameter (i) each time resulting in summing a different part of the array A, for example, thread 8 sums $A[80 : 89]$. For this example, we could have used user threads as they work independently without getting blocked, however, we used kernel threads as POSIX threads API allows the usage of kernel threads only. Solaris operating system allows both user and kernel threads and provide flexibility but this API is not a standard [15].

```
#include <stdio.h>
#include <pthread.h>
#define n 100
#define n_threads 10
int A[n]=2,1,...,8;      /* initialize */
pthread_mutex_t m1;
```

```
int total_sum;

void *worker(void *id)
{ int me=*((int *)id);
  int i, n_slice, my_sum;
  for(i=me*n_slice;i<me*n_slice+n_slice;i++)
    my_sum=my_sum+A[i];
  pthread_mutex_lock(&m1);
  total_sum=total_sum+my_sum;
  pthread_mutex_unlock(&m1);
}

main()
{ pthread_t threads[n];
  int i;
  pthread_mutex_init(&m1);
  for(i=1; i<=n_threads; i++)
    pthread_create(&threads[i],NULL,worker,i);
  for(i=1; i<=n_threads; i++)
    pthread_join(threads[i],NULL);
  printf("Total sum is = %d", total_sum);
}
```

We need to compile this program (sum.c) with the POSIX thread library as follows:

```
cc -o sum sum.c -lpthread
```

Threads can synchronize using locks, condition variables or semaphores which are data structures defined globally. A semaphore can be declared and initialized to be shared among threads with the initial value of 1 as follows:

```
#include <semaphore.h>
sem_t s1;
sem_init(&s1, 1, 1)
```

The wait and signal operations on semaphores are *sem_wait* and *sem_signal* respectively. In the following example, we will implement the *producer/consumer* example of Sect. 4.3.4 by using two threads and two semaphores, *full* and *empty*.

```
#include <stdio.h>
#include <pthread.h>
#include <semaphore.h>

sem_t full, empty;
int data;

void *producer(void *arg)
{ int i=1;
  while(i<10) {
    sem_wait(&empty);
```

```
    scanf("%d", data);
    sem_post(&full);
    i++;
  }
}

void *consumer(void *arg)
{ int i=1;
  while(i<10) {
    sem_wait(&full);
    printf("%d", output);
    sem_post(&empty);
    i++;
  }
}

main()
{ pthread_t prod, cons;
  int i;
  sem_init(&full,1,0);
  sem_init(&empty,1,1);
  pthread_create(&prod,NULL,producer,NULL);
  pthread_create(&cons,NULL,consumer,NULL);
  pthread_join(prod,NULL);
  pthread_join(cons,NULL);
}
```

The producer executes a *wait* on the *empty* semaphore which is initialized to 1, so it does not actually wait the first time. It then writes the input data from the keyboard to the global data location and signals the *full* semaphore where the consumer is waiting. The consumer then is activated and consumes data by printing it. It has to signal the waiting *producer* now so that the previous data is not overwritten. Note that two semaphores are needed as two processes wait on two different conditions to be signalled; the *consumer* waits for the availability of next data and the producer waits for the notification of the consumption of data written.

4.3.6 Parallel Processing in UNIX

UNIX is a multitasking operating system developed at Bell Labs first and at University of California at Berkeley with network interface extension, named Berkeley Software Distribution (BSD). It is written in C for the most part, has a relatively smaller size than other operating systems, is modular and distributed in source code which make UNIX as one of the most widely used operating systems. The user interface in UNIX is called *shell*, and the *kernel* which performs the core operating system functions such as process management and scheduling resides between the shell and the hardware as shown in Fig. 4.9.

Fig. 4.9 UNIX structure

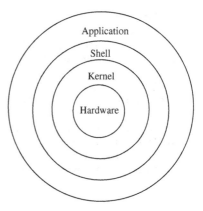

UNIX is based on processes and a number of system calls provide process management. A process can be created by a process using the system call *fork*. The caller
becomes the *parent* of the newly created process which becomes the *child* of the
parent. The *fork* system call returns the process identifier as an unsigned integer to
the parent, and 0 to the child. We can create a number of parallel processes using this
structure and provide concurrency where each child process is created with a copy
of the data area of the parent and it runs the same code as the parent. UNIX provides
various interprocess communication primitives and *pipes* are one of the simplest. A
pipe is created by the *pipe* system call and two identifiers, one for reading from a
pipe and one for writing to the pipe are returned. Reading and writing to a pipe are
performed by *read* and *write* system calls, similar to file read and write operations.
The following describes the process creation and interprocess communication using
pipes in UNIX. Our aim in this program to add the elements of an array in parallel
using two processes. The parent creates two pipes one for each direction and then
forks a child, sends the second half of the array to it by the pipe p_1 to have it calculate
the partial sum, and calculates the sum of the lower portion of the array itself. It then
reads the sum of the child from pipe p_2 and finds the total sum and displays it.

```
#include <stdio.h>
#define n   8
int c, *pt, p1[2], p2[2], A[n], sum=0, my_sum=0;

main()
{ pipe(p1); /* create two pipes one for each direction */
  pipe(p2); /* p1 is from parent to child,   */
  c=fork();  /* p2 is from child to parent */
  if(c!=0) { /* this is parent */
    close(p1[0]);  /* close the read end of p1 */
    close(p2[1]);  /* close the write end of p2 */
    for(i=0;i<n;i++)  /* initialize array */
      A[i]=i+1;
    write(p[1],&A[n/2],n/2);  /* send the second half of array */
```

```
   for(i=0;i<n/2;i++)
      my_sum=my_sum+A[i];
   while(n=read(p2[0], pt, 1));
   printf("Total sum is  = %d", total_sum);
 }
 else {  /* this is child */
   close(p2[0]);  /* close the read end of p2 */
   close(p1[1]);  /* close the write end of p1 */
   while(n=read(p2[0], pt, n));
   for(i=0;i<n/2;i++)
      my_sum=my_sum+A[i];
   write(p2[1],sizeof(int),sum);  /* send partial sum */
   }
 }
}
```

4.4 Distributed Computing

A distributed system is made of computational nodes that are connected over a communication network. The Internet, the grid, cluster of workstations and an airline reservation system are the examples of distributed systems. Two important benefits to be gained from distributed systems other than parallel processing are *resource sharing* and *fault tolerance*. Our aim in the context of this book will be to use a distributed system for parallel processing in order to solve computationally time consuming bioinformatics problems.

Nodes of a distributed systems use *message-passing* as the main method for communication. Each process may send a message to a single process (*unicast*), to a group of processes (*multicast*) or to all processes in the system (*broadcast*). Message passing can be performed synchronously or asynchronously. Two fundamental primitives of message-passing are the *send* and *receive* primitives. A synchronous *send* primitive will wait until the receiver has received its message and the synchronous *receive* will block the caller until a message arrives, on the other hand, asynchronous send and receive procedures do not block the caller. As the further actions of a receiver are frequently determined by the contents of the received message, the *receive* primitive is commonly used in blocking mode. However, the sender may assume its message has been delivered to the receiver over a reliable network and typically does not need to block. For this reason, the asynchronous *send* and synchronous *receive* combination which we will call *semi-synchronous* interprocess communication is mostly used in message-passing distributed computing systems.

4.4.1 Distributed Algorithm Design

A distributed algorithm runs at the computational nodes of a distributed system. Typically, the same algorithm code will run on different nodes with different data which is called the *single program multiple data* (SPMD) paradigm. At a higher level of operation than message passing, distributed algorithms can be specified as synchronous where a number of rounds are executed synchronously under the control of a central process; or asynchronous with no central control. The type of action to be performed in a distributed algorithm depends largely on the type and contents of the message received from neighbors. Algorithm 4.2 displays a commonly employed structure of a distributed algorithm where process i receives a message from a process j as in [3], and based on the type and contents of this message, it executes a specific procedure.

Algorithm 4.2 *General Distributed Algorithm Structure*

1: **int** i, j ▷ i is this node; j is the sender of the current message
2: **while** ¬*finished* **do**   ▷ all nodes execute the same code
3:     **receive** *msg(j)*
4:     **case** *msg(j).type* **of**
5:             *type_*1 : *Procedure_*1
6:             ... : ...
7:             *type_n* : *Procedure_n*
8: **end while**

4.4.2 Threads Re-visited

Although threads are mostly used for shared memory parallel processing, we can still have message-passing functionality using threads by additional routines at user level. One such library is presented in [2] where threads communicate by message passing primitives *write_fifo* and *read_fifo* which correspond to *send* and *receive* routines of a message passing system. The communication channels between processes are simulated by first-in-first-out (FIFO) data structures as shown below.

```
typedef struct {
sem_t send_sem;
sem_t receive_sem;
msgptr_t message_que[N_msgs];
} fifo_t;
```

Sending a message to a process is performed by first checking the availability of message space in the FIFO by executing a wait on the sending semaphore of the FIFO; then writing the message address to the FIFO of the receiving process and signalling its receiving semaphore so that it is activated. Using such a simulator provides a simple testbed to verify distributed algorithms and when the algorithm works correctly, performance tests can be performed using commonly used message-passing tools as described next.

4.4.3 Message Passing Interface

The Message Passing Interface Standard (MPI) provides a library of message passing primitives in C or Fortran programming languages for distributed processing over a number of computational nodes connected by a network [8, 10]. Its aim is to provide a portable, efficient and flexible standard of message passing. MPI code can be easily transferred to any parallel/distributed architecture that implements MPI. It can be used in parallel computers, clusters and heterogeneous networks. It is basically an application programming interface (API) to handle communication and synchronization among processes residing at various nodes of a paralell/distributed computing system. MPI provides point-to-point, blocking and non-blocking and collective communication modes. A previous commonly used tool for this purpose was Parallel Virtual Machine (PVM) [7]. The basic MPI instructions to run an MPI program are as follows.

- MPI_Init: Initializes the MPI library
- MPI_Comm_Size: Specifies the number of processes that will execute the code.
- MPI_Comm_Rank: The rank of a process which varies from 0 to $size - 1$ is obtained.
- MPI_Finalize: Cleans up the library and terminates the calling process.

Here is a simple program called *hello.c* in C showing how MPI works.

```
#include <stdio.h>
#include <mpi.h>

  main(int argc, char **argv)
  { MPI_Init(&argc, &argv);
    printf("Hello world");
    MPI_Finalize();
  }
```

The program is compiled and then run as follows:

```
mpicc -o hello hello.c
mpirun -np 8 hello
```

The program when run by the *mpirun* command takes the number of processes as an input which is 8 in this case. It then creates 7 other child processes by the MPI_Init command to have a total of 8 processes. Each of these processes then execute their program separately. We will have 8 "Hello world" outputs when this program is executed. Clearly, it would make more sense to have each child process execute on different data to perform parallel processing. In order to achieve this SPMD style of parallel programming, each process has a unique identifier which can then be mapped to the portion of data it should process. A process finds out its identifier by the MPI_Comm_rank command and based on this value, it can execute the same code on different data, achieving data partitioning. Two basic communication routines in MPI are the MPI_Send and MPI_Receive with the following syntax:

```
int MPI_Send(void *data, int n_data, MPI_Datatype type,
        int receiver, int tag, MPI_Comm comm);

int MPI_Recv(void *data, int n_data, MPI_Datatype type,
        int sender, int tag, MPI_Comm comm, MPI_Status *status);
```

where *data* is the memory location for data storage, *n_data* is the number of data items to be transferred with the given type, *receiver* and the *sender* are the identifiers of the receiving and the sending processes respectively, and *tag* is the type of message. The next example shows how to find the sum of an array in parallel using a number of processes. In this case, the root performs data partitioning by sending a different part of an array to each child process which calculate the partial sums and return it to the root which finds the total sum. The root first initializes the array and sends the size of the portion along with array data to each process. Message types for each communication direction are also defined.

```
#include <stdio.h>
#include <mpi.h>

#define n_data 1000
#define tag1 1   /* from root to workers */
#define tag2 2   /* from workers to root */
#define root 0
int array[n_data];

main(int argc, char **argv)
{  long int total_sum, partial_sum;
   MPI_Status status;
   int my_id, root, i, n_procs, n_portion;

   MPI_Init(&argc, &argv);

    /* find my id and number of processes */
    MPI_Comm_rank(MPI_COMM_WORLD, &my_id);
    MPI_Comm_size(MPI_COMM_WORLD, &n_procs);
    n_portion=n_data/n_procs;

    if(my_id == root) {   /* initialize array */
      for(i = 0; i < n_data; i++) {
```

```
        array[i]=i;
    }

    /* send a portion of the array to each worker */
    for(i= 1; i < _procs; i++) {
        MPI_Send( &array[i*n_portion], n_portion, MPI_INT,
            i, tag1, MPI_COMM_WORLD); }

    /* calculate the sum of my portion */
    for(i = 0; i < n_portion; i++)
        total_sum += array[i];

    /* collect the partial sums from workers */
    for(i= 1; i < n_procs; i++) {
        MPI_Recv( &partial_sum, 1, MPI_LONG, MPI_ANY_SOURCE,
            tag2, MPI_COMM_WORLD, &status);
        total_sum += partial_sum; }
    printf("The total sum is: %i", sum);
}

else { /* I am a worker, receive data from root */

    n_portion=n_data/n_procs;
    MPI_Recv( &array, n_portion, MPI_INT, root, tag2,
            MPI_COMM_WORLD, &status);

    /* Calculate the sum of my portion of the array */

    partial_sum = 0;
    for(i = 0; i < n_portion; i++)
        partial_sum += array[i];

    /* send my partial sum to the root */

    MPI_Send( &partial_sum, 1, MPI_LONG, root,
            tag2, MPI_COMM_WORLD);
}
MPI_Finalize();
}
```

MPI specification standard that can be implemented differently in various platforms. It is hardware and language independent providing ease in transporting code. It is by far the most commonly used parallel/distributed computing standard to develop applications. OpenMPI project provides open source MPI implementation which can also be used for shared memory parallel programming [9].

4.4.4 Distributed Processing in UNIX

BSD UNIX provides the socket-based communications over the Internet. A server is an endpoint of a communication between the two hosts over the Internet which performs some function as required by a *client* process in network communications using sockets. Communication over the Internet can be performed either as *connection-oriented* where a connection between a server and a client is established and maintained during data transfer. Connection oriented delivery of messages also guarantees the delivery of messages reliably by preserving the order of messages. Transmission Control Protocol (TCP) is a connection oriented protocol provided at layer 4 of the International Standards Organization (ISO) Open System Interconnect (OSI) 7-Layer model. In connectionless communication mode, there is no established or maintained connection and the delivery of messages is not guaranteed requiring further processing at higher levels of communication. Unreliable Datagram Protocol (UDP) is the standard connectionless protocol of the OSI model. Sockets may be used for connection-oriented communications in which case they are called *stream sockets* and when they are used for connectionless communication, they are called *datagram sockets*. The basic system calls provided by BSD UNIX are as follows:

- *socket*: This call creates a socket of required type, whether connection-oriented or connectionless, and the required domain.
- *listen*: A server specifies the maximum number of concurrent requests it can handle on a socket by this call.
- *connect*: A client initiates a connection to a server by this procedure.
- *accept*: A server accepts client requests by this call. It blocks the caller until a request is made. This call is used by a server for connection oriented communication.
- *read* and *write*: Data is read from and written to a socket using these calls.
- *close*: A connection is closed by this call on a socket.

Figure 4.10 displays a typical communication scenario of a connection-oriented server and a client. The server and the client are synchronized by the blocking system calls *connect* and *accept* after which a full connection is established using the TCP protocol. The server then reads the client request by the *read* and responds by the *write* calls. A server typically runs in an endless loop and in order to respond to further incoming requests, it spawns a child for each request so that it can return waiting for further client requests on the *accept* call. Threads may also be used for this purpose resulting in less management overheads than processes.

We can use socket-based communication over the Internet to perform distributed computations. Returning to the sum of an array example we have been elaborating, a server process can send portions of an array to a number of clients using stream or datagram sockets upon their requests and then can combine all of the partial results from the clients to find the final sum (see Exercise 4.9).

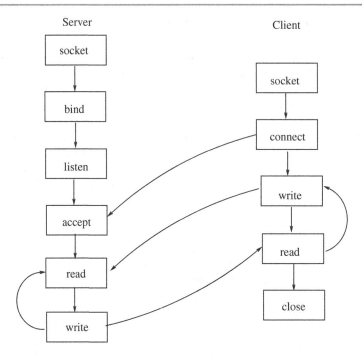

Fig. 4.10 A connection-oriented server and a client using socket-based communication

4.5 Chapter Notes

We have reviewed basic principles of parallel and distributed computing in this chapter. Parallel computing as we consider in the context of this book uses shared memory for data transfer among concurrent processes. Synchronization and communication are the two fundamental requirements for parallel and distributed processing. We have seen access to the shared memory should be controlled by mutual exclusion structures or semaphores, and processes synchronize on events by conditional variables to notify each other. Operating systems provide primitives for both mutual exclusion and conditional synchronization. Modern operating systems are based on processes which are instances of executing programs. Threads are lightweight processes with few data such as the program counter, stack and registers. Threads of a process share data within the process which must be protected as the shared data between the processes. POSIX threads are widely used for multi-threaded parallel processing applications.

There is no shared memory in distributed computing and message passing is the main method to provide synchronization and data transfer among processes. MPI which consist of various primitives for interprocess communication is a commonly used software to provide parallel and distributed processing. We can also employ basic network communication using UNIX BSD socket interface [13, 14]. This interface provides routines for creation, reading from and writing to data structures called *sockets*. However, MPI provides a neater interface to the application with the addition of routines for multicast and broadcast communication and is frequently employed for parallel and distributed processing. Threads are frequently used for shared memory parallel processing and MPI can be employed for both parallel and distributed applications.

Exercises

1. Provide a partitioning of the task dependency graph of Fig. 4.5 to three processors which considers evenly balancing the load among the processors and minimizing the IPC costs at the same time. Draw the Gant chart for this partitioning and work out the total IPC cost and the execution time of this partitioning.
2. Modify Algorithm 1.1 such that array A is first copied to an array B and all of the summations are done on array B to keep contents of A intact. Copying of A to B should also be done in parallel using n processors.
3. The *prefix sum* of an array A is an array B with the running sums of the elements of A such that $b_0 = a_0$; $b_1 = a_0 + a_1$; $b_2 = a_0 + a_1 + a_2$; ... For example if $A = \{2, 1, -4, 6, 3, 0, 5\}$ then $B = \{2, 3, -1, 5, 8, 8, 13\}$. Write the pseudocode of an algorithm using EREW PRAM model to find the prefix sum of an input array.
4. Provide an N-buffer version of the producer/consumer algorithm where we have N empty buffers initially and the producer process fills these buffers with input data as long as there are empty buffers. The producer reads filled buffer locations and consumes these data by printing them. Write the pseudocode of this algorithm with short comments.
5. A multi-threaded file server is to be designed which receives a message by the front end thread, and this thread invokes one of the *open*, *read* and *write* threads depending on the action required in the incoming message. The *read* thread reads the number of bytes from the file which is sent to the sender, the *write* thread writes the specified number of contained bytes in the message to the specified file. Write this program in C with POSIX threads with brief comments.
6. The PI number can be approximated by $\int_0^1 \frac{4}{(1+x^2)}$. Provide a multi-threaded C program using POSIX threads, with 10 threads each working with 20 slices to find PI by calculating the area under this curve.
7. Modify the program of Sect. 4.3.6 such that there are a total of 8 processes spawned by the parent to find the sum of the array in parallel.
8. Provide a distributed algorithm for 8 processes with process 0 as the root. The root process sends a request to learn the identifiers of the processes in the first round and then finds the maximum identifier among them and notifies each process of the maximum identifier in the second round.
9. Write an MPI program in C that sums elements of an integer array in a pipelined topology of four processes as depicted in Fig. 4.11. The root process $P0$ computes the sum of the first 16 elements and passes this sum along with the remaining 48 elements to $P1$ which sums the second 16 elements of the array and pass

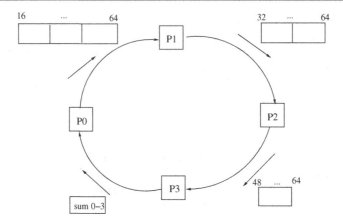

Fig. 4.11 MPI tasks for Exercise 9

the accumulated sum to *P2*. This process is repeated until message reaches *P0*
which receives the total sum and displays it in the final step. Note that the first
extra element of the message transferred can be used to store the partial sum
between processes.
10. Provide the pseudocode of a program using UNIX BSD sockets where we have
a server and *n* clients. Each client makes a connection-oriented call to the server,
obtains its portion of an integer array *A* and returns the partial sum to the server.
The server displays the total sum of array *A* when it receives all of the partial
sums from the clients.

References

1. Elsner U (1997) Graph partitioning, a survey. Technical Report, Technische Universitat Chemnitz
2. Erciyes K (2013) Distributed graph algorithms for computer networks. Computer communications and networks series. Chap. 18, Springer. ISBN 978-1-4471-5172-2
3. Erciyes K (2014) Complex networks: an algorithmic perspective. CRC Press, Taylor and Francis, pp 57–59. ISBN 978-1-4471-5172-2
4. Flynn MJ (1972) Some computer organizations and their effectiveness. IEEE Trans. Comput. C 21(9):948–960
5. Foster I (1995) Designing and building parallel programs: concepts and tools for parallel software engineering. Addison-Wesley
6. Gantt HL (1910) Work, wages and profit. The Engineering Magazine, New York
7. http://www.csm.ornl.gov/pvm/
8. http://www.mcs.anl.gov/research/projects/mpi/
9. http://www.open-mpi.org/
10. https://computing.llnl.gov/tutorials/mpi/
11. POSIX.1 FAQ. The open group. 5 Oct 2011

12. Quinn M (2003) Parallel programming in C with MPI and OpenMP. McGraw-Hill Science/Engineering/Math
13. Stevens WR (1998) UNIX network programming. In: Networking APIs: sockets and XTI, vol 1, 2nd edn. Prentice Hall
14. Stevens WR (1999) UNIX network programming. In: Interprocess communications, vol 2, 2nd edn. Prentice Hall
15. Sun Microsystems Inc. (1991) SunSoft introduces first shrink-wrapped distributed computing solution: Solaris. (Press release)
16. Tanenbaum A (2014) Modern operating systems 4th edn. Prentice-Hall

String Algorithms

5

5.1 Introduction

A *string S* consists of an ordered set of characters over a finite set of symbols called an *alphabet* Σ. The size of the alphabet, $|\Sigma|$, is the number of distinct characters in it and the size of a string, $|S|$, is the number of characters contained in it; also called the *length* of the string. For example, the DNA structure can be represented by a string over the alphabet of four nucleotides: Adenine (A), Cytosine (C), Guanine (G) and Thymine (T), hence $\Sigma = \{A, C, G, T\}$ for DNA; the RNA structure has $\Sigma = \{A, C, G, U\}$ replacing Thymine with Uracil (U), and a protein is a linear chain of amino acids over an alphabet of 20 amino acids $\Sigma = \{A, R, N, \ldots, V\}$.

String algorithms have numerous applications in molecular biology. Biologists often need to compare two DNA/RNA or protein sequences to find the similarity between them. This process is called *sequence alignment* and finding the relatedness of two or more biological sequences helps to deduce ancestral relationships among organisms as well as finding conserved segments which may have fundamental roles for the functioning of an organism. We may also need to search for a specific string pattern in a biological sequence as this pattern may indicate a location of a structure such as a gene in DNA. In some other cases, biologists need to find repeating substring patterns in DNA or protein sequences as these serve as signals in genome and also over representations of these may indicate complex diseases. Analysis of genome as substring rearrangements helps to find evolutionary relationships between organisms as well as to understand the mechanisms of diseases. In summary, DNA/RNA and protein functionalities are highly dependent on their sequence structures and we need efficient string manipulation algorithms to understand the underlying complex biological processes.

In this chapter, we start the analysis of strings by the string matching algorithms where we search the occurrences of a small string inside a larger string. We will see that there are a number of algorithms with varying time complexities. The next problem we will investigate is the *longest common subsequence* problem where our aim is to find the longest common subsequence in two strings. Longest increasing

© Springer International Publishing Switzerland 2015   81
K. Erciyes, *Distributed and Sequential Algorithms for Bioinformatics*,
Computational Biology 23, DOI 10.1007/978-3-319-24966-7_5

subsequence problem searches the longest increasing subsequence in a string and the longest common increasing subsequence is the search of such sequences in two strings. All of these problems are closely related to DNA sequence analysis. A suffix tree is a data structure that represents suffixes of a string. Suffix trees have many applications in bioinformatic sequence problems including exact string matching and other string algorithms such as the substring problems. We will investigate sequential and distributed construction of suffix trees and their applications in biological sequence problems. In all the topics described, we will first present sequential algorithms for these problems and then provide a review of contemporary parallel and distributed algorithms in these topics.

5.2   Exact String Matching

Given a string $S = s_1, \ldots, s_n$ with n elements, the string $S_P = s_i, \ldots, s_j \subset S$ is called a *substring* of S which starts at position i and ends at position j. A *prefix* $S_p = s_1, \ldots, s_i$ of a string S starts from the beginning of the string and ends at location i. A *suffix* $S = s_i, \ldots, s_n$ of a string S on the other hand, starts at location i and ends at the end of the string. As an example, let $S = madagaskar$, *mada* is a prefix and *skar* is a suffix of S.

In the string matching problem, we want to find the first but most commonly, all occurrences of a small string called *pattern* in a large string called *text* where symbols in the pattern and text are chosen from the alphabet Σ. Formally, we are given a text $T = t_1, \ldots, t_n$ of length n and a pattern $P = p_1, \ldots, p_m$ of length m and require to find the location i in T such that $t_i, \ldots, t_{i+m-1} = P$. For example, if $P = ccba$ and $T = abccbabaccbaab$ using the alphabet $\Sigma = \{a, b, c\}$, we need to find the locations $\{3, 9\}$ as $t_3, \ldots, t_6 = t_9, \ldots, t_{12} = ccba$. We will mostly use this notation, however, we will occasionally adopt the array notation as $S[i]$ to denote the ith element of a string S, for example, for convenience in representation in algorithms.

5.2.1   Sequential Algorithms

We will now review fundamental sequential algorithms for string matching starting with the naive algorithm which is a brute force algorithm that checks every possible combination. As this is a time consuming approach, various algorithms were proposed that have lower time complexities.

5.2.1.1   The Naive Algorithm

The brute force or the naive algorithm is the simplest method to search for a pattern in a text. The occurrences of the pattern can only start in the first $n - m + 1$ locations of the string. For the text $T = abccbabaccbaab$ and the pattern $P = ccba$ example, n is 14 and m is 4; we need to check the first 11 positions in T. Therefore, we start from the first locations in T and P and compare their elements. As long as we find a match, the indices are incremented and if the size of the pattern m is reached, a match is found and output as shown in Algorithm 5.1.

Algorithm 5.1 Naive algorithm

1: **Input**: text $T = t_1...t_n$, pattern $P = p_1...p_m$
2: **int** i, j, k
3: **for** $i \leftarrow 1$ to $(n - m + 1)$ **do**
4:     $k \leftarrow i; j \leftarrow 1$
5:     **while** $j \leq m$ **and** $t_k = p_j$ **do**
6:         $k \leftarrow k + 1$
7:     **end while**
8:     **if** $j > m$ **then**
9:         **output** match at position $i - j + 1$
10:    **end if**
11: **end for**

Let us consider the running of this algorithm until the first match, for the example above:

```
T =  a b c c b a b a c c b a a b
P =  c c b a
         c c b a
             c c b a   (a match found at location 3)
```

The time taken by this algorithm is $O(nm)$ since checking each occurrence of P is done in m comparisons and we need to check the first $n - m + 1$ locations of T. For very long strings such as the biological sequences, the total time taken will be significantly high. Baeza-Yates showed that the expected time taken in this algorithm for an alphabet of size c is [1]:

$$C = \frac{c}{c - 1}(1 - \frac{1}{c^m}(n - m + 1) + O(1) \tag{5.1}$$

5.2.1.2 Finite State Machine-Based Algorithm

A *finite automaton* or a *finite state machine* (FSM) (or simply a state machine) consists of a finite number of states, a finite number of transitions between these states, a start state and a number of states called accepting (final) states. An FSM can be in only one state at any time which is its current state and its next state is determined by this current state, and the input it receives. FSMs are used to model diverse problems such as communication protocols, language grammars, and sequence analysis. In a deterministic FSM, the next state is a single determined state and there may be a number of possible next states in the nondeterministic FSM. Formally, an FSM is a quintuple $(\Sigma, S, s_0, \delta, O)$ where:

- Σ is an input alphabet
- S is a finite nonempty set of states
- $s_0 \in S$ is the initial state
- $\delta : S \times \Sigma \rightarrow S$ is the state transition function

Let us illustrate these concepts by a string matching algorithm example. Given an input sequence T of length n over an alphabet $\Sigma = \{a, b, c\}$ of length m, we want to find all occurrences of the pattern $P = acab$ in S. The FSM diagram which consists of circles representing states with arcs showing the transitions between the states can be drawn as depicted in Fig. 5.1. Reaching the final state in the FSM means we have detected the pattern in the text.

An FSM table is an alternative way of showing the operation of the FSM. Typically, the rows of the table are indexed by the states and the inputs are the columns. In the example above, we have 5 states as 0 (the initial state), a, ac, aca, and $acab$. The possible inputs as read from the string T are a, b, and c. We can simply fill the entries of this table with the next state of the FSM based on the input and the current state as shown in Table 5.1.

For example, receiving input c at state aca will cause to return to initial state of 0. This way of representing an FSM provides a simple way of programming the algorithm. We will show how to implement this FSM using such a table in the programming language C. We first define the FSM table (*fsm_tab*) as a two-dimensional array of function pointers. Then each state changing action can be

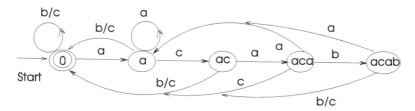

Fig. 5.1 FSM to find pattern *acab*

Table 5.1 Finite state machine table

States	Inputs		
	a	b	c
0	a	0	0
a	a	0	ac
ac	aca	0	0
aca	a	$acab$	0
$acab$	a	0	0

defined as a function address of which is placed in this table. The algorithm then simply directs the flow to the function specified by the current state and the current input in the table as shown below. The states are now labeled as 0, 1, 2, 3 and 4, and the inputs are 0, 1, and 2 corresponding to a, b, and c. The time to construct the FSM is $\Theta(m|\Sigma|)$ which is the size of the table, assuming there are no overlapping states. The time to search the input string is $\Theta(n)$, resulting in a total time complexity of $\Theta(m|\Sigma| + n)$ for this algorithm.

```
void (*fsm_tab[n_states][n_inputs])();    // fsm table declared
char S[n];         // input string S of length n
unsigned int curr_state=0; // initialize current state
act_00(){curr_state = 1; // action 00 changes current state to 1
         }            // when 'a' is received at state 0
         ...          // all functions are declared
/* initialize */
fsm_tab[0][0]=act_00;   // place function addresses in table
         ...            // do it for n_states * n_inputs times
main() { unsigned int count=0;
         while(count < n-m+1) {
         input=S[count];
         *fsm_tab[curr_state][input]; // go to the action
         if (curr_state == 4)  // state 4 is a match
             printf("pattern found at location %d", count);
         count++;
         }
      }
```

5.2.1.3 Knuth–Morris–Pratt Algorithm

Knuth and Morris, and around the same time Pratt discovered the first linear time string matching algorithm by analyzing the naive algorithm [2]. The general observation of the naive algorithm showed that there are unneeded searches in it and some of the comparisons may be skipped when previous comparisons are used. The main idea of the Knuth–Morris–Pratt (KMP) algorithm is to use a sliding window of size m and compare the characters in this window of text with the pattern. The window is slid from left to right until a mismatch is found. When this happens, the *prefix table Π*, which is an array of integers constructed prior to the execution of the KMP algorithm, is searched to determine where to move next in the text. This table basically shows how many characters to skip if there is a mismatch while seeking the pattern in the text, and this is the fundamental gain of this algorithm. In other words, when there is a mismatch, the window is shifted as many characters as specified in Π, saving redundant searches.

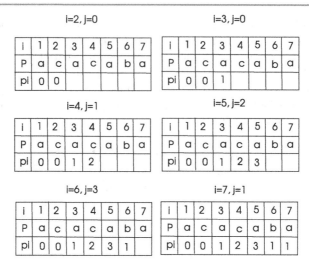

Fig. 5.2 The computation of the prefix array Π for the pattern $P = \{acacaba\}$

The formation of the prefix table Π is performed at initialization using the pattern P only. The idea here is to check incrementally whether any proper prefix of a pattern is also a proper suffix of it. If this is the case, the integer value is incremented. The entry $\Pi[i]$ is the largest integer smaller than i such that prefix $p_1, \ldots, p_{\pi_i}$ is also a suffix of $p_1, \ldots, p_i$. As an example, let us assume the pattern $P = \{acacaba\}$ and work out the values of Π as shown in Fig. 5.2.

The pseudocode of the prefix function to fill Π is shown in Algorithm 5.2.

Algorithm 5.2 Prefix Function

1: **procedure** *Prefix*($P[m]$)
2:     pattern $P[1..m]$ ▷ pattern of size m
3:     $\Pi[1] \leftarrow 0$ ▷ output array holding prefix values
4:     $j \leftarrow 0$
5:     **for** $i \leftarrow 2$ to m **do**
6:         **while** $j > 0$ **and** $P[j + 1] \neq P[i]$ **do**
7:             $j \leftarrow \Pi[j]$
8:         **end while**
9:         **if** $P[j + 1] = P[i]$ **then**
10:            $j \leftarrow j + 1$
11:        **end if**
12:        $\Pi[i] \leftarrow j$
13:    **end for**
14:    **return** Π
15: **end procedure**

Once this array is formed, we can use it when there is a mismatch to shift the indices in the text T. The operation of the KMP algorithm is described in pseudocode in Algorithm 5.3.

Algorithm 5.3 *KMP_Alg*

1: **Input** : text $T[1..n]$
2:     pattern $P[1..m]$
3: $\Pi \leftarrow Prefix(P)$ ▷ compute prefix function
4: $j \leftarrow 0$
5: **for** $i \leftarrow 1$ to n **do**
6:     **while** $j > 0$ **and** $P[j+1] \neq T[i]$ **do**
7:         $j \leftarrow \Pi[j]$
8:     **end while**
9:     **if** $P[j+1] = T[i]$ **then**
10:         $j \leftarrow j+1$
11:     **end if**
12:     **if** $j = m$ **then**
13:         **output** match at $i - m$
14:         $j \leftarrow \Pi[j]$
15:     **end if**
16: **end for**

Let us consider an example where $T = \{ccabababacbac\}$ and the pattern $P = \{acacaba\}$ described above so we can use the prefix table Π computed. As long as the characters in T match the characters in P, we increment the pointers and whenever there is a mismatch, the working window is shifted as many times as in the corresponding entry of Π shown below.

```
Prefix : 0 0 1 2 3 1 1
S =  c c a c a c a c a b a c a b
P =       a c a c a b a (first match found at location 3)
         a c a c a b a (second match)
         a c a c a b a (third match)
         a c a c a b a (fourth match)
         a c a c a b a (fifth match)
         a c a c a b a (mismatch at 6, shift 1)
           a c a c a b a  (mismatches at 1,2 and 3, shift 1 at 3)
             a c a c a b a (first match found at location 3)
             a c a c a b a (second match)
             a c a c a b a (third match)
             a c a c a b a (fourth match)
             a c a c a b a (fifth match)
             a c a c a b a (sixth match)
             a c a c a b a  (FULL MATCH)
```

This algorithm takes $O(m)$ time to compute the prefix values and $O(n)$ to compare the pattern to the text, resulting in a total of $O(m + n)$ time. The running time of this algorithm is optimal and can process large texts as it never needs to move backward. However, it is sensitive to the size of the alphabet $|\Sigma|$ as there will be more mismatches when this increases.

5.2.1.4 Boyer–Moore Algorithm

The Boyer–Moore (BM) algorithm is an efficient pattern matching algorithm that has usually a sub-linear computation time [3]. We have an input string T and a pattern P as before. The main idea of this algorithm is to search a pattern without looking at all characters of the text. Also, the matching process of the pattern with the text is performed from right to left since more information is gained this way. The processing of the whole pattern is still from left to right though, pattern $p_1, \ldots, p_m$ is compared with $t_{j+1}, \ldots, t_{j+m}$ where $1 \leq j \leq n - m$ is the current location in the text T. It implements two clever rules called *bad character rule* and *good suffix rule* to compute the number of right shifts whenever there is a mismatch, as described below.

Bad character rule

The pattern P is shifted right until the rightmost occurrence of the text character $T[j + i]$ in the pattern. Let us assume there is a mismatch at position $P[i] = a$ and $T[i+j] = b$. This implies $P[i+1..m] = T[i+j+1..j+m] = w$ since we are doing a right scan. If the pattern P does not contain b, then we can shift P until $T[i + j]$. Otherwise, we need to shift P until the rightmost incidence of b. We therefore need to do preprocessing for the bad suffix rule. For each character $x \in \Sigma$:

$$R(x) = \max \begin{cases} 0, \text{ if } x \notin P \\ \max\{i < m|\ P[i] = x\} \end{cases}$$

Informally, $R(x)$ is the position of the rightmost occurrence of x in P, and it is zero if x does not exist in P. The array R can be computed in $O(m)$ time for a pattern P of length m. Bad character rule is as follows. When a mismatch is found at position i with $T[k]$ being the mismatch character in T, shift the pattern right by $\max(1, i - R[T[k]]$. For example, given the pattern $P = $ a c b c b a, $R[a] = 6$, $R[b] = 5$, and $R[c] = 4$, let us see how this rule works for the text $T = $ a b c c b a a b b a c

```
T= a b c c b a a b b a c
P= a c b c b a    mismatch at position 3
                  i=3, k=3, T[3]=c, i-R[c]=-1, shift 1
```

Good suffix rule

The idea of the good suffix rule is to shift the pattern to the position of the next occurrence of the suffix $P[i+1], \ldots, P[m]$ in the pattern. It basically means we have matched a suffix of P and we know the following characters in T. The number of steps to shift when there is a mismatch is calculated according to both rules and the shift that has the greatest value is implemented. We need to preprocess the pattern to form arrays for both cases in order to implement these rules.

The Boyer–Moore algorithm shown in Algorithm 5.4 forms one array for the bad character rule and two arrays for the good suffix rule before the execution of the algorithm and uses them to shift the pattern when there is a mismatch. Algorithm 5.4 displays the simplified pseudocode of the algorithm where preprocessing forms the arrays.

Algorithm 5.4 *BM_Alg*

1: **Input** : text $T[1..n]$ ▷ text of size n
2:      pattern $P[1..m]$ ▷ pattern of size m
3: **Output** : Index of first substring of T matching P
4: preprocess to obtain arrays for bad character and good suffix rules
5: $i \leftarrow m - 1$
6: $j \leftarrow m - 1$
7: **repeat**
8:      **if** $P[j] = T[i]$ **then**
9:          **if** $j = 0$ **then**
10:             **return** i
11:          **else**
12:             $i \leftarrow i - 1$
13:             $j \leftarrow j - 1$
14:          **end if**
15:      **else**
16:          shift P by the maximum of the shifts obtained
17:          from the bad character rule and the good suffix rule
18:      **end if**
19: **until** $i > n - 1$

Its worst-case performance is $O(n+rm)$, where r is the number of occurrences of P in T. The average performance of BM algorithm is $O(n/m)$ meaning its performance gets better for longer patterns. A parallel implementation of BM algorithm is reported in [4].

5.2.2 Distributed String Matching

A general approach to provide distributed string matching would be to partition the string S into approximately equal segments and provide each processor with one segment. Each process p_i residing on a processor then would implement the string matching algorithm, whether KMP or BM algorithm, in its own segment. The partitioning of data is handled by a special process called the *supervisor* and the processes in the distributed system are called *workers*. This model is also called *master–slave* method of parallel processing and we will frequently adopt this approach for distributed processing to solve bioinformatics problems. We will generally assume there are k processors and k processes, each of which runs on a processor. Another distinction is whether the supervisor is involved in the actual algorithm to find results, for example, whether it also searches for a pattern as workers. We will show examples of both approaches, however, if there is high volume of data to be communicated by the supervisor or if the amount of work done by the workers can vary and cannot be determined beforehand, the supervisor may be assigned to perform management only.

Let us attempt to sketch a distributed algorithm to perform KMP algorithm in parallel over k processors. We have k processes $pr_0, \ldots, pr_{k-1}$ and pr_0 is the supervisor (master or the root process). As shown in Algorithm 5.5, pr_0 first works out the prefixes using the prefix function. As this initialization part of the algorithm takes $O(m)$ time and since the pattern is significantly shorter than the text T, this task can be handled by pr_0. It then partitions the input text T into approximately $k - 1$ partitions $T_1, \ldots, T_{k-1}$ and sends each worker a unique segment along with the prefix array values. Each worker then implements KMP algorithm in its segment and returns the results to the supervisor pr_0 which combines them and outputs the final result.

There is one issue to be handled though, the characters in the bordering regions of segments should be carefully considered. For example, let us assume we have the following text which is partitioned among three processes p_1, p_2, and p_3 and we search for the pattern *cba* in this text:

```
b c c b c | b a a c b | b c a b b
   p1              p2            p3
```

Neither the process p_2 nor p_1 will be aware of the match because they do not have the whole of the pattern *cba*. A possible remedy for this situation would be to provide the preceding process, which is p_1 in this case, with the first $m - 1$ characters of its proceeding process. Searching for matches in these bordering regions would then be the responsibility of the preceding process with a lower identifier. Assuming the communication costs are negligible, this approach which is sometimes called *embarrassingly parallel* will provide almost linear speedup.

Algorithm 5.5 *Dist_KMP*

1: **Input** : Text $T[1, n]$, pattern $P[1, m]$
2: *all_procs* $= \{pr_1, ..., pr_{k-1}\}$ ▷ process set
3: **Output** : *matched*   ▷ set of matched P values in T
4: *received* $\leftarrow \emptyset$
5: **if** $pr_i = pr_0$ **then**   ▷ if I am the supervisor process
6:     $\Pi \leftarrow Prefix(T)$ ▷ compute prefix values
7:     **for** $i = 1$ to $k - 1$ **do**
8:         **send** $(T[((i-1)n/k) + 1] \text{ - } T[in/k]) \cup \Pi$ to pr_i ▷ send a segment of T
9:     **end for**
10:    **while** *received* $\neq$ *all_procs* **do**   ▷ get matched minimum values from processes
11:        **receive** *match*($vals_i$) from pr_i
12:        *matched* $\leftarrow$ *matched* $\cup$ *vals*
13:        *received* $\leftarrow$ *received* $\cup$ pr_i
14:    **end while**
15:    **output** *matched*
16: **else**   ▷ I am a worker process p_i
17:    **receive** my segment M_i from pr_0
18:    *vals* $\leftarrow KMP(M_i)$ ▷ run KMP on my segment
19:    **send** *match*($vals_i$) to pr_0 ▷ send matches to supervisor
20: **end if**

5.3   Approximate String Matching

Given a text $T = t_1, \ldots, t_n$ of length n and a pattern $P = p_1, \ldots, p_m$, our aim in approximate string matching is to search for approximate matches of P in T. That is, a prespecified upperbound k on mismatches is allowed. When $k = 1$, this problem is commonly referred as the *1-mismatch problem*. The required output is the list of positions of approximate matches as in the exact matching problem. Let us illustrate this concept by an example; we have a text $T = acbbcabbacbc$ and the pattern is $P = cbb$ with the maximum allowed mismatch number $k = 1$. Intuitively, we can implement the naive algorithm that we used in exact matching by scanning the text T from left to right for the pattern P. This time, however, we allow for a maximum of one mismatch. The example below shows the operation of this algorithm for these two example sets.

```
T =  a c a b c b b b c b c
P =  c b b
         c b b  ( approximate match at position 2 )
           c b b
             c b b
               c b b ( exact match at position 5)
                 c b b
                   c b b
                     c b b
                       c b b ( approximate match at 9)
```

The pseudocode for this algorithm is shown in Algorithm 5.6. As in the naive exact algorithm, it has a time complexity of $O(nm)$.

Algorithm 5.6 Naive Approximate Matching

1: **Input**: text $T = t_1, ..., t_n$, pattern $P = p_1, ..., p_m$, upperbound k
2: **Output**: positions of approximate matches
3: **int** i,j
4: **for** $i \leftarrow 1$ to $(n - m + 1)$ **do**
5:      $d \leftarrow 0$
6:      **for** $j \leftarrow 1$ to m **do**
7:          **if** $T[i + j - 1] \neq P[j]$ **then**
8:              $d \leftarrow d + 1$
9:          **end if**
10:     **end for**
11:     **if** $d \leq k$ **then**
12:         **output** i
13:     **end if**
14: **end for**

5.4 Longest Subsequence Problems

The discovery of longest subsequences of a number of strings has many practical applications in bioinformatics such as finding the conserved regions of similarity and associating functionality. There are few variants of this problem as described next.

5.4.1 Longest Common Subsequence

Let us consider a string S which consists of symbols from a finite alphabet Σ. A *substring* $P \in S$ is a fragment of S which includes some contiguous symbols of S. A *subsequence* $Q \in S$ is found by deleting zero or more symbols from S. In other words, the symbols in Q need not be contiguous in S. An example of a substring A and subsequences B and C of a string S with alphabet $\Sigma = \{a, b, c\}$ is as follows:

```
S =  a b b a b c b a a b c c a
A =      b a b c
B =      a - c b a - - c
C =      b a b - b
```

The longest common subsequence problem can be stated as follows :

Definition 5.1 The Longest Common Subsequence (LCS) of two strings $A = (a_1, \ldots, a_n)$ and $B = (b_1, \ldots, b_m)$ is the longest sequence that is a subsequence of both A and B. LCS can be also be searched among a set of k strings $S = S_1, \ldots, S_k$.

The following example shows the LCS P of two strings A and B. In general, we are more interested in finding the length of LCS rather than its contents.

```
A =  a b b a b c b a b a
B =  b b a c c c a a c
P =  b b a c a a
```

Comparing two biological sequences is frequently needed to infer their relationship and understand their functionalities. The LCS between two such sequences is one step toward this goal and we will see how this problem can be solved for the more general case in the next chapter. The brute force algorithm to find LCS of strings A of size m and B of size n will consider all 2^m substrings of A and will search all of these for each element of B resulting in $n2^m$ computation time. Therefore, there is a need for algorithms with better time complexities.

5.4.1.1 Dynamic Programming Algorithm

The dynamic programming solution to this problem involves breaking it into smaller problems and using the solutions of the smaller problems to build larger solutions, as in the general approach of dynamic programming. In the search of LCS between two strings $A = a_1, \ldots, a_n$ and $B = b_1, \ldots, b_m$, we observe that for two prefixes $a_1, \ldots, a_i$ and $b_1, \ldots, b_j$ of these two strings; their LCS has a length of one more than the length of LCS of $a_1, \ldots, a_{i-1}$ and $b_1, \ldots, b_{j-1}$. This observation shows LCS problem can be solved using dynamic programming. Let us aim at finding the length of LCS between two strings A and B and consider two cases of this problem:

1. **Case 1**: Considering any two elements $A[i]$ and $B[j]$, the first case is they may be equal. In this case of $A[i] = B[j]$, the length of current LCS should be incremented. This relation can be stated recursively as follows:

$$LCS[i, j] = 1 + LCS[i - 1, j - 1] \qquad (5.2)$$

2. **Case 2**: When $A[i] \neq B[j]$, either $A[i]$ or $B[j]$ has to be discarded, therefore:

$$LCS[i, j] = \max(LCS[i - 1, j], LCS[i, j - 1]) \qquad (5.3)$$

Using these two cases, we can form the dynamic programming solution to this problem using two nested loops as shown in Algorithm 5.7. The array C holds the partial and final solutions to this problem.

Algorithm 5.7 LCS Dynamic Algorithm

```
1: Input : A = a₁, ..., aₘ, B = b₁, ..., bₙ
2: Output C[n, m]
3: int i,j
4: for i ← 1 to n do
5:     for j ← 1 to m do
6:         if aᵢ = bⱼ then
7:             C[i, j] ← C[i − 1, j − 1] + 1
8:         else
9:             C[i, j] ← max(C[i, j − 1], C[i − 1, j])
10:        end if
11:    end for
12: end for
```

For two strings $A = \{abacc\}$ and $B = \{babbc\}$, let us form the output matrix C as below. The time to fill this matrix is $\Theta(nm)$ with a total space $\Theta(nm)$, and the final length of LCS is in the last element of C, that is, $C[n, m] = |\text{LCS}(A, B)|$. In order to find the actual LCS sequence, we start moving up from the right corner of the matrix which holds the element $C[n, m]$ and backtrack the path we have followed while filling the matrix. The sequence elements discovered in this way are shown by arrows in the matrix and the final LCSes are $\{bacc\}$ and $\{abcc\}$ as there are two paths.

	B	b	a	b	c	c
A	0	0	0	0	0	0
a	0	0	1	1	1	1
b	0	1	1	2	2	2
a	0	1	2	← 2	2	2
c	0	1	2	2	3	3
c	0	1	2	2	3	4

5.4.1.2 Parallel and Distributed LCS Search

Several ways of parallelizing LCS using dynamic programming is described by Ukiyama and Imai [5]. The use of parallelization in diagonal direction of the matrix is implemented efficiently on the CM5 parallel computer which can provide parallel operations up to 16000 nodes. A bit-parallel algorithm for the LCS problem was proposed in [6] with $O(nm/w)$ time and $O(m/w)$ space complexities where w is the size of the machine word and n and m are the lengths of the two strings.

Dominant points are the minimal points of search and using them narrows the search space efficiently. Studies using dominant points to solve multiple LCS problem in parallel for more than two input strings have been reported in [7,8]. A parallel algorithm to solve LCS problem on graphics processing units (GPUs) was proposed by Yang et al. [9].

5.4.2 Longest Increasing Subsequence

As another example, let us consider the *longest increasing subsequence* (LIS) problem. Given a sequence $S = \{a_1 a_2, \ldots, a_n\}$ of numbers, L is the longest subsequence of S where $\forall a_i, a_j \in L, a_i \le a_j$ if $i \le j$. For example, given $S = \{7, 2, 4, 1, 6, 8, 5, 9\}$, $L = \{2, 4, 6, 8, 9\}$. A dynamic programming solution to the problem would again start from the subproblems and store the results to be used in future. Algorithm 5.8 shows how to find LIS using dynamic programming in $O(n^2)$ time [10]. The procedure *Max_Val* is used to find the maximum of numbers in an array in order to find the length of the longest path stored in the array L.

Algorithm 5.8 *LIS_dyn*

1: **Input** : $S[n]$ ▷ array of numbers
2: **Output** : length ▷ length of LIS
3: **for** $i \leftarrow 1$ to n **do**
4:     $L[i] \leftarrow 1$
5:     **for** $j \leftarrow 1$ to i **do**
6:         **if** $S[i] \ge S[j] \land L[i] \le L[j]$ **then**
7:             $L[i] \leftarrow L[j] + 1$
8:         **end if**
9:     **end for**
10: **end for**
11: **return** $Max\_Val(L)$

The longest common increasing subsequence (LCIS) is more general than LIS in which we search for an LIS of two or more sequences. Formally, given two strings $A = (a_1, \ldots, a_m)$ and $B = (b_1, \ldots, b_n)$ over an ordered alphabet $\Sigma = \{x_1 < x_2 < x_3\}$, the LCIS of A and B is the common subsequence of A and B of the longest length. An example LCIS L of two strings A and B is shown below:

```
A =  3 6 1 4 2 5 7 8 2
B =  2 3 8 4 1 9 5 1 8
P =  3 4 5 8
```

A variation of this problem is to consider equality such that each element x_i of the LCIS L is less or equal to the next element. The sequence formed in this manner where each $x_i \in L \le x_j \in L, \forall i < j$ is called the *longest common weakly increasing subsequence* (LCWIS).

5.5 Suffix Trees

A suffix tree introduced by Weiner [11] is a data structure that represents suffixes of a string. Suffix trees have numerous of applications in bioinformatics problems including exact string matching and pattern matching. Formally, a suffix tree can be defined as follows:

Definition 5.2 [12] A suffix tree T to represent an n character string S is a rooted directed tree with n leaves numbered 1 to n. Any intermediate node of this tree has at least two children and each edge has a label with a nonempty substring of S. For any leaf i of T, the concatenation of characters from the root to the leaf i shows a suffix $S[i, \ldots, n]$ of S. Edges that connect children of a node in T start with different characters.

This representation of a suffix tree, however, does not guarantee a valid suffix tree for any string S. If the prefix of a suffix of S is the same as its suffix, the path for the suffix will not be represented by a leaf. It is therefore general practice to place a special symbol such as $ at the end of S and hence every suffix of S ends with $ as shown in Fig. 5.3 which represents a suffix tree for the string $S = abbcbabc$.

A *generalized suffix tree* contains suffix trees for two or more strings. In this case, each leaf number of such a tree is represented by two integers that identify the string and the starting position of the suffix represented by the leaf in that string as shown in Fig. 5.4 where two strings $S_1 = acbab$ and $S_2 = bacb$ over an alphabet

Fig. 5.3 Suffix tree example for the string $S = abbcbabc$. The leaves are numbered by the indexes of the suffixes in S

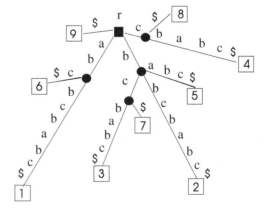

Fig. 5.4 A generalized
suffix tree example of two
strings, $S_1 = acbab$ and
$S_2 = bacb$. The leaves are
labeled by the string number
and the position and the
terminating character is "$"
for S_1 and "#" for S_2

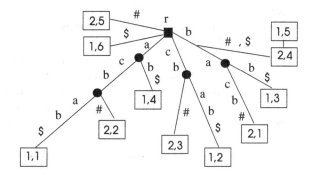

$\Sigma = \{a, b, c\}$ are represented by a generalized suffix tree. The terminating symbols
for the two distinct trees need to be different (Fig. 5.4).

For very long strings, the storage requirements for the suffix tree need to be
considered. For a string of length n over an alphabet Σ, the suffix tree T has n leaves
to represent n suffixes and it has a maximum of $2n$ edges. Each character of the
alphabet Σ can be represented by $\log |\Sigma|$ bits, total number of bits required to store
S would then be $O(n^2 \log \Sigma)$ bits. Rather than storing each edge e as a concatenation
of characters, we may represent an edge by two integers i and j where $S[i, \ldots, j]$ is
equal to the label of e. This way, the total storage required is reduced to $O(n \log n)$
since each integer can be represented by $\log n$ bits [13].

In this section, we first introduce suffix tree data structure and provide algorithms
for suffix tree construction. We then review parallel and distributed algorithms for
suffix tree construction. At the end of the section, suffix arrays which provide a more
compact way of storage are described.

5.5.1 Construction of Suffix Trees

There are various algorithms to construct a suffix tree T of a string S. Weiner was first
to provide a linear time algorithm to construct a suffix tree [11] which was modified
by McGreight to yield a space complexity of $O(n^2)$ [14]. Ukkonen proposed an
online algorithm which provided more space saving over McCreight's algorithm
[15] and Farach provided an algorithm to construct suffix trees using unbounded
alphabet sizes [16]. We will describe sequential and parallel algorithms to construct
suffix trees starting with a naive algorithm and then the algorithm due to Ukkonen.
We will then investigate methods for parallel construction of suffix trees.

5.5.1.1 The Naive Algorithm

The naive method to build a suffix tree of a string $S[n]$ would be to first form the
branch $S[1, n]$ in the tree and then insert $S[i, n]$ in sequence for $2 \leq i \leq n$, from
the longest to shortest one. Initially, the suffix $S[1, \ldots, n\$]$ is inserted in tree and

T_1 consists of this suffix only. At each step $i + 1$, the tree T_i is traversed starting at root r and a prefix of $S[i + 1, n]$ that matches a path of T_i starting from the root node is searched. If such a prefix is not found, a new leaf numbered $i + 1$ with edge label $S[i + 1, n\$]$ is formed as an edge coming out of the root. If such a prefix of $S[i + 1, n\$]$ exists as a prefix of a branch of T_i, we check symbols of the path in that branch until a mismatch occurs. If this mismatch is in the midst of an edge (u, v), we split (u, v) to form a new branch, otherwise if the mismatch occurs after a vertex w, the new branch is joined to w as shown in Algorithm 5.9.

Algorithm 5.9 *Naive_ST*

1: **Input** : String $S[1..n]$
2: **Output** : Suffix tree T_S representing S
3: **insert** $S[1, n\$]$ in T
4: **for** $i = 2$ to n **do**
5:    **traverse** T_i
6:    **if** there is a longest prefix $S[i..k]$ that matches a path in T_i **then**
7:       let $S[k + 1, n]$ be the non-matching substring
8:       (u, v) is the edge of the path where $S[k]$ is found
9:       **if** k is just before a vertex w in T_i **then**
10:          **form** a new edge joining w and leaf i
11:          **label** edge (w, i) with $S[i..k\$]$
12:       **else**
13:          **form** a new node w and a leaf numbered i
14:          **divide** edge (u, v) into edges (u, w) and (w, v) with w after k
15:          **label** edge (w, i) with $S[i..k\$]$
16:       **end if**
17:    **else**
18:       **form** a new leaf with edge label $S[i..n\$]$ and join it to the root
19:    **end if**
20: **end for**

The execution of Algorithm 5.9 is shown in Fig. 5.5 for the string $S = babcab$. We start with the longest suffix $S[1..n]\$$ in (a) and connect it to the root r to form T_1. The second suffix starting at location 2 does not have a matching path and therefore a new branch to T_1 is added to form T_2 as shown in (b). The suffix S[3..n] has the first character in common with the first suffix and therefore, we can split the first suffix branch as shown in (c). Continuing in this manner, the whole suffix tree is constructed.

The naive algorithm has a time complexity of $O(n^2)$ for a string S of length n and requires $O(n)$ space. The time needed for the ith suffix in ith iteration of this algorithm is $O(n - i + 1)$. Therefore, total time is:

$$\sum_{i=1}^{n} = O(n - i + 1) = \sum_{i=1}^{n} O(i) = O(n^2) \tag{5.4}$$

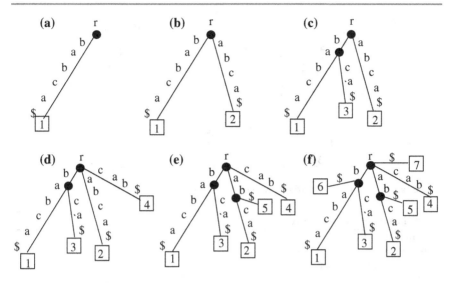

Fig. 5.5 Construction of a suffix tree for the string $S = \{babca\}$ by the naive algorithm

5.5.1.2   Ukkonenn's Algorithm

Ukkonen provided a suffix construction method that works online and has linear time complexity [15,17]. The online property means that the algorithm can process a string from left to right and provides a suffix tree for the current partial input without knowledge of the rest of the string. It is based on a simple approach which is improved using clever approaches. This method uses the concept of *implicit suffix trees* defined as follows:

Definition 5.3 [12] An implicit suffix tree T_{im} of a string S is obtained from the suffix tree T of S by deleting all terminal $ symbols from edges of T, then deleting all unlabeled edges and any nodes that have one child in T.

Ukkonen's algorithm basically constructs implicit suffix trees and extends them and it finally converts the implicit trees to explicit trees. This process is illustrated in Fig. 5.6 in which a suffix tree is converted to an implicit suffix tree. We will now briefly review Ukkonen's algorithm at high level as described in [12]. The algorithm has n phases and a tree I_{i+1} is constructed from the tree I_i of the previous phase. In the phase $i + 1$, there are $i + 1$ extensions and at extension j of phase $i + 1$, the suffix $S[1, j]$ is placed in the tree as shown in Algorithm 5.10.

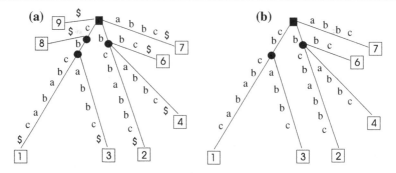

Fig. 5.6 a A suffix tree for the string $S = cbcbabcb$, **b** The implicit suffix tree of S

Algorithm 5.10 *Ukkonen_Alg*

1: **Input** : String $S[1, n]$
2: **Output** : Suffix tree T_S representing S
3: Construct the initial tree i_1
4: {**phase** $(i + 1)$}
5: **for** $i = 1$ to $n - 1$ **do**
6:     **for** $j = 1$ to $i + 1$ **do**
7:         **find** the path $S[j, i]$
8:         **if** there is not a path $S[j, i + 1]$ in i_i **then**
9:             **if** $S[i, j]$ is a leaf **then**   ▷ case 1
10:                 **append** $S[i + 1]$ to the leaf
11:             **else if** $S[i, j]$ is an internal node v **then**   ▷ case 2
12:                 **append** an edge labeled $S[i + 1]$ and a leaf labeled j to v
13:             **else if** $S[i, j]$ is inside an edge **then**   ▷ case 3
14:                 **add** a new node v to the end
15:                 **append** an edge labeled $S[i + 1]$ and a leaf labeled j to v
16:             **end if**
17:         **end if**
18:     **end for**
19: **end for**

In the first case, we find a prefix of the suffix and we append $S[i + 1]$ to this leaf. Otherwise, if $S[i, j]$ is an internal node v, we form a new edge labeled $S[i + 1]$ and attach a leaf labeled j to this edge. In case 3, $S[i, j]$ is not a leaf nor an internal node but it is inside an edge label. In this case, we form a new node v at the end of the $S[i, j]$ and insert a new edge labeled $S[i + 1]$ and a leaf j from v. Although in this naive form, the complexity of this algorithm is $O(n^3)$ which is unacceptable for long biological sequences, few tricks are used to reduce the time complexity to $O(n)$ which are the use of suffix links and edge-label compression.

5.5.1.3 Distributed Suffix Tree Construction

An early attempt to construct a suffix tree using the CRCW parallel RAM model was reported in 1976 in [18]. The algorithm consisted of two parts: in the first part, an approximate tree called the skeleton is built, this structure is then refined in the second part. In general, construction of suffix trees may be classified as *in-core*, *semi-disk-based*, and *out-of-core* algorithms [19]. The sequential algorithms of Weiner [11], McCreight [14], and Ukkonen [15] are basically in-core algorithms that provide linear time of suffix tree construction when the sizes of the string and associated the suffix tree are small enough to be accommodated in the main memory. For large strings, the main memory becomes a bottleneck. For example, the human genome has about 3G nucleotides and the suffix tree for human genome requires about 67 GB which is difficult to provide in many systems [19]. The problem is more challenging when constructing generalized suffix strings as the size now becomes the sum of the sizes of individual strings that make up the generalized tree.

Semi-disk-based methods divide the suffix tree into a number of subtrees and store these on the disk. In [20], the string is divided into substrings and a suffix tree for each substring is constructed and these are then merged to obtain the final suffix tree. The out-of-core algorithms are designed basically to process large data which cannot be stored in the main memory as a whole. B^2ST is one such algorithm which first builds a suffix array of a string, which is a compact way of representing a suffix tree as we will see, and then converts the suffix array to a suffix tree which can be performed in linear time [21]. This method provides a good locality of reference in practice. It partitions the input string S into k segments where $k = 2n/M$ with n being the size of the input string and M is the size of the main memory, and constructs a suffix array for each pair of the partitions. In the second phase of this algorithm, suffix arrays are combined and a suffix tree for the whole string from the combined suffix array is generated. The merge operation is similar to the merge phase of multi-way merge sort. This algorithm achieves the construction of the suffix tree in $O(n^2/M)$ time where M is the size of the main memory, and uses $O(n^2/M)$ temporary disk space. The authors reported this algorithm provides significant time gains when compared with the methods of TDD and TRELLIS. Although B^2ST is proposed as a sequential algorithm, it can easily be parallelized. We propose the sketch of the parallel algorithm based on this method as follows. We have k distributed processes, one of which is the supervisor and the rest are workers. The supervisor partitions the input string S into k segments and forms $k(k-1)/2$ segment pairs. It then distributes these pairs evenly to $k-1$ workers which form suffix array for these pairs in parallel. The partial k suffix arrays along with $k(k-1)/2$ order arrays can then be gathered by the supervisor which can form the final suffix tree.

Out-of-core or *disk-based* methods as sometimes called, typically partition the string into a number of substrings and the associated subtrees are constructed individually in the main memory. These subtrees are finally combined to form the final suffix tree. These steps are suitable for shared memory or distributed parallel processing in which each process can handle the building of a subtree. However, as noted in [22], parallel disk-based suffix tree construction algorithms are very scarce.

Earlier parallel suffix tree construction methods such as in [18] were mainly of theoretical in nature. Two relatively more recently reported parallel disk-based (out-of-core) suffix tree construction algorithms are the Wavefront method [23] and Elastic range (ERa) [19]. The wavefront method consists of a number of main steps to construct a suffix tree in parallel. In the initial network string caching step, a cache that consists of all main memories in the system is built. The following task generation step involves finding a set of variable length prefixes P of the input string S. The location of each prefix $p \in P$ is discovered in the third step called prefix location discovery. The subtrees are then constructed for each prefix p and finally are combined to yield the final suffix tree. The complexity of this algorithm is reported as $O(n^2/k)$ where k is the number of processors. The wavefront method has been experimented in distributed memory IBM BlueGene/L supercomputer with 1024 PowerPC 440 processors using MPI was for message passing and it was found to be scalable.

ERa takes a slightly different approach by employing vertical and horizontal partitions. The suffix tree is divided into a number of subtrees using variable length prefixes in the vertical partitioning. Moreover, the subtrees are grouped to share the input/output costs. Each subtree is further horizontally divided considering the currently processed number of paths in the subtree. The subtrees are then combined to yield the final suffix tree. The time complexity of this method is shown to be $O(n^2)$ where n is the length of the input string. The serial and shared memory–shared disk and shared-nothing parallel versions of this algorithm were presented. In the distributed construction, a 16-node Linux cluster was used as the testbed and significant speedups were reported when compared with the wavefront method [19]. A recent parallel suffix tree construction algorithm called parallel continuous flow is presented in [24] where a suffix tree is built by using the suffix array and longest common prefix structure.

5.5.2 Applications of Suffix Trees

There are numerous applications of suffix trees for string processing and biological sequence analysis. Two such applications, the exact string matching and LCS finding are relevant here and we will describe them next. We will see other suffix tree applications in the next chapters when they are used for sequence alignment and to find sequence repeats in DNA and proteins.

5.5.2.1 Exact String Matching

In exact string or pattern matching, given a text $S = s_1, \ldots, s_n$ and a pattern $P = p_1, \ldots, p_m$, we need to find all occurrences of P in S. When the suffix tree T of S is constructed, we observe that each occurrence of the pattern P in T should be a prefix of a suffix in T since any substring of a string S is a prefix of some suffix of S. Therefore, we can search the first character of each prefix from the root. If there is no match, we can discard the rest of that branch as pattern will not exist there. Based on this observation, we can search for pattern $P[1, m]$ in $S[1, n]$ by first forming the suffix tree T_S of S. We can then check each suffix starting from the root. Let $S[i, n]$ be a suffix under consideration. If $S[i] \neq P[1]$ we can abort searching that branch and continue with the next suffix. If there is match of P along a suffix branch, the number of tree edges lower than $P[m]$ will show the number of occurrences of the pattern in S. This procedure consists of the following steps:

1. Construct suffix tree T_S of the string S
2. Search edges of T_S (suffixes of S) starting from the root
3. If there is a match with P, record the number of edges of T_S below the last character of P. This is the number of patterns in S. The index numbers of the leaves are the positions of the patterns in S.

This method is illustrated for $S = baccacac$ and $P = ac$ in Fig. 5.7. Construction of the suffix tree can be performed in $O(n)$ time using Ukkonen's algorithm and finding an instance of a pattern takes $O(m)$ time. For z occurrences of a pattern in the text, total time is $O(n + z)$.

The same idea can be extended to check whether a given string Q is a suffix of a string S. We construct the suffix tree T of S as before and check for Q starting from the root. If we end at a leaf after all characters of Q are matched and the terminating character is reached, Q is a suffix of S. A biological sequence S may contain repeats and these have significance as we will see in Chap. 7. In order to find the longest repeat of S, we construct the suffix tree T of S and find the deepest level, level being the length of path from the root, node in T which has at least two leaves under it.

Suffix trees can also be used to solve the approximate string matching problem. For the case of 1-mismatch problem in which a maximum of one mismatch is allowed, we start by constructing the suffix tree T_S for the string. We then search for the pattern P in T_S as we did in the exact matching procedure but this time allowing for a maximum mismatch of 1.

Fig. 5.7 Exact pattern matching using a suffix tree. The suffix tree for the string $S = baccacac$ is first constructed and the pattern ac is searched in the suffixes. This pattern exists as a prefix in the middle branch and there are three lower edges of the tree after the match at indices 2, 5, and 7 which are the positions of the matches

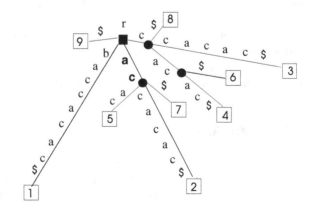

5.5.2.2   Longest Common Substring

In the longest common substring problem, we search for the longest common sub-strings of a number of strings $S_1, \ldots, S_m$. This problem is slightly different than the LCS problem as noted before, since the substrings consist of consecutive elements, whereas in LCS problem the characters of the subsequences may and in general will not be adjacent to each other. For example, given the strings $S_1 = \{accbca\}$ and $S_2 = \{bcbcba\}$; cbc is a common substring and a common subsequence of S_1 and S_2; however, cba is a common subsequence but not a common substring of these two substrings. In the commonly used terminology, substrings consist of consecutive characters but subsequences may not consist of consecutive characters.

We can use the generalized suffix trees to find longest common substrings of two or more strings. The general idea of this approach is the observation that a common substring of the input strings should also be a common prefix of one of their not necessarily common suffixes. Since we are inspecting the longest of such substrings, we should go as deep as possible in the suffix tree for these common prefixes. The algorithm then can be sketched to consist of the following steps:

1. *Input*: $S = \{S_1, \ldots, S_n\}$: A set of strings to search
2. Construct the generalized suffix tree T for S.
3. Label each internal node v of T by $(i_1, \ldots, i_k)$ if for each j there is a leaf reachable from v labeled by the character i_j. In other words, if v is on the path from the root to a leaf for a suffix of string S_k, label v with k.
4. An internal node labeled by all the strings represents a common substring, the common substring consisting of characters between the root and the node v.
5. We then provide another label L_v for such a node v which shows its depth in T. The node v that is labeled with all strings and has the highest depth is marked and the longest common substring consists of characters between the root and this node.

We will illustrate the operation of this algorithm by a simple example of two input strings $S_1 = ccbac$ and $S_2 = bccba$. The generalized suffix tree is first constructed and the nodes are labeled as shown in Fig. 5.8. The longest common substring found is $ccba$.

If n is the total length of strings under consideration, construction of the gener-alized suffix tree takes $O(n)$ time and labeling of the internal nodes also takes $O(n)$ time. Finally, the deepest node is determined in $O(n)$ steps, and hence the total time taken for this suffix tree implementation is $O(n)$.

5.5.3   Suffix Arrays

The suffix array data structure proposed in [25] is a compact representation of a string. A suffix array SA of a string S stores the suffixes of S in lexicographic order. For example, the string $S = cbbacb$ can be represented as a suffix array as shown in Table 5.2.

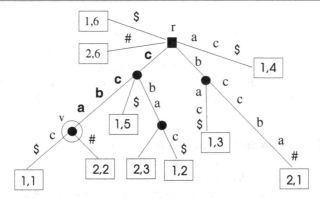

Fig. 5.8 Longest common substring discovery using a suffix tree. The generalized suffix tree T for the strings $S_1 = ccbac$ and $S_2 = bccba$ is first constructed and the internal node that has leaves of both strings in its subtree are marked which are all of the internal nodes. The internal node v shown in circle is the deepest one and the prefix $ccba$ leading to v is the longest common substring of S_1 and S_2

In a simple and straightforward way of constructing a suffix array, we can enumerate all of the suffixes of the string S and sort them which can be done in $O(n^2 \log n)$ time. Alternatively, a suffix array of a suffix tree can be formed by first constructing the suffix tree T_S of the string S and then traversing T_S using lexicographical depthfirst search. This results in $O(n + m)$ time, where n is the number of nodes and m is the number of edges of the suffix tree, which can be considered to be $O(n)$. However, we need to provide memory space to store the suffix tree first which can be discarded after forming the suffix array.

A suffix array can be used for exact string matching by performing a binary search on it for the pattern. In order to find pattern P in S, we need to find a suffix that begins with P in T_S. We need to compare each suffix of S with T in $O(m)$ time and the time taken by the binary search is dependent on the logarithm of the size of search space. Therefore, total time taken to find a match of T in S, or conclude a match does not exist is $O(m \log n)$.

Table 5.2 Suffix array for the string $cbbacb$

Index	Suffix	SA[i]	Sorted
1	cbbacb$	6	$
2	bbac$	4	ac$
3	bac$	3	bac$
4	ac$	2	bbac$
5	c$	5	c$
6	$	1	cbbacb$

In the original suffix array construction algorithm in [25], a method called *prefix doubling* is used in which suffixes that have the same prefixes of length k are grouped into *buckets* and the suffixes in a bucket are sorted using their first $2k$ characters. The bucket numbers are updated and this process is repeated for a maximum of $\log n$ rounds until all suffixes are in buckets of length 1 resulting in $O(n \log n)$ time [26]. A survey of suffix array construction algorithms can be found in [27].

5.5.3.1 Distributed Suffix Array Construction

The main operation of the suffix array construction is the sorting of the suffixes which can be performed in parallel. An early attempt to provide a parallel in-core suffix array construction using distributed memory computers was proposed by Futamura [28] using this idea. The suffix tree of the text is first constructed and the suffixes are grouped into buckets such that suffixes in the same bucket all start with the same prefix of length w. These buckets are distributed to distributed processes for forming the partial suffix arrays. The results are gathered in one process which combines all of the partial arrays to form the whole suffix array. Algorithm 5.11 shows one way of implementing this distributed algorithm using the supervisor–worker model. Out of core algorithms to construct suffix arrays can be found in [29–31].

Algorithm 5.11 *Dist_SA*

1: **Input** : text $T[1, n]$
2: $all\_procs = \{pr_0, ..., pr_{k-1}\}$ ▷ process set
3: **Output** : $SA[1, n]$ ▷ suffix array of T
4: $received \leftarrow \emptyset$
5: **if** $pr_i = pr_0$ **then**
6:      **compute** suffix tree S of T ▷ I am the supervisor process
7:      **partition** suffixes of S into buckets $b_1, ..., b_m$ of same prefixes of equal length
8:      **distribute** evenly buckets $b_1, ..., b_m$ to $pr_1, ..., pr_k$
9:      **while** $received \neq all\_procs$ **do**   ▷ get partial suffix arrays from processes
10:          **receive** $sufarray(sa_i)$ from pr_i
11:          $m\_suffarray \leftarrow m\_suffarray \cup sa_i$
12:          $received \leftarrow received \cup pr_i$
13:      **end while**
14:      **merge** the sorted partial arrays $sa_1, ..., sa_{k-1}$ to obtain sorted SA
15:      **output** SA
16: **else**   ▷ I am a worker process
17:      **receive** my buckets $B_i = b_j, ..., b_{j+l}$ from pr_0
18:      **sort** the suffixes in B_i into sa_i ▷ construct partial suffix array
19:      **send** $sufarray(sa_i)$ to pr_0 ▷ send partial suffix array to supervisor
20: **end if**

5.6 Chapter Notes

We reviewed fundamental algorithms for strings starting with exact string matching in this chapter. Approximate string matching refers to the case of matching when a number of mismatches to the pattern are allowed. The first algorithm for exact string matching called the naive algorithm checked all possible occurrences of the pattern in the text at the expense of high time complexity of $O(nm)$ which prohibits its use with large strings such as biological sequences. We then described algorithms with better time complexities. Knuth–Morris–Pratt algorithm is deterministic as it provides asymptotic bounds but Boyer–Moore algorithm is probably the most favorable algorithm for string matching due to its low complexity and also as its performance gets better with increased alphabet size. Both of these algorithms require preprocessing of tables to be used during pattern search. Table 5.3 displays the comparison of the performances algorithms.

Rabin–Karp algorithm is a randomized algorithm that has a linear time complexity in many cases [32]. It uses hashing to compute a *signature* for each m character substring of the text T and then checks whether the signature of the pattern P is equal to a signature of T. When there is a match, all substrings are searched to find the actual match using the naive algorithm. Aho–Corasick algorithm constructs a finite state machine of input keywords and uses it to find the pattern among the keywords [33].

The distributed implementations of these algorithms are straightforward, we need to partition the input string into a number of processors and each process should then search for the pattern in its partition. The supervisor which is a designated process, gathers all of the results obtained and combines these to obtain the final output. However, there are only few studies on the parallelization of these algorithms on distributed memory computing systems.

Suffix trees provide a convenient method of representing strings. Construction of suffix trees can be performed by various algorithms including the naive method. The space complexity of such an algorithm should also be considered as long strings

Table 5.3 String matching algorithm comparison

Algorithm	Preprocessing	Average search	Worst search
Naive algorithm	0	$O(n+m)$	$O(nm)$
FSM-based algorithm	$\Theta(m\lvert\Sigma\rvert)$	$O(n)$	
Knuth–Morris–Pratt algorithm	$O(m)$	$O(n)$	
Boyer–Moore algorithm	$O(m+\lvert\Sigma\rvert)$	$O(n/m)$	$O(n)$
Rabin–Karp algorithm	$O(n+m)$	$O(nm)$	
Aho–Corasick algorithm	$O(n)$	$O(m+z)$	

and their suffix trees will require significant memory. Hence, the algorithms can be basically divided into in-core and out-of-core methods. We described two in-core algorithms and the distributed algorithms to construct suffix trees are generally out-of-core approaches which provide construction of subtrees in individual processors. The string is typically partitioned into a number of segments and a subtree is formed for each segment. The properties of a segment to be distributed varies for different algorithms, typically a segment pair or prefixes can be used. The main schemes partition the data into a number of processes and the supervisor gathers and combines the results as before. Suffix arrays provide a compact way of storing suffix trees in memory. They basically sort the suffixes of the tree and store the indexes and the string only. The price paid, however, is the higher time complexity in various applications such as pattern matching.

Exercises

1. Find the pattern ATTCT in the following DNA sequence using the naive algorithm. Show all iterations of the algorithm.

   A T G G C T A T T C T A T G G C T A G

2. Construct an FSM to find the pattern $P = babc$ with alphabet $\Sigma = \{a, b, c\}$ in any input string. Build the FSM table displaying the states and the inputs. Show the state transitions for the input string $S = cabbabccb$ using this FSM.

3. It is required to find the occurrences of the pattern $P = cbacb$ in the string $S = acbacccbacbb$ using the KMP algorithm. First, construct the prefix table Π for the pattern P. Then, show step-by-step execution of the KMP algorithm to search P in S using Π.

4. Give qualitative reasons as to why the BM algorithm is more efficient than other pattern matching algorithms.

5. Find the suffix tree of the string $S = accbacba$ using the naive algorithm. What is the time complexity of this approach and why?

6. Search for the pattern bac in string $S = ccbacbbc$ by constructing a suffix tree. How many steps did it take to find this pattern in S?

7. Construct a generalized suffix tree for two strings $S_1 = cacaca$ and $S_2 = ccbaca$. Show that the time taken to construct a generalized suffix tree for a set S of strings is the total length of the strings in this set. Work out also the longest common substring of these two strings using the method described in Sect. 5.5.2.

8. Construct the suffix array for the string $S = cabbcabbc$. Find the occurrences of the pattern $P = cab$ in S using the suffix array of this string.

References

1. Baeza-Yates R (1989) String searching algorithms revisited. In: Dehne F, Sack JR, Santoro N (eds) Workshop in algorithms and data structures, Lecture notes on computer science, vol 382. Springer, Ottawa, Canada, pp 75–96
2. Knuth DE, Morris JH, Pratt VR (1977) Fast pattern matching in strings. SIAM J Comput 6(2):323–350
3. Boyer RS, Moore JS (1977) A fast string searching algorithm. Commun ACM 20(10):761–772
4. Breslauer D, Galil Z (1993) A parallel implementation of Boyer-Moore string searching algorithm. Sequences II:121–142
5. Ukiyama N, Imai H (1993) Parallel multiple alignments and their implementation on CM5. In: Proceedings of genome informatics workshop, pp 103–108
6. Crochemore M, Iliopoulos CS, Pinzon YJ, Reid JF (2001) A fast and practical bit vector algorithm for the longest common subsequence problem. Inform Process Lett 80(6):279–285
7. Korkin D, Wang Q, Shang Y (2008) An efficient parallel algorithm for the multiple longest common subsequence (MLCS) problem. In: Proceedings 37th international conference on parallel processing, pp 354–363
8. Liu W, Chen L (2006) A parallel algorithm for solving LCS of multiple bioseqences. In Proceedings fifth international conference on machine learning and cybernetics. Dalian, China, pp 4316–4321
9. Yang J, Xu Y, Shang Y (2010) An efficient parallel algorithm for longest common subsequence problem on GPUs. In: Proceedings of world congress on engineering 2010, vol I, WCE 2010, June 30-July 2, 2010, London, U.K
10. Erciyes K (2014) Complex networks: an algorithmic perspective. CRC Press Taylor and Francis, pp 38–39, ISBN 9781466571662
11. Weiner P (1973) Linear pattern matching algorithms. In: Proceedings of 14th IEEE symposium on switching and automata theory, pp 1–11
12. Gusfield D (1997) Algorithms on strings, trees and sequences. Computer science and computational biology. Cambridge University Press
13. Sung W-K (2009) Algorithms in Bioinformatics: a practical Introduction. Chapman & Hall/CRC Mathematical and computational biology. Chap 3, November 24
14. McCreight E (1976) A space-economical suffix tree construction algorithm. J ACM 23(2):262–272
15. Ukkonen E (1993) Approximate string-matching over suffix trees. Combinatorial pattern matching, Springer LNCS vol 684, pp 228–242
16. Farach-Colton M, Ferragina P, Muthukrishnan S (2000) On the sorting-complexity of suffix tree construction. J ACM 47(6):987–1011
17. Ukkonen E (1995) On-line construction of suffix trees. Algorithmica 14(3):249–260
18. Apostolico A, Iliopoulos C, Landau G, Schieber B, Vishkin U (1988) Parallel construction of a suffix tree with application. Algorithmica 3:347–365
19. Mansour E, Allam A, Skiadopoulos S, Kalnis P (2011) Era: efficient serial and parallel suffix tree construction for very long strings. Proc VLDB Endowment 5(1):49–60
20. Phoophakdee B, Zaki MJ (2007) Genome-scale disk-based suffix tree indexing. In: Proceedings of ACM SIGMOD, pp 833–844
21. Barsky M, Stege U, Thomo A, Upton C (2009) Suffix trees for very large genomic sequences. In: Proceedings of ACM CIKM, pp 1417–1420
22. Ghoting A, Makarychev K (2009) Serial and parallel methods for I/O efficient suffix tree construction. In: Proceedings of ACM SIGMOD, pp 827–840
23. Ghoting A, Makarychev K (2009) Indexing genomic sequences on the IBM Blue Gene. In: Proceedings of conference on high performance computing networking, storage and analysis (SC), pp 1–11

24. Matteo Comin M, Montse Farreras M (2014) Parallel continuous flow: a parallel suffix tree construction tool for whole genomes. J Comput Biol 21(4):330–344
25. Manber U, Myers G (1993) Suffix arrays: a new method for on-line string searches. SIAM J Comput 25(5):935–948
26. Rajasekaran S, Nicolae M (2014) An elegant algorithm for the construction of suffix arrays. J Discrete Algorithms 27:21–28
27. Puglisi S, Smyth W, Turpin A (2007) A taxonomy of suffix array construction algorithms. ACM Comput Surv 39(2)
28. Futamura N, Aluru S, Kurtz S (2001) Parallel suffix sorting. In: Proceedings of 9th international conference on advanced computing and communications, pp 76–81
29. Dementiev R, Karkkainen J, Mehnert J, Sanders P (2008) Better external memory suffix array construction. J Exp Algorithmics 12:3–4
30. Karkkainen J, Sanders P, Burkhardt S (2006) Linear work suffix array construction. J ACM 53(6):918–936
31. Kurtz S, Choudhuri J, Ohlebusch E, Schleiermacher C, Stoye J, Giegerich R (2001) Reputer: the manifold applications of repeat analysis on a genome scale. Nucleic Acids Res 29(22):4633–4642
32. Karp RM, Rabin MO (1987) Efficient randomized pattern-matching algorithms. IBM J Res Dev 31(2):249–260
33. Aho AV, Corasick MJ (1975) Efficient string matching: an aid to bibliographic search. Comm ACM 18(6):333–340

Sequence Alignment

6.1 Introduction

Sequence alignment is the process of comparing two or more sequences by searching for a series of characters that appear in the same order in these sequences. In DNA sequence alignment, we would be searching for an alignment of nucleotides, whereas amino acid sequences are aligned in proteins. Using sequence alignment, similar segments of DNA, RNA, or proteins can be identified which may indicate functional, structural, or evolutionary relationships between these sequences [23].

The general aim of any sequence comparison method in bioinformatics is to determine whether the similarities between two or more sequences is incidental or they are derived from a common ancestral sequence in which case they are *homologous*. Homology indicating a common ancestor may reveal a common function or structure. Homologous genes, for example, are derived from the same ancestral gene which is altered due to a number of mutations. Homology is displayed more easily in protein amino acid sequences than the DNA nucleotide sequences. This is due to the smaller alphabet of 4 nucleotides (A, C, G, T) in DNA sequences compared to 20 amino acids in proteins. That is, the chances of finding matches in DNA sequences are greater than finding amino acid matches in protein sequences. Also, different codons in DNA encode for the same amino acid and the 3D structure of a protein is determined by its amino acid sequence. In order to have the same functionality of a protein which is based on its 3D shape, the evolutionary process in a protein sequence is slower. In general, we would be interested in both the alignment of DNA/RNA nucleotide sequences and protein amino acid sequences.

Sequence comparison also allows finding evolutionary relationships among organisms. This association is commonly used to construct *phylogenetic trees* and *networks* which display ancestor–descendant affinities and these structures can be used for a wide range of applications including disease analysis as we will see in Chap. 15. Also, the distances between the sequences as computed by sequence alignment algorithms are frequently input to sequence clustering algorithms which groups biological sequences based on their similarities as we review in Chap. 7. A *sequence*

© Springer International Publishing Switzerland 2015

K. Erciyes, *Distributed and Sequential Algorithms for Bioinformatics*,
Computational Biology 23, DOI 10.1007/978-3-319-24966-7_6

motif is a repeating DNA nucleotide or protein amino acid sequence which has a biological significance. Sequence comparison methods are used to discover these motifs as we will analyze in Chap. 8. In summary, the distances and similarities between biological sequences is required in various sequence analysis methods and the alignment methods are usually the first step that provides the needed input to all of these methods.

Two sequences are aligned in *pairwise alignment* and multiple sequences are aligned in *multiple alignment*. In *global alignment*, two homologous sequences of similar lengths are compared over their entire sequence. This method is used to find similarities of two closely related sequences. In many cases, however, only certain segments of two sequences may be similar but the rest of the sequences may be completely unrelated. For example, two proteins may consist of a number of domains and only one or two of these domains may be similar. The global alignment will not display a high similarity between these two proteins in this case. *Local alignment* refers to the method of finding similar regions in two sequences which may have very different lengths. Multiple sequence alignment can be performed by global alignment if the input sequences are closely related and we are searching for the similarity of these sequences as a whole; or local alignment in which case we are interested to find similar subsequences of otherwise not related sequences. Different types of alignment methods need different algorithms; however, they can be coarsely classified as dynamic programming based, heuristic or a combination of both in general.

In this chapter, we first state the sequence alignment problem, describe ways of evaluating goodness of any alignment method and then analyze representative sequential global, local and multiple sequence alignment algorithms in detail. We then provide parallel/distributed algorithms aimed to solve these problems and review current research in this area.

6.2 Problem Statement

Sequence alignment is the basic and most fundamental method of comparing two biological sequences. In the very common application of such alignment, we have an input sequence called the *query* that needs to be identified since it is newly discovered or not aligned before; and this query is typically aligned with each sequence in a database of sequences. The sequences in the database that have the highest scores are then identified as the ones having highest similarity and therefore relatedness to the query sequence. This affinity in base structures may imply phylogenetic relationships and also similar functionality to aid the analysis of the newly discovered sequence.

6.2.1 The Objective Function

We need to asses the quality of an alignment which reflects its goodness. The cost-benefit approach identifies three scores during alignment:

- The benefit of aligning two identical characters (*match*)
- The cost of aligning two different characters (*mismatch* or *substitution*)
- The cost of aligning a character in a sequence with a gap in the other sequence (insertion or deletion-*indel*). The first sequence has a gap in the related column in insertion, and the second sequence has a gap in the corresponding column in deletion.

Insertion and deletion of gaps refer to the operations on the first sequence, that is, insertion/deletion means inserting/deleting a gap to/from the first sequence. An alignment between two DNA sequences X and Y is shown below with matches, mismatches, insertions, and deletions.

```
position      1  2  3  4  5  6  7  8  9  10  11
X:            G  A  G  T  A  -  C  -  G  C   T
Y:            A  A  G  -  A  G  C  T  -  G   T

positions of comparison :
matches      : 2, 3, 5, 7 and 11
mismatches   : 1 and 10
insertions   : 6 and 8
deletions    : 4 and 9
```

A positive score is associated with a match and negative scores are used to penalize a mismatch and an indel. The negative scores or penalties are based on observed statistical occurrences of an indel and a mismatch and typically, the indels are penalized more, reflecting their relatively less prevalence in the genome alignment. As an example, let us use the scores +2 for a match, –1 for a mismatch and –2 for an indel. Given the two DNA sequences X = ATGGCTACAC and Y = GTGTACTAC, we can have various alignments four of which are shown with mismatches (m) and indels (i) marked. Among these four options, the alignment in (a) or (b) should be chosen as they both have the highest scores.

```
A T G G C T A C - A C         A T G G C T A C -   A C
G T - G - T A C T A C         G T G - - T A C T   A C
m   i i     i                 m     i i       i
     (a) Score = 7                 (b) Score = 7


A T G - G C T A C - A C       A T G - G C T A G A C
- - G T G - T A C T A C       G T G T A C T - - A C
i i   i i       i             m     i m     i i
     (c) Score = 4                 (d) Score = 4
```

The aim of any alignment method is to maximize the total score. However, there are exponential number of combinations to check and if we can find an alignment that has a higher score than others, then using it should be preferred. Formally, alignment of two sequences can be defined as follows:

Definition 6.1 (*sequence alignment*) Let $\sum_{\text{org}}$ be an alphabet and $X = x_1 \ldots x_n$ and $Y = y_1 \ldots y_m$ be two sequences over this alphabet and let $\sum \leftarrow \sum_{\text{org}} \cup \{-\}$, that is, the space character added to the original alphabet. An alignment of these two sequences is a two-row matrix where the first row are the elements of X and the second row are the elements of Y and each row contains at least one element of Σ_{org}.

A related parameter between two sequences is the *edit distance* between them which is defined as follows:

Definition 6.2 (*edit distance*) Edit distance, or the Levenshtein distance, is the minimum number of substitutions, insertions, and deletions between two sequences. Hamming distance is an upper bound on edit distance.

For the above example, the edit distance between the two sequences is 4 which occurs in (a). The procedure for sequence alignment is very similar to finding LCS between them; however, we now have costs associated with matches, mismatches, and indels and search for the highest scoring alignment.

6.2.2 Scoring Matrices for Proteins

Proteins consist of a sequence of amino acids from a 20-letter alphabet. Some mismatches are more likely to occur in proteins and they are more frequently encountered than other substitutions. This fact necessitates the use of a weighting scheme for each amino acid substitution. Scoring matrices for mismatches in protein amino acid sequences define the scores for each substitution in these sequences. The two widely used matrices for this purpose are the point accepted mutation (PAM) matrix [11] and the blocks substitution matrix (BLOSUM) [17]. They both use statistical methods and are based on counting the observed substitution frequency and comparison of this value with the expected substitution frequency.

A positive score in the entry m_{ij} of a PAM matrix M means that the probability of the substitution between i and j is more than its expected value; therefore, it bears some significance. The entry m_{ij} is formed by considering the expected frequencies of i and j, and the frequency of alignment between i and j in the global alignment of homologous sequences [4]. The nth power of M is then taken to form the PAM-n matrix such as PAM-80, PAM-120, or PAM-250. A large value of n should be used to align proteins that are not closely related.

PAM may not provide realistic values for remotely related protein sequences as it uses extrapolation of values. BLOSUM matrix structure proposed by Henikoff and Henikoff [17] overcomes this difficulty by analyzing segments of proteins rather than the whole. If two segments of proteins under consideration have similarity over a threshold value, they are clustered. The threshold value t is specified as BLOSUM-p which means the matrix is generated by combining sequences which have at least t % similarity. The BLOSUM62 matrix which which is formed by clustering sequences that have at least 62 % identity level is shown below. A small value of t is used for distantly related protein sequences and more closely related ones can be aligned using a larger value.

```
    A   R   N   D   C   Q   E   G   H   I   L   K   M   F   P   S   T   W   Y   V   B   Z   X   *
A   4  -1  -2  -2   0  -1  -1   0  -2  -1  -1  -1  -1  -2  -1   1   0  -3  -2   0  -2  -1   0  -4
R  -1   5   0  -2  -3   1   0  -2   0  -3  -2   2  -1  -3  -2  -1  -1  -3  -2  -3  -1   0  -1  -4
N  -2   0   6   1  -3   0   0   0   1  -3  -3   0  -2  -3  -2   1   0  -4  -2  -3   3   0  -1  -4
D  -2  -2   1   6  -3   0   2  -1  -1  -3  -4  -1  -3  -3  -1   0  -1  -4  -3  -3   4   1  -1  -4
C   0  -3  -3  -3   9  -3  -4  -3  -3  -1  -1  -3  -1  -2  -3  -1  -1  -2  -2  -1  -3  -3  -2  -4
Q  -1   1   0   0  -3   5   2  -2   0  -3  -2   1   0  -3  -1   0  -1  -2  -1  -2   0   3  -1  -4
E  -1   0   0   2  -4   2   5  -2   0  -3  -3   1  -2   0  -1   0  -1  -3  -2  -2   1   4  -1  -4
G   0  -2   0  -1  -3  -2  -2   6  -2  -4  -4  -2  -3  -3  -2   0  -2  -2  -3  -3  -1  -2  -1  -4
H  -2   0   1  -1  -3   0   0  -2   8  -3  -3  -1  -2  -1  -2  -1  -2  -2   2  -3   0   0  -1  -4
I  -1  -3  -3  -3  -1  -3  -3  -4  -3   4   2  -3   1   0  -3  -2  -1  -3  -1   3  -3  -3  -1  -4
L  -1  -2  -3  -4  -1  -2  -3  -4  -3   2   4  -2   2   0  -3  -2  -1  -2  -1   1  -4  -3  -1  -4
K  -1   2   0  -1  -3   1   1  -2  -1  -3  -2   5  -1  -3  -1   0  -1  -3  -2  -2   0   1  -1  -4
M  -1  -1  -2  -3  -1   0  -2  -3  -2   1   2  -1   5   0  -2  -1  -1  -1  -1  -1   1  -3  -1  -1  -4
F  -2  -3  -3  -3  -2  -3  -3  -3  -1   0   0  -3   0   6  -4  -2  -2   1   3  -1  -3  -3  -1  -4
P  -1  -2  -2  -1  -3  -1  -1  -2  -2  -3  -3  -1  -2  -4   7  -1  -1  -4  -3  -2  -2  -1  -2  -4
S   1  -1   1   0  -1   0   0   0  -1  -2  -2   0  -1  -2  -1   4   1  -3  -2  -2   0   0   0  -4
T   0  -1   0  -1  -1  -1  -1  -2  -2  -1  -1  -1  -1  -2  -1   1   5  -2  -2   0  -1  -1   0  -4
W  -3  -3  -4  -4  -2  -2  -3  -2  -2  -3  -2  -3  -1   1  -4  -3  -2  11   2  -3  -4  -3  -2  -4
Y  -2  -2  -2  -3  -2  -1  -2  -3   2  -1  -1  -2  -1   3  -3  -2  -2   2   7  -1  -3  -2  -1  -4
V   0  -3  -3  -3  -1  -2  -2  -3  -3   3   1  -2   1  -1  -2  -2   0  -3  -1   4  -3  -2  -1  -4
B  -2  -1   3   4  -3   0   1  -1   0  -3  -4   0  -3  -3  -2   0  -1  -4  -3  -3   4   1  -1  -4
Z  -1   0   0   1  -3   3   4  -2   0  -3  -3   1  -1  -3  -1   0  -1  -3  -2  -2   1   4  -1  -4
X   0  -1  -1  -1  -2  -1  -1  -1  -1  -1  -1  -1  -1  -1  -2   0   0  -2  -1  -1  -1  -1  -1  -4
*  -4  -4  -4  -4  -4  -4  -4  -4  -4  -4  -4  -4  -4  -4  -4  -4  -4  -4  -4  -4  -4  -4  -4   1
```

6.3   Pairwise Alignment

Pairwise alignment is the comparison of two distinct DNA or protein sequences. It can be performed globally to compare two sequences as a whole or locally to detect similar subsequences in the two sequences as outlined below.

6.3.1   Global Alignment

Global alignment assumes that the sequences to be compared are sequentially homologous and attempts to align all of the sites optimally within the sequences.

Fig. 6.1 An entry $S[i, j]$ of
the alignment matrix

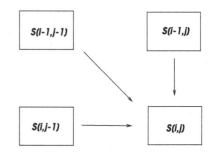

An *alignment matrix* S is a convenient way of displaying the alignment between two sequences. In order to represent two sequences X and Y of lengths n and m, S contains n rows and m columns. When aligning two sequences, we can have four options. The characters match; they do not match; a gap is inserted in the first sequence; or a gap is inserted in the second sequence. The filling of the alignment matrix is based on selecting the option which gives the highest score. An entry $S[i, j]$ of an alignment matrix depends on the values of the entries just before it in the preceding column, row, and diagonal as shown in Fig. 6.1. The first row and column of this graph are initialized with the gap penalties when these occur at the beginning of sequences X and Y.

We can therefore compute the value of the element (i, j) by checking the three previous entries at $(i - 1, j)$, $(i - 1, j - 1)$, and $(i, j - 1)$. The first value of the array is initialized to 0. Finding all other entries can be done using the dynamic programming approach where subsolutions are used to find the solution. Needleman and Wunsch provided the first dynamic algorithm for this purpose as described next.

6.3.1.1 Needleman–Wunsch Algorithm

Let us specify the scores for each character comparison. A match is given a score α, a mismatch a β and an indel γ when two characters a_i and b_i are compared, as shown below. The match score is commonly positive as this is what we require and the other scores are negative.

$$\text{score}(a_i, b_i) = \begin{cases} \alpha, a_i = b_i \\ \beta, a_i \neq b_i \\ \gamma, a_i =' -' \text{ or } b_i =' -' \end{cases}$$

Let us assume an optimal alignment A between two sequences X and Y, and $A' \subset A$ as an alignment of $X' \subset X$ and $Y' \subset Y$. If A is optimal, A' is also an optimal alignment which leads to the following dynamic algorithm solution. We consider the prefixes $X_i = x_1 \ldots x_i$ and $Y_j = y_1 \ldots y_j$ of two sequences. In order to find the optimal alignment, we need to select the one that gives the highest score from the following:

1. Align x_i with y_i and X_{i-1} with B_{j-1}
2. Align x_i with a gap and X_{i-1} with B_j
3. Align y_j with a gap and X_i with B_{j-1}.

The first case shows whether there is a match or a mismatch and δ is equal to α when there is a match, and β when a mismatch occurs. Aligning a prefix X_i of X with no element of Y is the product of the length of X_i with the gap penalty and the same is valid for Y. Therefore, $M[i, 0] = \gamma i$ and $M[0, j] = \gamma j$. Needleman–Wunsch algorithm brings together all of the concepts we have discussed until now in the dynamic programming based algorithm as shown in Algorithm 6.1.

Algorithm 6.1 NW_Alg

1: **Input** : Sequences $X = \{x_1, ..., x_n\}$ and $Y = \{y_1, ..., y_m\}$
2: **Output** : Array $S[n, m]$
3: $S[0, 0] \leftarrow 0$
4: **for** $j = 1$ to n **do**
5:     $S[0, j] \leftarrow \gamma \times j$
6: **end for**
7: **for** $i = 1$ to m **do**
8:     $S[i, 0] \leftarrow \gamma \times i$
9: **end for**
10: **for** $j = 1$ to n **do**

$$M[i, j] = max \begin{cases} M[i-1, j] + \gamma \\ M[i, j-1] + \gamma \\ M[i-1, j-1] + \delta(a_i, b_j) \end{cases}$$

11: **end for**

Let us form the alignment array M for two given DNA sequences $X = \text{ACCGT}$ and $Y = \text{AGCCTC}$ with $n = 5$ and $m = 6$; and we select $\alpha = 2$, $\beta = -1$ and $\gamma = -1$ for simplicity. We first fill the first row and column of M with gap penalties and assign 0 value for the first entry. We then form each entry using the dynamic programming relation in the NW algorithm to obtain the final array as shown below.

		A	G	C	C	T	C
	0	-1	-2	-3	-4	-5	-6
A	-1	2	← 1	0	-1	-2	-3
C	-2	1	1	3	2	1	0
C	-3	0	0	3	5	4	3
G	-4	-1	2	2	4	4	3
T	-5	-2	1	1	3	6	← 5

The global alignment problem can now be reduced to finding the best scoring path between vertices $M[0, 0]$ and $M[n, m]$. We can now start from the lowest right corner of M and work our way upwards until we reach $M[0, 0]$ by following in reverse direction of the path we have chosen while filling the array. An up arrow means a gap in the top sequence, a left arrow represents a gap in the second sequence on the lefthandside, and a diagonal arrow shows a match or a mismatch without any gap in that position. The alternative paths result in alternative alignments with the same score. Implementing this procedure for the above example yields the following alignment with a score of 5 (4 matches and 3 indels).

```
A G C C - T C
A - C C G T -
```

The time to fill the array is $O(nm)$ which is the size of the array and hence the space requirement is the same. At the end of the algorithm, best alignment score is stored in $M[n, m]$.

6.3.2 Local Alignment

The global alignment may not provide the correct results because of the genome shuffling and rearrangements. A segment of a sequence inversion may happen in a sequence causing a subsequence to look radically different. Local alignment provides us information about the conserved subsequences within organisms. A local alignment between two sequences with four matches and a mismatch is shown in bold below.

```
A T G C T A G T G C C
G C A C T T G T A A T
```

As a general rule, subsequences of the sequences are aligned separately without considering the general order of the global sequences in local alignment. Given two sequences $X = x_1 \ldots x_n$ and $Y = y_1 \ldots y_m$, some parts of X and Y may be aligned with high scores by local alignment but the remaining subsequences may be very different. One way of tackling this problem is to identify all possible subsequences of X and Y and then compute all global alignments between every pair of these subsequences and select the one with the highest score. Unfortunately, this brute force algorithm can find subsequences of X and Y in $O(n^2)$ and $O(m^2)$ times, and performing global alignment results in complexity of $O(n^2m^2nm)$ or $O(n^3m^3)$ which is unacceptable for large sequences. A commonly used algorithm for local alignment was proposed by Smith and Waterman [31] which is an adaptation of the NW algorithm for local alignment. The main differences between these two algorithms are as follows. First, a fourth value of zero is allowed in addition to the three possible values in the NW algorithm to prevent negative values in the alignment graph. The first row and column of the array M contain zeros now to discard gaps occurring in the beginning of sequences. We still attempt to find a path with maximum value but we do not have to start from the beginning to allow for local alignments. Instead, we start with the maximum value of the array and stop when a zero is encountered which signals the end of the regional alignment. Algorithm 6.2 shows the pseudocode of this algorithm.

Algorithm 6.2 SW_Alg

1: **Input** : Sequences $X = \{x_1, ..., x_n\}$ and $Y = \{y_1, ..., y_m\}$
2: **Output** : Array $S[n, m]$
3: $S[0, 0] \leftarrow 0$
4: **for** $j = 1$ to n **do**
5:     $S[0, j] \leftarrow 0$
6: **end for**
7: **for** $i = 1$ to n **do**
8:     $S[i, 0] \leftarrow 0$
9: **end for**
10: **for** $j = 1$ to n **do**

$$M[i, j] = max \begin{cases} M[i-1, j] + \gamma \\ M[i, j-1] + \gamma \\ M[i-1, j-1] + \delta(a_i, b_j) \\ 0 \end{cases}$$

11: **end for**

As an example, given two DNA sequences $X = $ GGATACGTA and $Y = $ TCATACT with $n = 9$ and $m = 7$, and scoring as before with $\alpha = 2$, $\beta = -1$ and $\gamma = -1$; we proceed similar to the global alignment algorithm by first initializing the first row and column of the similarity matrix M. We then form each entry using the SW algorithm and fill the entries of M, keeping the track of the selected path as in NW algorithm. Backtracking starts from the highest value element of M this time, stopping whenever a zero is encountered as shown below.

We start with the maximum value entry 9 and backtrack the path until a 0 is encountered. The obtained maximum local alignment between these two sequences is shown below in bold.

$$\begin{aligned} &\text{G G } \textbf{A T A C G T} \text{ A} \\ &\text{T C } \textbf{A T A C} \text{ - } \textbf{T} \end{aligned}$$

	G	G	A	T	A	C	G	T	A
	0	0	0	0	0	0	0	0	0
T	0	0	0	2	← 1	0	0	2	1
C	0	0	0	1	1	3	2	1	1
A	0	0	2	1	3	2	2	1	3
T	0	0	1	4	3	2	1	4	3
A	0	0	2	3	6	5	4	3	6
C	0	0	1	2	5	8	← 7	6	5
T	0	0	0	3	4	7	6	9	8

If we start with 6 shown in the upper path, a local alignment with one mismatch and one gap, and a score of 6 is obtained as follows.

G G A T A C G T A
T - C A T A C T

We can have various local alignments between these two sequences using alternative paths, starting with the next largest value and continuing until a 0. The time and space complexities of this algorithm are $O(nm)$ as in NW algorithm since we need to fill the alignment matrix as before.

6.4 Multiple Sequence Alignment

Multiple sequence alignment (MSA) aims at aligning more than two biological sequences. We have a set of k input sequences $S = S_1, \ldots, S_k$ and our aim is to provide alignment of these k sequences. We can have global and local multiple alignment in MSA. In theory, we can use NW algorithm for the global alignment of k sequences, for each possible pairs of a set of k sequences $S = S_1, \ldots, S_k$ each with a length of n, invoking this algorithm $k(k-1)/2$ times. For large k, this method is inefficient due to its increased time complexity of $O(n^2k^2)$. For this reason, heuristics are widely used for MSA. The main methods of MSA can be stated as follows:

- *Exact methods*: These algorithms typically use dynamic programming outlined and have high running times and can be effective for only 3 or 4 sequences.
- *Approximation algorithms*: They have polynomial run times but only approximate the solution. However, performance is guaranteed to be within the approximation ratio.

- *Heuristic methods*: The algorithms based on heuristics typically search only a subset of the possible alignments and find an alignment that is suboptimal. There is no performance guarantee but they are widely used in practice.
- *Probabilistic methods*: They assume a probabilistic model and search alignment that best fits this model.

We will now take a closer look at representative algorithms for these methods starting with an approximation algorithm.

6.4.1 Center Star Method

The center star method is an approximation algorithm with a ratio of 2. The main idea of this algorithm is to identify a sequence which is closest to all others as the center and then work out the alignments of all sequences with respect to this center. There are various methods to measure the distances between the sequences. As a simple approach, we can find the consensus sequence S_{cs} of $S = \{S_1, \dots, S_n\}$ of n input sequences which is the sequence containing the most frequent symbols of each column to be matched in each sequence. We can then work out the distance of each sequence S_i to S_{cs} and mark the sequence with shortest distance to S_{cs} as the central sequence S_c. Alternatively, we can compute pairwise distances between all pairs of sequences which is called the *sum of pairs distance* which is used in the center star method. The center star algorithm specifically consists of the following steps:

1. *Input*: A set $S = \{S_1, \dots, S_k\}$ of k sequences of length n each.
2. *Output*: MSA of sequences in S.
3. Work out the distance matrix D between sequences such that d_{ij} entry of D is equal to the distance between S_i and S_j.
4. Find the center sequence S_c that has the minimum value of sum of pairs, $\sum_{i=1}^{k} d_{ij}$.
5. For each $S_i \in S \setminus S_c$, find an optimal global alignment between S_i and S_c using Needleman–Wunsch algorithm.
6. Insert gaps in S_c to complete MSA.

The center sequence is the one that is most similar to all other sequences. For example, given the four sequences below, we can find their pairwise similarities as shown.

```
S1: A C C G T G G C    S1: A C C G T G A T    S1: A C C G T G T T
S2: C G C C T C T T    S3: C A G G T C T G    S4: C G T A A T A G
        d=3                    d=3                    d=5

S2: C G C C T C G A    S2: C G C C T C A G    S3: C G C C T C A G
S3: C A G G T C T A    S4: C G T A A T T A    S4: C G T A A T T A
        d=2                    d=4                    d=1
```

Fig. 6.2 The star tree
formation for four sequences

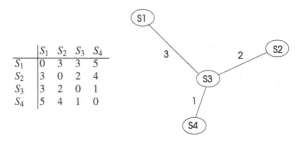

	S_1	S_2	S_3	S_4
S_1	0	3	3	5
S_2	3	0	2	4
S_3	3	2	0	1
S_4	5	4	1	0

The total number of comparisons is $k(k-1)/2$ times, 6 in this case, resulting in a total time complexity of $O(k^2n^2)$ for this step. We can then form the distance matrix D with these values and find the sequence that has the greatest similarity to all others as shown in Fig. 6.2. This step involves summing rows of the matrix D and detecting the sequence with the lowest sum, which is S_3 in this case, in $O(k^2n^2)$ time.

We now need to align sequences S_1, S_3, S_4 to the central sequence S_2 by the Needleman–Wunsch algorithm in $O(kn^2)$ time. Finally, gaps are inserted in the aligned sequences to complete the multiple sequence alignment $O(k^2n)$ resulting in a total time of $O(k^2n^2)$ since the first step dominates. It can be shown using the triangle inequality between three sequences that the approximation ratio of this algorithm is 2 [33].

6.4.2   Progressive Alignment

Progressive alignment methods employ heuristic algorithms to compute the MSA of a set of sequences. A general approach is to to align two closely related sequences and then progressively align other sequences. Typically in the first step, all possible pairwise alignments of k sequences for a total of $k(k-1)/2$ pairs are computed. A phylogenetic tree (see Chap. 14) that shows the evolutionary relationships based on their distances is then estimated and used as a guide to perform alignment. The most similar sequences that are close to each other in the phylogenetic tree are then pairwise aligned. CLUSTALW is one such widely used global progressive alignment tool that can be used for both DNA and protein sequence analysis [32]. It performs the following steps:

1. Computation of all pairwise alignment scores and forming the distance matrix D based on these scores.
2. Construction of a phylogenetic tree T using D by the neighbor-joining (NJ) method (see Sect. 14.3).
3. Perform MSA with the sequences starting with the closely related ones in the tree.

The NJ algorithm proposed by Saitou and Nei [29] is basically a hierarchical clustering algorithm that iteratively groups closely related input data which are also farthest to the rest of inputs. The CLUSTALW algorithm then iteratively performs

Fig. 6.3 A phylogenetic tree which has the five input sequences *A, B, C, D,* and *E* as its leaves. Sequences *A* and *B* are first aligned as they are closest, and then sequences *C* and *D* are aligned followed by the alignment of the resulting sequence with *E*. Finally, the resulting two sequences are aligned

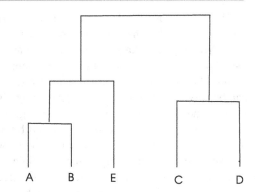

pairwise sequence alignment with the closest sequences in the tree. Figure 6.3 displays an example phylogenetic tree where the input sequences are the leaves of this tree.

The NJ algorithm in general will produce unrooted trees where the input sequences may not be equidistant to their ancestors. The root in this tree is placed in a location from which the average distances on its both sides are equal. The CLUSTALW algorithm will use this tree to pairwise align the closest sequences as guided by the tree. Starting from the leaves, the closest leaves are aligned iteratively to form larger clusters at each step. Progressive alignment methods introduce significant errors when sequences are distantly related. Also, the guide tree is formed using pairwise alignments which may not reflect the evolutionary process accurately.

6.5 Alignment with Suffix Trees

Suffix trees can be used for global sequence alignment. A fundamental method used for this purpose is called *anchoring* in which similar regions called *anchors* in two sequences are first identified using suffix trees. The segments between the anchors are then aligned using dynamic programming or using the same method recursively or a combination of both approaches.

MUMmer is one such algorithm that uses suffix trees for extracting maximal unique matches (MUMs) that are used for anchoring [12]. Given two sequences X and Y of lengths n and m, a MUM is a subsequence of both X and Y of length greater than a given threshold d. A MUM of X and Y has to be unique in both of them. A brute force algorithm needs to search all possible prefixes of both strings in $O(nm)$ time. However, this problem can be simplified by the aid of a generalized suffix tree. We need to build a generalized suffix tree for the two strings and search for internal nodes that have exactly two leaves, one from each sequence. We then check whether the node representing the substring is maximal. If this condition is satisfied, the prefix starting from the root and ending at the internal node represents a MUM. This algorithm takes $O(n + m)$ steps which is the time to construct the generalized suffix tree and also the time for other steps.

The order of MUMs is also conserved between related genomes, and therefore we can predict that the conserved regions in two biological sequences contain ordered MUMs rather than randomly distributed MUMs. The idea behind the MUMmer algorithm is this observation and it attempts to find these conserved regions by finding the longest common subsequence (LCS) of them [12]. The LCS problem can be solved by the dynamic programming algorithm we reviewed in Sect. 5.4 in $O(n^2)$ time and $O(n^2)$ space or by using generalized suffix trees. However, since each MUM is unique, it can be replaced by a special character allowing a solution in $O(n \log n)$ time [33]. The regions between the anchors are aligned using the Needleman–Wunsch algorithm. Multiple genome aligner (MGA) is a tool for multiple sequence alignment based on suffix trees [18]. The longest nonoverlapping sequence of maximal multiple exact matches (multiMEMs) are computed and then used to guide the multiple alignments in this algorithm. The LAGAN [5] is another tool based on anchoring; however, it uses the CHAOS local alignment algorithm and uses the local alignments produced by CHAOS as anchors limiting the search area of the Needleman–Wunsch algorithm around these anchors [6]. LAGAN provides the visual display of alignment results.

6.6 Database Search

It is of interest to compare a newly discovered biological sequence against many other existing ones in databases to find its affinity to them, and therefore to predict and compare the functionality of the new sequence. The sequences deciphered using modern sequencing techniques have increasingly large sizes making it difficult to align them using the SW or NW algorithms which have $O(nm)$ time complexities. The focus of the research studies have then been the design of algorithms that use heuristics and provide approximate but fast solutions. There are many sequence alignment tools for this purpose two of which are more commonly used than others and we will describe them briefly.

6.6.1 FASTA

FASTA is an early local pairwise sequence alignment tool for database comparison of a biological sequences [27]. Its predecessor was called FASTAP [21] and handled protein sequence alignment only, and since FASTA can search for both protein and DNA sequences, it was called FASTA (Fast-All). It is a heuristic algorithm that compares a given input query sequence against the sequences in a database. Its operation can be summarized as follows:

1. Given a query sequence Q and a set of sequences $S = S_1, \ldots, S_n$ in the database, it searches for exact matches of length l between the query and a database sequence $S_i \in S$. These matches are called *hotspots*. Commonly used values of values of

l are 2 for protein amino acid sequences and between 4–6 for DNA sequence comparisons.
2. The hotspots are combined into a long sequence called *initial regions*. These regions are scored using the similarity matrix M. Only a small part of M is aligned and the best scoring 10 alignments are considered for the next step.
3. Using dynamic programming, the ten best partial alignments are combined to give a longer alignment.
4. SW algorithm is used to align these sequences.

The main idea of this program is to find subsequence matches between the query sequence and each of the sequences in the database, enlarge them and compute local alignment in these regions using dynamic programming. There are few efforts on parallelizing FASTA such as [19,30] on a cluster of workstations.

6.6.2 BLAST

Basic local alignment search tool (BLAST) developed at the National Center for Biotechnology Information by Altchul and colleagues [1] is a popular tool for local sequence alignment. BLAST and its derivative algorithms are one of the most widely used tools for sequence alignment. The main idea of BLAST is to search only a subspace of the sequences. In its basic version, gaps are not allowed during alignment which simplifies the alignment procedure greatly. The assumption here is if there is a similarity between two sequences, it will show even if the gaps are not allowed.

A *segment pair* in BLAST is defined as a pair of equal-length subsequences between two sequences S_1 and S_2 which are aligned without gaps. A maximal segment pair (MSP) of S_1 and S_2 is the highest scoring segment between them. As the first step, BLAST searches all sequences with length l in the database that have an MSP score higher than a threshold τ with the input query Q [9]. It searches the short sequences first and then extends them. The found subsequences are called *hits* which are then extended in both directions to find if the score is higher than τ. In detail, BLAST performs the following steps:

1. We are again given a query sequence Q and a set of sequences $\mathcal{S} = S_1, \ldots, S_n$ in the database. BLAST searches hits of length l that have an MSP of score higher than τ between the query and the database sequence $S_i \in \mathcal{S}$. Typical values of l are 3 for protein amino acid sequences and 11 for DNA sequences. The threshold τ is dependent on the scoring matrix used.
2. It searches for pairs of hits which have a maximum distance of d between them.
3. The hit pairs are extended in both directions and the alignment score is checked at each extension. This process is stopped when the score does not change. The pair of hits scoring above a threshold after the extension are called high scoring pairs (HSPs).
4. The consistent HSPs are combined into local alignment that gives the highest score.

The newer versions of BLAST alow gaps [2,26]. The BLAST algorithm also provides an estimate of the statistical significance of the output. This tool is available for free usage at www.ncbi.nlm.nih.gov/blast/.

6.7 Parallel and Distributed Sequence Alignment

We have reviewed basic global and local exact alignment algorithms and the commonly used database alignment tools which use heuristic methods that provide approximate results. The database tools are simpler to parallelize on a distributed memory computer system as we can easily partition the database across the machines or duplicate it if its size is not very large. We will first look at ways of parallelizing the exact algorithms and then review existing methods for distributed alignment using the database tools.

6.7.1 Parallel and Distributed SW Algorithm

The SW algorithm is a dynamic programming method to provide local alignment of two sequences as we have reviewed. We can have fine-grain or coarse-grain parallel running of this algorithm on a number of processors [7]. In fine-grain parallel computing, we have small tasks that cooperate more frequently for small data sizes.

Forming the alignment matrix is the most time-consuming part of the algorithm. We can have a fine-grain parallel mode of SW algorithm by assigning each cell of the alignment matrix to a process. The value of each cell $S[i, j]$ in this matrix is dependent on the values of the preceding row, column, and diagonal values. We can therefore employ a scheme in which every process responsible for the cell $S[i, j]$ that calculates its value sends it to the cells $S[i + 1, j + 1]$, $S[i, j + 1]$, and $S[i + 1, j]$ for further processing as shown in Fig. 6.4. As the computation progresses along waves which increase in size until diagonal and then decrease, this scheme is called the *wavefront* method [4]. We would need nm processes to fill the matrix for two sequences of lengths n and m. Hence, specific architectures such as array processors are suitable for this method. An early attempt that used this approach was reported in [14] which used 12 processors with shared memory, and another implementation was described in [28]. The same technique can be used to find global alignment between two sequences using the NW algorithm.

The coarse-grain distributed sequence alignment is basically based on parallel database operations in which sequences from a database are searched in parallel. In a typical supervisor–worker parallel computation model, the supervisor process sends a number of sequences to each worker to align. The workers send the results to the supervisor which ranks them and keeps the best alignments. As the processing times of workers will be of varying lengths, a dynamic load balancing strategy is usually needed in this mode of operation to keep processes busy at all times [7].

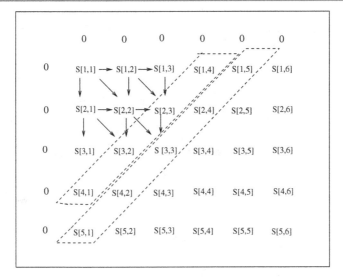

Fig. 6.4 The wavefront method for parallel SW algorithm

6.7.2 Distributed BLAST

When the database to be searched is comparatively small, a simple way to provide parallelism is to replicate it on a number of machines. We can then divide the batched queries to k processors which implement the BLAST algorithm in parallel. The supervisor–worker model can be adopted in this case in which the supervisor process gathers all of the results obtained from the individual workers and outputs the final result. Figure 6.5 shows this process visually. This method has been the subject of various studies including BeoBLAST [15] and Hi-per BLAST [24].

Recent sequencing techniques allow discovery and provision of many biological sequences which constitute large databases. These databases cannot usually be accommodated in a single computer and a convenient way of providing parallelism using such large databases is to partition the data. We can again implement the supervisor–worker model where each process implements BLAST on partial data and sends the partial results to the supervisor which combines them to get the final output. The general approach of a distributed BLAST algorithm employing partitioned database is shown in Fig. 6.6.

The TurboBLAST tool implements database segmentation along with load balancing and scheduling algorithms to run BLAST on a cluster of workstations [3]. This approach has also been applied in mpiBLAST [10] which uses the message passing interface (MPI) parallel programming environment [16]. Its claimed benefits are first the decreased disk I/O operations due to partitioned database and the reduction of interprocess communications between processes as each worker uses data in its partition only. At the start of mpiBLAST, each worker process notifies the supervisor of the database segments it has. The supervisor then inputs the query

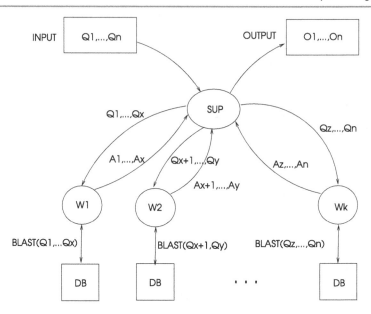

Fig. 6.5 Distributed BLAST using replicated database. The input query batch is $Q_1, \ldots, Q_n$. Each worker W_i has a copy of the database and receives a portion of the input query batch from the supervisor process (SUP). Each worker then runs part of the BLAST query in its local database, obtains the results $(A_p, \ldots, A_q)$ and sends them to the supervisor which combines the partial results and outputs them

batch and broadcasts the batch to all workers. Upon a worker W_i announcing it is idle and can start working, the supervisor assigns W_i a database segment. The worker W_i then performs alignment in the segment it is allocated and reports the result to the supervisor process. The operation of the algorithm is very similar to what is depicted in Fig. 6.6 with the additional enhanced load balancing in which a worker that has finished searching a database can be assigned another search in a different database segment as assigned by the supervisor. The authors report that mpiBLAST achieves super-linear speedup in all tests [10].

The multithreaded versions of BLAST are also available to run on shared memory multiprocessor systems. This mode of operation is similar to partitioned database approach; however, the database is loaded to shared memory now and each thread can work in its partition. Thread and shared memory management may incur overheads and cause scalability issues. NCBI BLAST [25] and WU BLAST [34] are the examples of multithreaded BLAST systems [35]. The UMD–BLAST is an interface that enables to use the most suitable parallel/distributed BLAST algorithm. It inputs the database size, query batch size, and query length and determines which algorithm to use. For large databases which cannot be accommodated in the memory of a single computer, UMD–BLAST uses mpiBLAST; for long query batches with not very large query lengths, BLAST++ which employs replicated database is used. Otherwise, the multithreaded BLAST is employed and the outputs are combined [35].

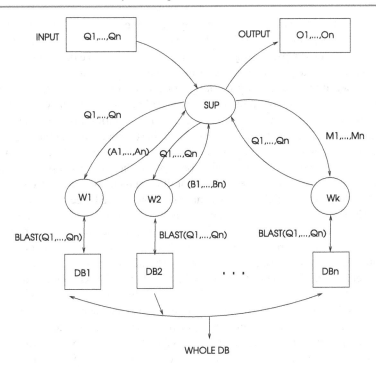

Fig. 6.6 The operation of distributed BLAST using partitioned database. Each worker W_i now has a segment of the database and receives the full input query batch from the supervisor process (SUP). It then runs all of the BLAST query in its database partition, finds the results, and sends them to the supervisor which combines the partial results and outputs them

6.7.3   Parallel/Distributed CLUSTALW

Let us review the main three steps of the CLUSTALW algorithm [32]. In the first step, it computes pairwise distances followed by the construction of the guide tree using the neighbor-joining algorithm in the second step. This tree is used as a guide to perform alignment in the last step where the leaves are first aligned followed by the alignment of close nodes in the tree in sequence.

Parallelization whether using shared memory or distributed memory computers involves implementing these three steps in parallel. The first step requires calculation of distance between k sequences with $k(k-1)/2$ comparisons. This step is trivial to parallelize again using the supervisor–worker model of parallel computation. We can have the supervisor send groups of sequences to each worker process which compute the distances and the results are then gathered at the supervisor. The parallel implementation of the second and third steps is not so straightforward due to the data dependencies involved.

ClustalW-MPI [20] is the distributed implementation of the CLUSTALW method using the MPI parallel programming environment based on the described approach above. The distance matrix is first formed by allocating *chunks* of independent tasks to processes. Large batches result in decreased interprocess communication times but may have poor load balancing. On the other hand, small size of batches provides balanced process loads at the expense of increased communication overheads.

The guided tree is formed by the neighbor-joining method as in CLUSTALW; however, few modifications to the original algorithm resulted in the complexity of $O(n^2)$ time for constructing the guided tree. As this algorithm searches sequences that are closer to each other but also have the highest distance to all other clusters, a parallel search method was designed to search for such sequence clusters. However, details of this method are not described in the paper. In the final progressive alignment step, a mixture of fine and coarse-grained parallelism methods is used. Coarse-grain parallelism involves aligning external nodes of the guided tree and the speedup obtained is reported as $n / \log n$ where n is the number of nodes of the tree. The authors also implemented recursive parallelism and calculated the forward and backward steps of the dynamic programming in parallel. They showed experimentally the speedup achieved for aligning 500-sequence test data as 15.8 using 16 processors.

Another parallel version of CLUSTALW, called pCLUSTAL, which can run on various hardware from parallel multiprocessors to distributed memory parallel computers using MPI was described in [8]. This study also uses supervisor–worker paradigm in which the supervisor process p_0 maps the sequence-pairs to processes and each process then performs sequential CLUSTALW algorithm on its own data set. The results are then gathered at p_0 which builds the guided tree T. It then examines T for independently executable alignments and assigns these to processes. The final step involves gathering of all the alignment results at p_0. The experiments were carried on protein sequences of average length of 300 amino acids. They showed the time-consuming pairwise alignment step takes time proportional to $1/k$ where k is the number of processors.

A shared memory implementation CLUSTALW in SGI multicomputers was described in [22] using OpenMP and speedups of 10 on 16 processors was reported, and a comparison of various implementations is presented in [13].

6.8 Chapter Notes

Comparison of biological sequences using alignment is needed as the first step of various analysis methods in bioinformatics, for example, alignment of sequences provides their affinities which can be used to infer pyhlogenetic relationships between them. Global alignment refers to comparing two or more sequences as a whole, and local alignment methods attempt to align subsequences of the sequences under consideration. The alignment methods, whether global or local, can be broadly classified as exact and heuristic approaches. Exact methods typically use dynamic programming and have favorable performances as we have seen in SW and NW algorithms. However, even these linear times become problematic when the size of the sequences is very large. Heuristic methods do not search exact solutions and typically narrow the search space by sampling of the data which results in favorable run times for large sequences. FASTA and widely used BLAST are two commonly used tools which adopt heuristic tools.

However, even the heuristic methods in sequential form are increasingly becoming more inadequate as the sizes of databases increase due to the expansion in the number of discovered sequences as a result of high volume efficient sequencing technologies. A possible way to speedup the heuristic alignment methods is to employ parallel and distributed processing. This can be achieved typically either by replicating the database if this is not relatively large, or partitioning it. We described these two approaches as implemented in various BLAST versions.

Sequence alignment is probably one of the most investigated and studied topic in bioinformatics and the tools for this purpose are among the mostly publicly used software in bioinformatics. There are books devoted solely to this topic as general alignment or multiple sequence alignment, and this topic is treated in detail in many contemporary bioinformatics books. Our approach in this chapter was to briefly review the fundamental methods of alignment only, with emphasis on distributed alignment.

Exercises

1. Work out the global alignment between the two DNA sequences below using the dynamic programming approach of NW algorithm. Show all matrix iterations.

    A T G G C T A G T A C C
    G T G C T T G T A C C

2. Find the local alignment between the two protein sequences below using the SW algorithm. Show all matrix iterations.

    B N Q R S T U R V Y A C K
    A N Q T T V T U R X E A C

3. For the following four DNA sequences, implement center star method of multiple sequence alignment by first finding the distances between them and forming the distance matrix D. Find the central sequence and align all of the sequences to the central sequence using NW algorithm and finally insert gaps in sequences to complete the alignment.

    S1: A C C G A A C
    S2: A G C G C T G
    S3: C C C T A T G
    S4: A T C G A T G

4. Compare FASTA and BLAST in terms of method used and the accuracy achieved.

5. Given the following two DNA sequences, draw the general suffix tree for them and find the LCS of these two sequences using this generalized suffix tree.

$$G \; T \; A \; C \; C \; T \; A \; A \; G \; T \; C \; A$$
$$A \; G \; T \; C \; T \; G \; A \; A \; C \; T \; G$$

6. Provide the pseudocode of a distributed BLAST algorithm based on supervisor–worker model. Assume the input queries are distributed to k processes and each worker returns the results to the supervisor.

References

1. Altschul SF, Gish W, Miller W, Myers EW et al (1990) Basic local alignment search tool. J Mol Biol 215(3):403–410
2. Altschul SF, Madden TL, Schaffer AA, Zhang J, Zhang Z, Miller W, Lipman DJ (1997) Gapped BLAST and PSI-BLAST: a new generation of protein database search programs. Nucleic Acids Res 25(17):3389–3402
3. Bjornson R, Sherman A, Weston S, Willard N, Wing J (2002) Turboblast: a parallel implementation of blast based on the turbohub process integration architecture. In: IPDPS 2002 Workshops
4. Boukerche A (2006) Computational molecular biology. In: Albert YZ (ed) Parallel computing for bioinformatics and computational biology models, enabling technologies, and case studies. Wiley series on parallel and distributed computing, Chap, 6
5. Brudno M, Do C, Cooper G, Kim M, Davydov E, Green ED, Sidow A, Batzoglou S (2003) LAGAN and multi-LAGAN: efficient tools for large-scale multiple alignment of genomic DNA. Genome Res 13:721–731
6. Brudno M, Chapman M, Gottgens B, Batzoglou S, Morgenstern B (2003) Fast and sensitive multiple alignment of large genomic sequences. BMC Bioinform 4:66
7. Chaudhary V, Liu F, Matta V, Yang LT (2006) Parallel implementations of local sequence alignment: hardware and software. In: Albert YZ (ed) Parallel computing for bioinformatics and computational biology models, enabling technologies, and case studies. Wiley series on parallel and distributed computing, Chap, 10
8. Cheetham JJ et al (2003) Parallel CLUSTALW for PC clusters, computational science and its applications. In: ICCSA 2003, pp 300–309. LNCS. Springer
9. Cristianini N, Hahn MW (2006) Introduction to computational genomics, a case studies approach. Cambridge University Press, Cambridge
10. Darling A, Carey L, Feng W-C (2003) The design, implementation, and evaluation of mpi-BLAST. In: Cluster world conference and expo and the 4th international conference on linux clusters: the HPC revolution, San Jose, CA
11. Dayhoff MO, Schwartz RM, Orcutt BC (1978) A model of evolutionary change in proteins. Atlas Protein Seq Struct 5(3):345–352
12. Delcher AL, Phillippy A, Carlton J, Salzberg SL (2002) Fast algorithms for large-scale Genome alignment and comparison. Nucleic Acids Res 30(11):2478–2483
13. Duzlevski O (2002) SMP version of ClustalW 1.82, unpublished. http://bioinfo.pbi.nrc.ca/clustalw-smp/

14. Galper AR, Brutlag DR (1990) Parallel similarity search and alignment with the dynamic programming method. Technical report KSL 90–74, Stanford University
15. Grant J, Dunbrack R, Manion F, Ochs M (2002) BeoBLAST: distributed BLAST and PSI-BLAST on a Beowulf cluster. Bioinformatics 18(5):765–766
16. Gropp W, Lusk E, Skjellum A (2014) Using MPI: portable parallel programming with the message passing interface, 3rd edn. MIT Press, ISBN: 9780262527392
17. Henikoff S, Henikoff JG (1992) Amino acid substitution matrices from protein blocks. Proc Nat Acad Sci USA 89(22):10915–10919
18. Hohl M, Kurtz S, Ohlebusch E (2002) Efficient multiple genome alignment. Bioinformatics 18:312–320
19. Janaki C, Joshi RR (2003) Accelerating comparative genomics using parallel computing. Silico Biol 3(4):429–440
20. Li Kuo-Ben (2003) ClustalW-MPI: ClustalW analysis using distributed and parallel computing. Bioinformatics 19(12):1585–1586
21. Lipman DJ, Pearson WR (1985) Rapid and sensitive protein similarity searches. Science 227(4693):1435–1441
22. Mikhailov D, Cofer H, Gomperts R (2001) Performance optimization of Clustal W: parallel Clustal W, HT Clustal, and MULTICLUSTAL. White papers, Silicon Graphics, Mountain View, CA
23. Mount DM (2004) Bioinformatics: sequence and genome analysis, 2nd edn. Cold Spring Harbor Laboratory Press, Cold Spring Harbor. ISBN 0-87969-608-7
24. Naruse A, Nishinomiya N (2002) Hi-per BLAST: high performance BLAST on PC cluster system. Genome Inform 13:254–255
25. National Center for Biotechnology Information, NCBIBLAST. http://www.ncbi.nih.gov/BLAST/
26. Pearson WR (1990) Rapid and sensitive sequence comparison with FASTP and FASTA. Method Enzymol 183:63–98
27. Pearson WR, Lipman DJ (1988) Improved tools for biological sequence comparison. Proc Nat Acad Sci USA 85:2444–2448
28. Rognes T, SeeBerg E (2000) Six-fold speedup of Smith Waterman sequence database searches using parallel processing on common microprocessors. Bioinformatics 16(8):699–706
29. Saitou N, Nei M (1987) The neighbor-joining method: a new method for reconstructing phylogenetic trees. Mol Biol Evol 4(4):406–425
30. Sharapov I (2001) Computational applications for life sciences on Sun platforms: performance overview, Whitepaper
31. Smith TF, Waterman MS (1981) Identification of common molecular subsequences. J Mol Biol 147:195–197
32. Thompson JD, Higgins DG, Gibson TJ (1994) CLUSTALW: improving the sensitivity of progressive multiple sequence alignment through sequence weighting, positions-specific gap penalties and weight matrix choice. Nucleic Acids Res 22:4673–4680
33. Wing-Kin S (2009) Algorithms in bioinformatics: a practical introduction. CRC Press (Taylor & Francis Group), Chap. 5
34. WU blast http://blast.wustl.edu/blast/README.html. Washington University School of Medicine
35. Wu X, Tseng C-W (2006) Searching sequence databases using high-performance BLASTs. In: Albert YZ (ed) Parallel computing for bioinformatics and computational biology models, enabling technologies, and case Studies. Wiley series on parallel and distributed computing, Chap, 9

Clustering of Biological Sequences

7.1 Introduction

Clustering is the process of grouping objects based on some similarity measure. The aim of any clustering method therefore is that the objects belonging to a cluster should be more similar to each other than to the rest of the objects analyzed. Clustering is one of the most studied topics in computer science as it has numerous applications such as in bioinformatics, data mining, image processing, and complex networks such as social networks. Recent technologies provide vast amounts of biological data and clustering of this data to obtain meaningful groups is one of the fundamental research areas in bioinformatics. Clustering of biological sequences commonly involve grouping of three types of data; the genome, protein amino acid sequences, and expressed sequence tags (ESTs) which are small segments from complementary DNA (cDNA) sequences [1]. ESTs are used for gene structure prediction and discovery, and a cDNA is formed by transcription from the mRNA.

We will make a distinction between clustering data points which we will call *data clustering*, and clustering objects which are represented as vertices of a graph in which case we will use the term *graph clustering*. Graph clustering methods for biological networks will be reviewed in Chap. 11, and our emphasis in this chapter is data clustering as applied to biological sequences.

The two classical clustering methods of data are the *hierarchical clustering* and *partitional clustering*. Hierarchical clustering algorithms iteratively build clusters either starting with each data point as a cluster and merge them to form larger clusters in each step, or start with whole data as one cluster and iteratively divide the clusters to form smaller clusters. The first method is called *agglomerative* and the second is the *divisive* hierarchical clustering. The partitional clustering methods on the other hand, attempt to directly group data points, and start with some initial clustering to be modified at each step to yield better clusters.

Since our aim is to group similar DNA/RNA protein sequences, their similarities can be evaluated using the sequence alignment methods and tools such as BLAST in the first step of clustering. The edit distance which is the minimum number of indel

© Springer International Publishing Switzerland 2015

K. Erciyes, *Distributed and Sequential Algorithms for Bioinformatics*, Computational Biology 23, DOI 10.1007/978-3-319-24966-7_7

and substitution operations between two sequences can then be found and the hierarchical or partitional methods can be employed to find clustering of the biological sequences. Nevertheless, there are methods specifically tailored and optimized for biological sequences other than these fundamental approaches. These methods can be broadly categorized as *sequence alignment-based* approaches and *alignment-free* approaches.

In this chapter, we first describe the clustering problem formally by reviewing basic similarity and validation parameters. We will then inspect the two main methods first and then investigate the clustering algorithms aiming at biological sequences. The parallel and distributed algorithms for various approaches will be discussed to conclude the chapter.

7.2 Analysis

Our aim in clustering is the grouping of similar objects. We therefore need to asses the similarities or dissimilarities between the data points under consideration. The methods to accomplish this assessment can be grouped under distance measures and similarity measures.

7.2.1 Distance and Similarity Measures

Distance measures show the dissimilarity between two data points; the more distant they are, naturally, the more dissimilar they are. The *Minskowski distance* between two data points $p_i = p_{i1}, ..., p_{ir}$ and $p_j = p_{j1}, ..., x_{jr}$ each having r dimensions is defined as below [11].

$$d(p_i, p_j) = \left(\sum_{k=1}^{r} (d(p_{ik}, p_{jk})^m) \right)^{1/m} \tag{7.1}$$

When $m = 2$, this distance is called the *Euclidian distance* as widely used in practice. The weighted Euclidian distance is defined as:

$$d(p_i, p_j, W) = \sqrt{\left(\sum_{k=1}^{r} (w_k \cdot d(p_{ik}, p_{jk}))^2 \right)} \tag{7.2}$$

where w_j is the weight assigned to the jth component. The *cosine similarity* is defined as the vector dot product of two vectors x and y characterizing the two data points as follows:

$$s(x, y) = \frac{x \cdot y}{\| x \| \| y \|} \tag{7.3}$$

7.2.2 Validation of Cluster Quality

Once we have the clusters formed, we need to check the quality of the groups formed. Our aim in any clustering method is to group similar objects into the same cluster which should be well separated from other clusters. Validation of a clustering method therefore involves checking how similar the objects in a cluster are and how dissimilar they are to the members of other clusters. Statistical methods can be employed to find mean, variance, and standard error of the members of a cluster to find intra-cluster similarity. Separation of clusters can be evaluated by computing their distances and the similarity/separation ratio may display the quality of the clustering method [15]. Moreover, this ratio of partitioning can be compared with the ratio of a random partitioning method to validate its performance.

Formally, the internal cluster dissimilarity parameter IS_{C_k} of a cluster C_k can be evaluated by computing the average distance between its members to its centroid c_k as follows:

$$IS_{C_k} = \frac{1}{n_k} \sum_{p_i \in C_k} d(p_i, c_k) \tag{7.4}$$

where n_k is the number of data objects in cluster C_k, p_i is a member object of cluster C_k, and c_k is its centroid. We may be interested to find the average value of the intra-cluster similarity values which can be calculated as:

$$IS_C = \frac{1}{N} \sum_{i \in C_k} n_k I_{C_k} \tag{7.5}$$

where N is the total number of clusters. The inter-cluster distance ES_{C_p, C_q} between two clusters C_p and C_q can be approximated by the distance between their centroids c_p, c_q as:

$$ES_{(C_p, C_q)} = d(c_p, c_q) \tag{7.6}$$

and the average value of this parameter for the whole data is:

$$ES_C = \frac{1}{\sum_{p \neq} n_p n_q} \sum_{i \in C_k} n_p n_q E_{C_p} E_{C_q} \tag{7.7}$$

It would be sensible to compare the average value of intra-cluster similarity to the average value of inter-cluster distance $R_C = IS_C / ES_C$, and we want this ratio to be as high as possible for a good clustering method. We can therefore compare the R_{C_i} values for t clustering methods $i = 1, \ldots, t$ and determine their relative performances. There are many other approaches to compare clustering methods other than this simple approach and the reader is referred to [11] for a detailed analysis.

7.3   Classical Methods

Fundamental approaches in classical clustering are the hierarchical, partitional, density-based, and grid-based methods.

7.3.1   Hierarchical Algorithms

Hierarchical clustering is a widely used method for clustering objects due to its simplicity and acceptable performance for data of medium size. This method of clustering can be classified as *agglomerative* or *divisive*. In agglomerative clustering, each data point is considered as a cluster initially. The two clusters that are closest to each other using some metric are then merged to form a new cluster at each step. This process is repeated until a single cluster is obtained. For individual points, Euclidian distance suffices to determine the smallest distances, and in the general case, distance between two clusters C_i and C_j can be defined as follows:

- *Single link*: The distance $d(C_i, C_j)$ is the distance between the two closest points $x \in C_i$ and $y \in C_j$.
- *Complete link*: $d(C_i, C_j)$ is between the two farthest points in C_i and C_j.
- *Average link*: The distance is calculated by finding the average of the distances between every pair of points in C_i and C_j.

Figure 7.1 displays these measures. The divisive algorithms on the other hand start with one cluster that contains all of the objects and then divide the clusters iteratively into smaller ones. The output of a hierarchical clustering is a structure called a *dendogram* which is a tree showing the clusters merged at each step and the dendogram can be divided into the required number of clusters by a horizontal line.

Fig. 7.1 Hierarchical clustering example. Single-link distance between the clusters C_1 and C_2 is 4 as it is the shortest distance, the complete-link distance is 14 as it is the longest distance and the average-link distance is 8.11

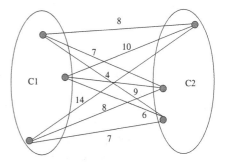

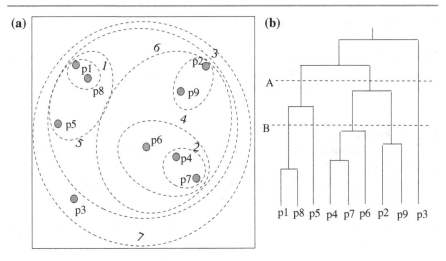

Fig. 7.2 A single-link agglomerative hierarchical clustering example. The steps of the algorithm in (**a**) is shown by italic numbers outside the clusters. The output dendogram is shown in (**b**) is divided into clusters $\{p_1, p_8, p_5\}$, $\{p_4, p_7, p_6, p_2, p_9\}$, and $\{p_3\}$ by horizontal line A; and $\{p_1, p_8\}$, $\{p_5\}$, $\{p_4, p_7, p_6\}$, $\{p_2, p_9\}$, and $\{p_3\}$ by B

A single-link agglomerative hierarchical clustering example is shown in Fig. 7.2a where 7 clusters are formed and the dendogram shown in (b) shows the nesting of these clusters. Algorithm 7.1 shows the pseudocode of the agglomerative hierarchical clustering algorithm.

Algorithm 7.1 *Single-Link Hierarchical Clustering*

1: **Input**: $P = \{p_1, p_2, ..., p_n\}$ ▷ n data points to be clustered
2: **Output**: A dendogram showing the cluster structure
3: **compute** the proximity matrix for data points
4: **repeat**
5:      **combine** the two closest clusters into one cluster
6:      **update** the proximity matrix
7: **until** there is only one cluster

Each iteration of the agglomerative clustering algorithm requires $O(n^2)$ time to find the shortest distance in matrix D, and this process is repeated for $n - 1$ times resulting in a time complexity of $O(n^3)$. Using suitable data structures, time complexity can be reduced to $O(n^2 \log n)$. The hierarchical algorithms are easy to implement and may provide the best results in some cases. Another advantage of these algorithms is that the number of clusters is not needed beforehand. However, these algorithms are not sensitive to noise data points and have difficulty in discovering arbitrary-shaped clusters. They also do not directly optimize an objective function.

7.3.2 Partitional Algorithms

Partitional algorithms attempt to find clustering directly by assigning some arbitrary cluster reference points initially. The most widely used algorithm in this category is the k-means algorithm and its derivatives.

7.3.2.1 k-means Algorithm

The k-means algorithm starts with initial cluster centers called *centroids*. Each data point is then assigned to its closest distance centroid. Assuming we start with k centroids c_1, ..., c_k, the clusters will be formed around these reference points. The new centroids for the formed clusters in the previous step are then re-computed, and each data point is again assigned to one of the new centroids. This process continues until some objective function is achieved, for example, when the coordinates of the centroids do not change significantly meaning we have stable clusters. Algorithm 7.2 displays the pseudo code for the k-means algorithm.

Algorithm 7.2 *k-means Algorithm*

1: **Input**: $\mathcal{P} = \{p_1, p_2, ..., p_n\}$ ▷ n data points to be clustered
2: **Output**: $\mathcal{C} = \{C_1, C_2, ..., C_k\}$ ▷ k clusters
3: **choose** k initial centroid points c_1, ..., c_k
4: **repeat**
5:      **assign** each data point p_i to its closest centroid
6:      **find** the new centroids for each cluster formed around old centroids
7: **until** an objective function is achieved

Three important points to be considered in this algorithm are as follows:

1. **Estimating the number of clusters initially**: Determining the value of k is a fundamental issue in the k-means algorithm. In the simplest case, we can use a number of k values iteratively and select the one that gives the best results. Using a simple heuristic, k can be selected to be approximately equal to $\sqrt{n/2}$, where n is the number of data points [16]. In another approach called the *elbow method*, the percentage of variance explained by the clusters is plotted against the number of clusters; and the k value corresponding to the point where the gain decreases is selected [14]. A survey of methods used for this purpose can be found in [10].

2. **Assigning the initial centroid points**: The initial centroid locations affect the clustering structure obtained significantly. In the simplest case, we can assign the initial centroids randomly and run the k-means algorithm for a number of times, say 10 times, and select the initial placement that gives the best results. In another approach, the hierarchical clustering algorithm can be executed on data and the centroids can be selected as the centers of the obtained clusters. For other methods, selecting centroids that are well separated is a favorable heuristic

since the formed clusters will eventually be separate groups of data points. There is not a single method of selecting initial centroids that works for all input data combinations as reported in [17].

3. **Definition of the objective function**: As the objective function, we need to ensure the clusters obtained are stable, in other words, their centroids remain almost constant. A widely used parameter for the objective function is the sum of squared errors (SSE) defined as follows:

$$SSE(\mathcal{C}) = \sum_{j=1}^{N} \sum_{p_i \in C_k} d(p_i, c_k)^2 \tag{7.8}$$

where C_k is the cluster k and c_k is the centroid of this cluster. The SSE parameter displays the sum of the square of distances of each data point to its centroid, accumulated over all clusters. The centroids can be selected using this criteria and the objective function can then be stated as to minimize the SSE parameter.

Figure 7.3 shows the execution of the k-means algorithm in a small dataset. In order to obtain k clusters of n data objects with each having r dimensions, the k-means algorithm runs in $O(nmkr)$ time where m is the number of iterations. Typically, all of these parameters are much smaller than data size n providing several steps of convergence time. The main advantage of k-means algorithm is that it has low time complexity and the convergence is obtained in several number of steps in practice. It has some limitations though; it works fine when the size of the clusters are similar, and their densities are in comparable ranges. It may not produce the right clusters if these conditions are not valid or the data contains outliers and noise. Also, another disadvantage of this algorithm is that the initial placement of the centroids affect the quality of clustering obtained and therefore additional sophisticated methods are needed for this purpose. However, k-means algorithm and its derivatives have found wide range of applications including bioinformatics due to their simplicities and low running times.

7.3.2.2 *k*-medoid Algorithm

There are few algorithm that are derived from the main k-means algorithm. In the k-medoids algorithm, centroids which are called *centrotypes* now, are real data points instead of coordinates. The initial positioning of the centrotypes should be handled with care as this affects the clusters as in the k-means algorithm. The discovery of the centrotypes is also different than the k-means algorithm for example as in the partitioning around medoids (PAM) algorithm. Each data point in a cluster is swapped with the centrotype and the cost function for this configuration is determined. The replacement that results in the lowest cost function is decided as the centrotype and the clusters are computed using this new centrotype. The cost function can be

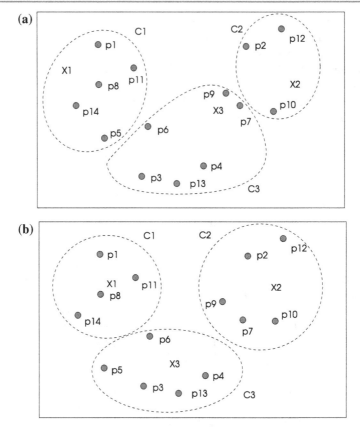

Fig. 7.3 *k*-means algorithm example, **a** There are 14 data points p_1, ..., p_{14} and the initial centroids for 3 clusters are X1, X2, and X3. Each data point is assigned to its nearest centroid resulting in clusters C1, C2, and C3. **b** The new centroids are computed and each data point is assigned to its nearest centroid again

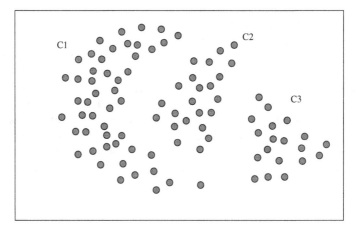

Fig. 7.4 Irregular clusters obtained by a density-based clustering algorithm

implemented as the average distance to the centrotype or the sum of distances to it. Algorithm 7.3 shows the pseudocode for this algorithm which has a similar structure to the k-means algorithm. The k-modes algorithm is very similar to the k-medoids algorithm which can efficiently determine clusters of categorical data.

Algorithm 7.3 *k-medoids Algorithm*

1: **Input**: $P = \{p_1, p_2, ..., p_n\}$ ▷ n data points to be clustered
2: **Output**: $C = \{C_1, C_2, ..., C_k\}$ ▷ k clusters
3: **select** k initial centrotype points $c_1, ..., c_k$
4: **repeat**
5:     **assign** each data point to its closest medoid and form clusters
6:     **for all** $C_i \in C$ **do**
7:         **for all** $p_j \in C_i$ **do**
8:             **swap** c_i and p_j
9:             **evaluate** the total cost function
10:         **end for**
11:         **select** p_j with the total lowest cost as the new centrotype c_i
12:     **end for**
13: **until** an objective function is achieved

7.3.3 Other Methods

Some datasets have nonspherical shapes, and clusters in these data cannot be easily discovered by the hierarchical or partitional algorithms as these assume data obey a probability distribution. *Density-based clustering* algorithms aim to discover clusters of such arbitrary-shaped data. Density-based spatial clustering of applications with noise (DBSCAN) is one such algorithm where density inside a fixed radius area around a data point is considered [8]. A *core* point p has at least *MinPts* number of data points within the radius of *Eps* from itself. A point q within the *Eps* radius of a point p is called *density-reachable* from p. If there is another point that is both density-reachable from both p and q; p and q are said to be *density-connected*. In DBSCAN, clusters are formed with density-connected points. *Noise* data points are not contained in any cluster and *border* data points are included in clusters but they do not have dense regions around them, that is, they are not core points. The irregular clusters obtained by a density-based clustering algorithm are shown in Fig. 7.4. In grid-based clustering, the data space is partitioned into a number of grid cells. Data densities in each of these cells is evaluated and neighboring cells with high densities are combined to form clusters.

7.4 Clustering Algorithms Targeting Biological Sequences

Our aim in the clustering of biological sequences is to arrange them in groups to asses their functions and evaluate phylogenetic relationships between them which can be used for many applications such as understanding diseases and design of therapies.

7.4.1 Alignment-Based Clustering

As stated before, the edit distance between two sequences is the number of indels (insertions-deletions) and substitutions to transform one to another. For example, two DNA sequences S_1 and S_2 and their alignment are given below:

```
S1: A T T C G T T G G A C C   S1: A T T C G T T G G A C C
S2: G T T C T T A G A C       S2: G T T C - T T A G A - -
                                  m       i       m   i i
```

There are two substitutions and 3 indels resulting in an edit distance of 5 between these two sequences. The lengths of the sequences to be aligned need not be the same as in this example. For global alignment where we searched for the alignment of two or more whole sequences, we can use the dynamic programming approach of the Smith–Waterman algorithm of Sect. 6.4 with time complexity of $O(nm)$, where n and m are the lengths of the two sequences. For the clustering of biological sequences, using global alignment tools seems as the right choice at first glance as we are interested in the clustering of the whole sequences. However, proteins which are chains of amino acids consist of groups of these molecules called *domains*. It was found that two globally very much different proteins may have similar domains. Surprisingly, these proteins may have similar functions [4]. Global alignment of such proteins would result in high edit distances, and they would be placed in different clusters although they should be included in the same cluster as they have similar functions. Under these circumstances, local alignment algorithms such as Needleman–Wunsch algorithm or tools like FASTA and BLAST should be used to find the distances between the sequences. All of these alignment methods, whether global or local, provide the edit distances between the sequences and once these are obtained, we can use any of the well-known clustering methods that consider distances between data points such as hierarchical or partitional algorithms. Although these classical algorithms have polynomial time complexities, the biological sequences are huge making even these polynomial times unacceptable in practice. For this reason, significant research has been oriented toward clustering aiming at biological sequences making use of some property of these sequences.

7.4.2 Other Similarity-Based Methods

In *keyword-based* clustering, a biological sequence S is divided into a number of equal-length short segments, and the frequencies of these segments are computed.

Then, a vector V_S which has an element v_i showing the frequency of the segment $s_i \in S$ is formed. The similarity search process between two segments S_1 and S_2 represented by vectors V_1 and V_2 can then be performed by vector operations such as dot product or Euclidian distance calculation [4]. The subsequence-based clustering is another method that aims biological sequences. The frequent subsequences called *motifs*, as we will describe in the next chapter, are searched in a sequence and once these are discovered, sequences with similar motifs can be clustered.

7.4.3   Graph-Based Clustering

The general idea of the graph-based sequence clustering method is to represent the sequences as nodes of a graph G. The edges of the graph represent the similarities between the sequences. The similarities or dissimilarities in the form of distances can be obtained using a sequence alignment method output of which is the distance matrix D. We can then check this matrix for distance $d(u, v)$ between each pair of sequences (u, v) in $O(n^2)$ time, and if $d(u, v)$ is less than a defined threshold τ, edge (u, v) is added to the edges of the graph G. Once graph G is formed, clustering of sequences is reduced to dividing G into locally dense regions which are the sequences with high similarities. Algorithm 7.4 displays the pseudocode for this algorithm.

Algorithm 7.4 *Graph-based Sequence Clustering*

1: **Input**: $S = \{S_1, S_2, ..., S_n\}$ ▷ set of n sequences clustered
2:          and D : distance matrix between all sequences
3:          τ : threshold for distance between sequences
4: **Output**: $C = \{C_1, C_2, ..., C_k\}$ ▷ set of k clusters
5: $V \leftarrow S$ ▷ initialize $G(V, E)$
6: $E \leftarrow \emptyset$
7: **for all** $u \in V$ **do**   ▷ construct $G(V, E)$
8:      **for all** $v \in V$ **do**
9:          **if** $d(u, v) \leq \tau$ **and** $u \neq v$ **then**
10:             $E \leftarrow E \cup (u, v)$
11:          **end if**
12:      **end for**
13: **end for**
14: $C \leftarrow Graph\_Cluster(G, k)$ ▷ partition $G(V, E)$ into k clusters

---

Figure 7.5 displays a small sample graph formed and the clustering of this graph into dense local regions. There are many graph clustering algorithms as we will investigate in Chap. 11.

Fig. 7.5 Clustering of a
small graph

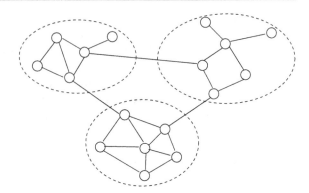

7.5 Distributed Clustering

The parallel and distributed algorithms for clustering large biological sequences can
be broadly classified as memory-based algorithms and disk-based algorithms [4]. If
data can be accommodated in the memories of the distributed memory computers,
memory-based algorithms can be used. Otherwise, data on multiple disks should
be managed. *MapReduce* is a software environment which handles disk resident
data in parallel [6]. The distributed algorithms for biological sequence clustering in
general follow the data partitioning approach. The distances between n sequences are
initially distributed by a supervisor process to k processes each of which performs
local clustering of sequences it is assigned. The results are then typically sent to
the supervisor which merges the clusters and do some postprocessing. It may also
send the partial results to the worker processes for further refinement. Algorithm
7.5 displays the pseudocode for a generic distributed algorithm which works as
described.

7.5.1 Hierarchical Clustering

The hierarchical clustering algorithms provide favorable results superior to the results
obtained from k-means algorithm, and also they have the advantage that the number
of clusters and the initial centroid locations is not required beforehand. However, the
fact that they have quadratic time complexity renders their use in biological sequences
which have large data sizes. Therefore, parallelization of hierarchical algorithms has
been a pursued topic of study for many researchers. However, the main difficulty
encountered is the need for global data access in these algorithms. We need to find
globally, out of all existing clusters, the closest clusters in each step which means
we need to check each cluster-to-cluster distance.

Algorithm 7.5 *Generic Distributed Clustering Algorithm*

```
1: Input : S = {S₁, S₂, ..., Sₙ}                          ▷ set of n sequences to cluster
2:          D : distance matrix between all sequences
3: Output : C = {C₁, C₂, ..., Cₘ}                                    ▷ set of m clusters
4:          P = {p₀, ..., p_{k-1}}                                    ▷ set of k processes
5: if pᵢ = p₀ then                                              ▷ if I am the supervisor
6:    for i = 1 to k − 1 do                                  ▷ send sequences to workers
7:        send rows ((i − 1)n/k)) + 1 to in/k of D to pᵢ
8:    end for
9:    for all pᵢ ∈ P do                                ▷ receive partial clusters from workers
10:       receive clusters(C_{pᵢ}) from pᵢ
11:       C' ← C' ∪ C_{pᵢ}
12:   end for
13:   merge clusters in C' to get C and do postprocessing
14: else                                                          ▷ I am a worker
15:   receive my rows
16:   cluster data to get C_{pᵢ}
17:   send C_{pᵢ} to p₀
18: end if
```

We will describe two ways of parallelizing the agglomerative hierarchical clustering (AHC) algorithm in a distributed memory computing system. Our first approach is based on finding the minimum distance between the clusters in parallel and the second method makes use of the observation that the dendogram output of an AHC algorithm is in fact the MST of a full connected graph showing the distances between the data points.

7.5.1.1 A Distributed AHC Algorithm Proposal

We can sketch a simple distributed algorithm to run AHC algorithm using a number of processes. A close look at the algorithm reveals that the two main time-consuming tasks in AHC algorithm are finding the minimum distance value between the clusters and updating the distances to the newly formed clusters. A simple approach to provide parallel processing using distributed memory computers is then to row-wise 1-D partition the distance matrix D among k processes. There is a central process p_0 and each process p_i gets $\lceil n/k \rceil$ rows of the original matrix from p_0 initially. Each p_i then iteratively finds the minimum distance value d_{p_i} in the row values it owns after which it sends this value to the root p_0. The root process collects all these values and finds the global minimum, and broadcasts the global minimum value to all processes. The process(es) which have the merged clusters update their data structures and all processes compute distances to the newly formed cluster. The pseudocode for this algorithm is shown in Algorithm 7.6 where supervisor is also involved in finding the closest clusters.

We will describe the operation of the distributed algorithm for a small set of 6 biological sequences $S_1, ..., S_6$ to be processed by 3 processes p_0, p_1 and p_2.

Algorithm 7.6 *Dist_AHC*

1: **Input** : $S = \{S_1, S_2, ..., S_n\}$ ▷ set of n sequences to cluster
2:            $D[n, n]$: distance matrix between all sequences
3:            $P = \{p_0, p_1, ..., p_{k-1}\}$ ▷ set of k processes
4: **Output** : $C = \{C_1, C_2, ..., C_m\}, T$ ▷ set of m clusters and a phylogenetic tree T
5: **if** $p_i = p_0$ **then**   ▷ if I am the root process
6:     **for** $i = 1$ to $k - 1$ **do**
7:         **send** rows $((i - 1)n/k)) + 1$ to in/k of D to p_i
8:     **end for**
9:     **for** $round = 1$ to $n - 1$ **do**   ▷ loop for n-1 rounds
10:        $count \leftarrow 0$; $vals \leftarrow 0$
11:        **find** $d_{min}(u, v)$ in my partition
12:        $vals \leftarrow vals \cup \{(u, v, d_{min})\}$
13:        **while** $count < k$ **do**   ▷ get local minimum values from processes
14:            **receive** $proc\_min(u, v, val)$ from p_i
15:            $vals \leftarrow vals \cup \{(u, v, val)\}$
16:            $count \leftarrow count + 1$
17:        **end while**
18:        $min\_val(u, v) \leftarrow min(vals)$ ▷ find the global minimum distance
19:        **broadcast** $new\_min(u, v, min\_val)$ to all processes
20:    **end for**
21: **else**   ▷ I am a worker process
22:    $round \leftarrow 0$
23:    **receive** my row partition $D[my\_rows]$ from p_0
24:    **while** $round < k$ **do**   ▷ get local minimum values from processes
25:        $d_{p_i}(u, v) \leftarrow$ minimum distance between u and v in $D[my\_rows]$
26:        **send** $proc\_min(u, v, d_{p_i})$ to p_0
27:        **receive** $new\_min(p, q, val)$ from p_0
28:        **if** p or $q \in my\_rows$ **then**
29:            **if** $p \leq q$ **then**
30:                **form** new cluster C_p by merging rows p and q into row p
31:            **end if**
32:            **remove** row q
33:        **end if**
34:        **compute** distances to the new cluster C_p
35:        $round \leftarrow round + 1$
36:    **end while**
37: **end if**

The distance matrix D is already formed using a sequence alignment tool like
BLAST. This matrix as row-partitioned between these processes is given as below.
In the first step, the global minimum distance is 1 between the sequences S_1 and S_4.

These two sequences are clustered and the distances to this new cluster are computed as shown in Fig. 7.6. This processs continues until there is one cluster that
contains all of the clusters as shown in Figs. 7.7 and 7.8 and the final tree constructed
is shown in Fig. 7.9.

	S_1	S_2	S_3	S_4	S_5	S_6	
S_1	0	12	15	1	9	5	p_0
S_2	12	0	4	5	8	14	
S_3	15	4	0	10	2	11	p_1
S_4	1	5	10	0	7	13	
S_5	9	8	2	7	0	6	p_2
S_6	5	14	11	13	6	0	

$\rightarrow$

	(S_1,S_4)	S_2	S_3	S_5	S_6	
(S_1,S_4)	0	5	10	7	5	p_0
S_2	5	0	4	8	14	
S_3	10	4	0	2	11	p_1
S_5	7	8	2	0	6	p_2
S_6	5	14	11	6	0	

Fig. 7.6 Distributed AHC algorithm first and second rounds. The minimum distance is between sequences S_1 and S_4 which are clustered to form the matrix on the *right*. The minimum distance in the right matrix is between S_3 and S_5 which are clustered

	(S_1,S_4)	S_2	(S_3,S_5)	S_6	
(S_1,S_4)	0	5	7	5	p_0
S_2	5	0	4	14	
(S_3,S_5)	7	4	0	6	p_1
S_6	5	14	6	0	p_2

$\rightarrow$

	(S_1,S_4)	(S_2,S_3,S_5)	(S_6)	
(S_1,S_4)	0	5	5	p_0
(S_2,S_3,S_5)	5	0	6	p_1
S_6	5	6	0	p_2

Fig. 7.7 Distributed AHC algorithm third and fourth rounds. The minimum distance in the *left* matrix above is between clusters (S_2) and (S_3, S_5) which are clustered. The minimum distance now is selected arbitrarily to be between clusters (S_1, S_4) and (S_6) as shown in the *right* matrix

	(S_1,S_4,S_6)	(S_2,S_3,S_5)	
(S_1,S_4,S_6)	0	5	p_0
(S_2,S_3,S_5)	5	0	p_1

Fig. 7.8 Distributed AHC algorithm last iteration. The minimum distance now is between clusters (S_1, S_4, S_6) and (S_2, S_3, S_5) which are clustered to form the last cluster which contains all of the clusters

Analysis

There are k processes as $p_0, \ldots, p_{k-1}$, and each process p_i has n/k rows to work with, each p_i finds the smallest distance in its group in $O(n^2/k)$ time. There are $n - 1$ rounds of the distributed algorithm, and the total time taken is therefore $O(n^3/k)$. The sequential algorithm had $O(n^3)$ time complexity, hence, the speedup obtained is the processor number k ignoring the communication overheads.

7.5.1.2 A Distributed Graph-Based AHC Algorithm

An alternative and promising approach in the parallelization of the hierarchical clustering algorithms is based on the following. Finding the dendogram output using the single-link proximity metric is equivalent to finding the MST of a fully connected weighted graph in which vertices are the data points and the weights display the dissimilarities (distances) between the vertices. We will illustrate this concept by an example where our aim is again to cluster the same 6 biological sequences $S_1, \ldots, S_6$ as in the example of previous section. Let us assume the edit distances for these 6 sequences are available from an alignment program and are included in the distance

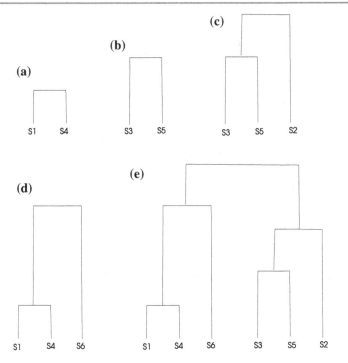

Fig. 7.9 The partial trees constructed with the distributed algorithm in each round. The trees in (a),…, (e) correspond to the rounds 1,…, 5. The tree in (e) is final (Not to the scale)

matrix D with an entry d_{ij} showing the distance between sequences i and j as before. We will repeat this matrix here as follows.

	S_1	S_2	S_3	S_4	S_5	S_6
S_1	0	12	15	1	9	5
S_2	12	0	4	5	8	14
S_3	15	4	0	10	2	11
S_4	1	5	10	0	7	13
S_5	9	8	2	7	0	6
S_6	3	14	11	13	6	0

The first step of the proposed parallel method involves constructing a full connected graph K_6 using these sequences as vertices and the edge weights denoting their distances in D. We then search for an MST of this graph using a suitable algorithm. The three algorithms to build an MST of a weighted graph are the Boruvka's, Prim's, and Kruskal's algorithms. Boruvka's algorithm tests each node of the graph and works by always selecting the lightest edges. Let us use Kruskal's algorithm and sort edges in increasing weights. We include edges in MST starting from the lightest edge as long as they do not create any cycles with the existing edges of the partial MST and continue this process for $(n - 1)$ times. The MST obtained from K_6 is shown in Fig. 7.10.

The output dendogram using the single-link agglomerative hierarchical clustering algorithm for the same dataset is displayed in Fig. 7.11. We can see that these are equivalent. In fact, the hierarchical clustering algorithm and the Kruskal's algorithms work using a similar principle, we always combine the closest pairs of clusters in the clustering algorithm and we choose the lightest weight edge in the MST algorithm. Cycle creating edges are discarded in the MST algorithm, similarly, two existing nodes of a cluster are never processed in the clustering algorithm. The construction of the MST dominates the time taken for the clustering algorithm and hence, we will look at ways of parallelizing this step.

MST Construction

Considering the three algorithms to build MST which are Prim's, Kruskal's, and Boruvka's algorithms, it can be seen at first glance that Prim's and Kruskal's algo-

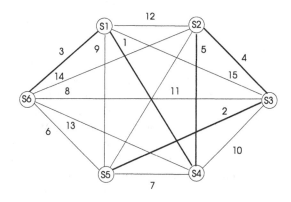

Fig. 7.10 MST of the K_6 graph representing sequence distances shown in *bold lines*

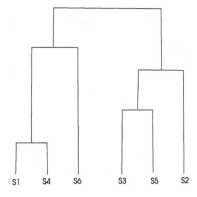

Fig. 7.11 Dendogram obtained from Fig. 7.10

rithms process only one edge at a time and therefore are not suitable for independent operations on processors, hence are difficult to parallelize. Boruvka's algorithm on the other hand provides independent operations on vertices and hence can be parallelized more conveniently. For this reason, we will elaborate on Boruvka's algorithm which consists of the following steps:

1. Find the lightest weight edge (u, v) from each vertex u in G. The edge (u, v) is included in MST.
2. Combine u and v to have a component which contains vertices u and v and the edge (u, v) between them.
3. Combine components with their neighbor components using the lightest edge (u, v) between them. The edge (u, v) is included in MST.
4. Repeat Step 3 until there is only one component.

Combining the components is called the *contraction* process and the edges used in contraction constitute the MST. The operation of this algorithm is depicted in Fig. 7.12 where the MST is formed only in two steps. The MST is unique as the edge weights are distinct, and Kruskal's or Prim's algorithm will provide the same MST.

We can form a simple distributed version of this algorithm by distributing the partitions of graph G to p processors each of which performs Boruvka's algorithm on its subgraph until they all have one component. They can then exchange messages to perform contraction between them.

7.5.2 *k*-means Clustering

The k-means algorithm can be parallelized in a simple manner to run in a distributed memory computing system. We implement the supervisor–worker paradigm again in which one of the processes is assigned as the root or the supervisor process that controls the overall running of the algorithm. We assume the distance matrix $D[n, n]$ for n sequences which has an entry d_{ij} showing the distance between sequences i and j is already available as output from a sequence alignment tool. The general idea of the distributed k-means algorithm is to have each process form the partial clusters for the points it owns around the current centroids distributed by the supervisor at each step. The rows of D are evenly distributed to all processes using row-wise 1D partitioning initially as in the generic distributed clustering algorithm.

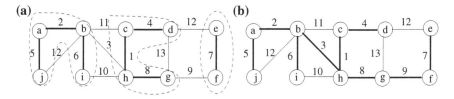

Fig. 7.12 Execution of Boruvka's algorithm in a sample graph. **a** The first step. **b** The second step. *Bold lines* are included in the MST

Algorithm 7.7 shows the operation of the distributed k-means using k processes with p_0 serving as the supervisor. Different than the generic algorithm, the supervisor itself is also involved in the clustering process. In the initialization phase, p_0 assigns the initial m centroids $c_1, ..., c_m$ to the set M and sends the partitions rows along with M to each worker process. Each process $p_i \neq p_0$ then works out the clusters of the data points in its rows by computing their distances to the centroids, and sends the partial cluster sets to the root at the end of the round, the root merges the partial clusters from all workers to find the full clusters, calculates the new centroids and sends these to the workers for the next round. This process continues until the centroids are stable in which case p_0 sends *stop* message to notify workers that the process is over. It should be noted that data points are sent once in the first round and only the newly calculated centroids are transferred in each round. Each process should then assign the data points in its rows to the clusters based on these centroids and send the partial clusters to the root at the end of a round. This process continues until an objective function is met.

Algorithm 7.7 *Distributed k-means Clustering Algorithm*

1: **Input**: $S = \{S_1, S_2, ..., S_n\}$ ▷ set of n sequences to cluster
2:         $D[n, n]$: distance matrix between all sequences
3:         $P = \{p_0, p_1, ..., p_{k-1}\}$ ▷ set of k processes
4: **Output**: $C = \{C_1, C_2, ..., C_m\}$ ▷ set of m clusters
5: **if** $p_i = p_0$ **then**   ▷ I am the supervisor
6:     **for** $i = 1$ to $k - 1$ **do**   ▷ send rows to workers
7:        **send** rows $((i - 1)n/k) + 1$ to in/k of D to p_i
8:     **end for**
9:     $M \leftarrow$ initial centroids $c_1, ..., c_m$
10:     **while** centroids are not stable **do**
11:        $C' \leftarrow \emptyset$
12:        **broadcast** M to all processes
13:        $C' \leftarrow$ clusters in my rows 1 to n/k
14:        **for all** $p_i \in P$ **do**   ▷ receive partial clusters from workers
15:           **receive** $cluster(C_{p_i})$ from p_i
16:           $C' \leftarrow C' \cup C_{p_i}$
17:        **end for**
18:        **merge** clusters in C' to get C
19:        $M \leftarrow$ new centroids $c_1, ..., c_m$ of C
20:     **end while**
21:     **broadcast** *stop* to all processes
22: **else**   ▷ I am a worker
23:     **receive** my rows
24:     **while** *stop* not received from p_0 **do**
25:        **receive** centroids M
26:        **cluster** data in my rows to get C_{p_i}
27:        **send** $cluster(C_{p_i})$ to p_0
28:     **end while**
29: **end if**

7.5.3 Graph-Based Clustering

The similarity graph for a set of biological sequences can be constructed by connecting two similar sequences by an edge. Our aim is now to find dense regions in this graph to build clusters. There exists various algorithms for this purpose as we will see in Chap. 11 when we investigate clustering in biological networks. For now, we need to find a method such that any sequential graph algorithm can be performed in parallel if we can partition the graph among the processes of the distributed computing system. There are numerous algorithms for this purpose, for example, we may require to divide the graph into partitions such that the number of vertices in each partition is approximately equal. We may have a different objective such as to have a minimum number of edges between the partitions or minimum sum of weights of edges if the graph is weighted. In this case, our aim is basically to seek clusters of the graph. We can even pursue both of these requirements, that is, partitions should have similar number of vertices with few edges between the partitions. If our aim is simply load balancing to provide each process with nearly equal workload, then partitioning for similar number of vertices in each partition would be sensible, especially if the clustering algorithm employed has a time complexity dependent on the vertex number rather than edges, as frequently encountered in practice. We will describe a simple method for this purpose next.

7.5.3.1 Graph Partitioning by BFS

Graph partitioning of a graph is different than clustering of it in few aspects. First, the partitions do not overlap, and we usually determine the number of partitions of a graph beforehand and more importantly, our aim is to divide a graph $G(V, E)$ into k subgraphs V_1, V_2, .., V_k such that $|V_1| \approx |V_2| \approx \ldots |V_k|$ with the minimal number of edges between the partitions. As can be seen, our aim here is different than clustering where we try to discover existing dense parts of a graph. For edge weighted graphs, the partitioning method should provide a minimal total weight of edges between the partitions. If vertex also have weights, we have an additional requirement to have a similar total weight of vertices in each partition.

We search for clusters in biological networks to understand their behavior, rather than partitioning them. However, we will frequently need to partition a graph representing such a network for load balancing in a distributed memory computing system. Our aim is to send a partition of a graph to a process in such a system so that parallel processing can be achieved. We will now describe an efficient yet very simple method of partitioning an unweighted, undirected simple graph into two similar-sized partitions.

We can partition such a graph $G(V, E)$ using the BFS algorithm of Sect. 3.3 as follows as in [7]. When the BFS algorithm is executed in G for a root vertex v, we obtain a BFS tree T rooted at v where each node other than the root has a parent and each node other than the leaf nodes have at least one child, and there are n-1 tree edges in this graph, n being the number of vertices. Other than the BFS tree edges, there are edges that exist between the vertices at the same level in the tree called *horizontal* edges and also edges between the vertices of adjacent levels called *inter-level* edges which are not included in the tree as shown in Fig. 7.13. As we iteratively run the BFS algorithm, we can record the total number of nodes s searched up to that point, and when $s \geq n/2$, we place all of the nodes discovered so far in partition 1 and the rest in partition 2. This will require a minor modification to the BFS algorithm and will provide almost equal-sized partitions. As the line that represents this partition will be just between the adjacent levels and not cross horizontal edges, we will have fewer edges between the two partitions and hence satisfy the second requirement of partitioning. Figure. 7.13 displays such a partitioning.

The time complexity of this algorithm is $O(n+m)$ as the ordinary BFS algorithm. As we frequently need to partition G into k partitions called *k-way partitioning*, we need to run this algorithm recursively for each partition such that at $\log k$ steps, we will achieve the required partitions. The total complexity will then be $O(m \log k)$ assuming $m \geq n$ in a connected graph.

7.5.4 Review of Existing Algorithms

There are various available algorithms and software for the parallel clustering of data, some aiming the biological sequences. These algorithms can be broadly categorized as memory-based and disk-based as was noted before.

Density-based distributed clustering (DBDC) [3] algorithm aims to discover arbitrary-shaped clusters. Data points are partitioned and sent to a number of processors which find the local clusters and send these back to the supervisor process which uses density-based spatial clustering of applications with noise (DBSCAN) [8] to combine the local clusters into global ones. The supervisor then sends the global clusters to the workers for postprocessing. This process is very similar to the procedure described in our generic distributed clustering algorithm of Algorithm 7.5. The speedup obtained by DBDC is favorable with good quality results.

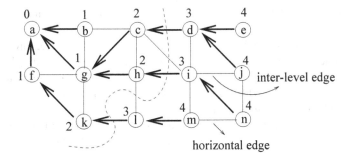

Fig. 7.13 BFS-based partitioning of a graph. The tree rooted at vertex a have edges shown by *bold arrows*, levels of vertices are shown next to them, and the dividing line is between levels 2 and 3 providing equal partitions of size 7

Clustering through MST in parallel (CLUMP) is a clustering method designed to detect dense regions of biological data [19]. It can be used to detect clusters in any network including biological networks but as its main parallel operation is the construction of the MST of the graph, we can also use it to cluster biological sequences. MST-based clustering is used in various applications including biological networks [23]. A review of parallel MST construction algorithms is provided in [18]. Similar to the CLUMP algorithm, a distributed single-link AHC algorithm, this time using MapReduce is proposed in [12] in which the MST of the complete graph is built in parallel to yield the dendogram output. The distance matrix D is partitioned into s splits, and subgraphs are formed by each of these two splits. The MST for each subgraph is then computed using Prim's MST algorithm in parallel, and the resulting partial MSTs are merged to give the final MST using Kruskal's algorithm which eliminates redundant edges. The authors showed the proposed algorithm is scalable and achieves a speedup of 160 on 190 computer cores.

PKMeans is a parallel k-means algorithm that uses MapReduce for disk-based data management [25]. Data is partitioned; local clustering is handled by the *mapper*, and the global clustering is performed by the *combiner* and *reducer* modules of the MapReduce. DisCo [20] and BoW [5] both use MapReduce and data partitioning for distributed clustering of data points. A recent work by Zhang et al., provided a parallel k-means algorithm called MKmeans using MPI [24] using data partitioning. Wang et al., provided a shared memory and a distributed clustering algorithm for large-scale biological sets [22] using affinity propagation [9]. Surveys of clustering methods for gene expression data can be found in [2,13,21].

7.6 Chapter Notes

A cluster of objects have some measure of similarity between them compared to other objects under consideration. We need to cluster biological sequences with the aim of deducing their functionalities better. If a number of sequences are in the same cluster, they have similar sequence structures and presumably their functionalities are similar.

We started this chapter by defining basic distance and similarity measures. There are various other and more sophisticated metrics of distance and similarity between objects other than the Euclidian distance, Minkowski distance, and Cosine similarity we described. We then provided a brief review of the fundamental classical data clustering methods of hierarchical and partitional clustering and other methods of density-based clustering in the first part. There is not a single best method that can be implemented for all applications since each have merits and demerits as outlined.

For clustering biological sequences, the edit distance output by sequence alignment algorithms would provide us the similarities between the sequences. We can therefore use any classical method with these calculated distances. However, biological sequences have large sizes making it difficult to implement even the polynomial time classical algorithms for this purpose. There are a number of clustering methods targeting the biological sequences. Keyword-based clustering divides the sequence into fixed-size segments and forms a vector with elements showing the frequencies of these sequences. Various vector comparisons can then be employed to find the similarities of the sequences and cluster them. Subsequence clustering attempts to find common subsequences among the input and tries to cluster them using these subsequences. A graph with nodes representing sequences and edges denoting their similarities can be built and clustering is then reduced to finding dense regions of this graph as these indicate closely related sequences. This process is called graph-based clustering, and there are various algorithms to detect these dense regions as we will see in Chap. 11 when we investigate clustering in biological networks.

In the final part of the chapter, we reviewed parallel and distributed clustering algorithms for classical data clustering and biological sequence clustering. We proposed a distributed single-link hierarchical clustering algorithm in which the distances between the clusters are computed in parallel. The second approach of this clustering method is based on constructing the MST of the complete graph with edges showing the distances. The popular k-means algorithm can be easily parallelized on distributed memory computers as pursued in various research studies. Providing a distributed version of graph-based clustering of biological sequences is more difficult due to the dependencies of subtasks involved. More research is needed in this area, namely, distributed graph-based clustering of biological sequences.

Exercises

1. Compare the hierarchical and partitional clustering algorithms in terms of accuracy, time complexity, and additional requirements.
2. For the data points shown in Fig. 7.14, work out the dendogram output using the agglomerative hierarchical clustering algorithm.
3. Workout the clusters of data points in Fig. 7.14, this time with divisive hierarchical clustering algorithm.
4. Use k-means algorithm to cluster the data points displayed in Fig. 7.15. The objective function is specified as stable cluster membership, that is, members of each cluster do not change.

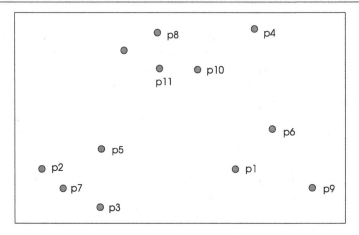

Fig. 7.14 Data points for Exercises 2 and 3

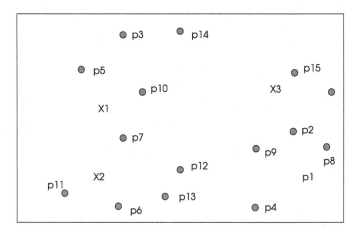

Fig. 7.15 Data points for Exercise 4

5. The distance matrix for 6 data points is given as below. Assuming three processes are available, work out clusters formed using the distributed AHC algorithm executed in parallel by the three processes. Show every iteration of the distributed AHC algorithm.

Fig. 7.16 Example graph for Exercise 6

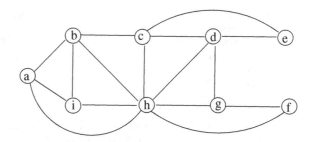

	S_1	S_2	S_3	S_4	S_5	S_6
S_1	0	4	1	8	3	11
S_2	4	0	11	14	5	6
S_3	1	11	0	9	10	7
S_4	8	4	9	0	13	4
S_5	3	5	10	13	0	2
S_6	11	6	7	4	2	0

6. Work out the two partitions formed by the BFS-based graph partitioning algorithm of Sect. 7.5.3 for the graph shown in Fig. 7.16 for the source vertex a.

References

1. Adams MD, Kelley JM, Gocayne JD, Dubnick M, Polymeropoulos MH, Xiao H, Merril CR, Wu A, Olde B, Moreno RF, Kerlavage AR, McConbie WR, Venter JC (1991) Complementary DNA sequencing: expressed sequence tags and human genome project. Science 252:1651–1656
2. Asyali MH, Colak D, Demirkaya O, Inan MS (2006) Gene expression profile classification: a review. Curr Bioinform 1:55–73
3. Cao F, Ester M, Qian W, Zhou A (2006) Density-based clustering over an evolving data stream with noise. In: SDM Conference, pp 326–337
4. Charu C Aggarwal, Chandan K Reddy (ed) (2014) Data clustering algorithms and applications. CRC Press, Taylor and Francis
5. Cordeiro RLF, Traina Jr C, Traina AJM, Lopez J, Kang U, Faloutsos C (2011) Clustering very large multi-dimensional datasets with MapReduce. In: Proceedings of KDD, pp 690–698
6. Dean J, Ghemawat S (2004) MapReduce: simplified data processing on large clusters. In: Proceedings of OSDI, pp 139–149
7. Erciyes K (2014) Complex networks: an algorithmic perspective, pp 145–147. CRC Press, Taylor and Francis, ISBN 978-1-4471-5172-2
8. Ester M, Kriegel H-P, Sander J, Xu X (1996) A density-based algorithm for discovering clusters in large spatial databases with noise. In: Proceedings of the second international conference on knowledge discovery and data mining, KDD, pp 226–231
9. Frey BJ, Dueck D (2007) Clustering by passing messages between data points. Science 315:972–976
10. Gordon A (1996) Null models in cluster validation. In: Gaul W, Pfeifer D (eds) From data to knowledge. Springer, New York, pp 32–44

11. Han J, Kamber M (2001) Data mining: concepts and techniques. Morgan Kaufmann Publishers, San Francisco
12. Jin C, Patwary Md MA, Agrawal A, Hendrix W, Liao W-K, Choudhary A (2013) DiSC: a distributed single-linkage hierarchical clustering algorithm using MapReduce. In: Proceedings of SC workshops, the fourth international workshop on data intensive computing in the clouds (DataCloud 2013)
13. Kerr G, Ruskin HJ, Crane M, Doolan P (2008) Techniques for clustering gene expression data. Comput Biol Med 38:283–293
14. Ketchen DJ Jr, Shook CL (1996) The application of cluster analysis in strategic management research: an analysis and critique. Strateg Manag J 17(6):441–458
15. MacCuish JD, MacCuish NE (2010) Clustering in bioinformatics and drug discovery. CRC Press, Taylor and Francis
16. Mardia K et al (1979) Multivariate analysis. Academic Press, London
17. Meila M, Heckerman D (1998) An experimental comparison of several clustering methods. In Microsoft research report MSR-TR-98-06, Redmond WA
18. Murtagh F (2002) Clustering in massive data sets. In: Handbook of massive data sets, pp 501–543
19. Olman V, Mao F, Wu H, Xu Y (2009) Parallel clustering algorithm for large data sets with applications in bioinformatics. IEEE/ACM Trans Comput Biol Bioinform 6:344–352
20. Papadimitriou S, Sun J (2008) DisCo: Distributed co-clustering with Map-Reduce: a case study towards petabyte-scale end-to-end mining. Proc ICDM 2008:512–521
21. Pham TD, Wells C, Crane DI (2006) Analysis of microarray gene expression data. Curr Bioinform 1:37–53
22. Wang M, Zhang W, Ding W, Dai D, Zhang H, Xie H, Chen L, Guo Y, Xie J (2014) Parallel clustering algorithm for large-scale biological data sets. PLoS One. doi:10.1371/journal.pone.0091315
23. Xu R, Wunsch D II (2005) Survey of clustering algorithms. IEEE Trans Neural Networks 16(3):645–678
24. Zhang J, Wu G, Hu X, Li S, Hao S (2013) A Parallel clustering algorithm with MPI-MKmeans. J Comput 8(1):10–17
25. Zhao W, Ma H, He Q (2009) Parallel k-means clustering based on MapReduce. In: Proceedings of CloudCom 2009, pp 674–679

Sequence Repeats

8

8.1 Introduction

The repeat analysis of DNA/RNA or proteins involves finding the location and determining the characteristics of the repetitive subsequences in these cellular structures. Given a number of DNA sequences, we can concatenate them to form a single long sequence and then perform repeat analysis. In this form, the repeat analysis problem is similar to the problem of sequence alignment since the repeats in such a sequence are the matching parts of the individual sequences. Repeating nucleotide sequences in human DNA occur frequently and various other organisms have also repeating subsequences.

The repeating patterns in human genome are basically of two types; *tandem repeats* and *dispersed repeats*. Tandem repeats are repeating adjacent subsequences such as GATAGATAGATA where the subsequence GATA is repeated 3 times consecutively. They usually reside in the non-coding region of DNA but also in genes and there is a high rate of change in the number of some repeats. Tandem repeats of amino acids in proteins also exist and these vary from a single amino acid to 100 or more residues. Tandem repeats in DNA also have varying sizes, from 1 bp to a complete gene which is possible. Short tandem repeats (STRs), also called *microsatellites*, are typically between 1 and 6 bp long and are considered as genetical markers of individuals [9]. Finding tandem repeats has many implications; they signal the location of genes and they are assumed to have some function in gene regulation. They can be used for identification of an individual in forensics and the increased number of tandem repeats is associated with a number of genetically inherited diseases such as fragile-X mental retardation, central nervous system deficiency, myotonic dystrophy, Huntington's disease, and also with diabetes and epilepsy [8]. They can also be used to identify complex diseases such as cancer [23,24]. Additionally, they are also needed in population genetic studies where the number of repeats is similar between related individuals.

© Springer International Publishing Switzerland 2015

K. Erciyes, *Distributed and Sequential Algorithms for Bioinformatics*, Computational Biology 23, DOI 10.1007/978-3-319-24966-7_8

The dispersed repeats are noncontiguous and these patterns are commonly called *sequence motifs*. Such a sequence motif is a nucleotide sequence in a DNA or RNA structure, typically in genes, or amino acid sequence in a protein that is found statistically more frequent than any other sequences. For example, a sequence motif of the sequence CG**ATC**AGCT**ATC**CCT**ATC**GGA is ATC which is repeated 3 times. Such motifs can be searched in two ways: over-represented sequence repeats in a single DNA or protein; or conserved motifs in a set of orthologous DNA sequences. A *structural motif* of a protein on the other hand is usually formed by a certain sequence of amino acids to result in a three-dimensional physical shape of the protein. A sequence motif in a gene may encode for a structural motif in a protein.

DNA sequence motifs are usually present in transcription factor binding sites (TFBS) where proteins such as *transcription factors* bind to these sites to regulate the expression of genes. Hence, discovery of such motifs helps to understand the mechanisms of gene expression [17,34,35]. *Motif discovery* or *motif extraction* is the process of finding recurring patterns in biological data. Motif discovery algorithms can be classified as statistical and combinatorial. Statistical algorithms are based on the computation of the probability of a motif existence and although they provide suboptimal solutions, they are usually faster than exact algorithms due to restricted search space. Combinatorial motif discovery algorithms on the other hand are exact algorithms in the general sense. A subclass of combinatorial algorithms based on graph theory has become popular recently as we will investigate. If the motif to be searched is known beforehand, the motif discovery problem in this form is reduced to pattern matching where we searched all occurrences of a pattern P in a text T as we reviewed in Chap. 5.

In the general case of repeat analysis, a number of mutations in repeats are permissable as observed in practice. In this case, the search is for approximate repeats, whether tandem repeats or sequence motifs, rather than exact ones. Searching repeating patterns is more directed toward nucleotide sequence repeats as amino acid sequence repeats in proteins are not as dynamic as the DNA/RNA sequences. Yet, another point to consider is whether the search of a motif is targeted toward a single long sequence or we need to search conserved motifs in a set of relatively short sequences. In summary, we have various facets of the sequence repeat problem as stated below:

1. *Tandem or dispersed repeats*: Search is for consecutive patterns or a pattern that is dispersed in noncontiguous locations in the sequence.
2. *Repeat structure is identified before the search or not*: If the characters in the repeat are known before the search, the problem is reduced to pattern matching in which several exact algorithms exist as we reviewed in Sect. 5.2.
3. *Pattern-based or profile-based*: We search the pattern of the motif in the first approach and the locations of the motif in the latter.
4. *Single or multiple sequence search*: We may search a repeat sequence in a single sequence or conserved repeats in a number of sequences. In the first case, the increased number of repeats may have some biological significance such as an indication of a disease state. The search for conserved segments in a number of

sequences is commonly performed to compare them and find their relatedness to infer phylogenetic relationships among them in the latter.

5. *Exact or approximate repeats*: We may search for an exact repeat or repeats that have at most a specified Hamming or edit distance between them.
6. *Probabilistic or combinatorial methods*: Probabilistic methods sample data based on a probability model and provide inexact solutions. The combinatorial methods, however, typically provide all repeats at the expense of increased computation time.
7. *Simple or structured repeats*: Whether the search is for a simple motif or a structured DNA motif which consists of a central motif and one or two *satellites* at the left and right of the central motif.

Naturally, the search method employed can have a possible combination of some of these characteristics. For example, our aim may be to find pattern discoveries of structured and dispersed motifs in a set of sequences using probabilistic methods. In this chapter, we will first describe methods for finding tandem repeats mainly in DNA sequences. We will then investigate motif discovery problem and then describe sequential and distributed algorithms with focus on graph-theoretic motif search.

8.2 Tandem Repeats

Tandem repeats are observed in both the genome and proteins. A tandem repeat that is repeated more than twice is sometimes called a *tandem array* [32], we will, however, use the term *repeat* to mean both. A tandem repeat is called *primitive* if it does not contain any other tandem repeats, otherwise it is *non-primitive*. For example, GTAC is a primitive tandem repeat and TATA is a non-primitive tandem repeat as it contains another repeating pattern of TA. A repeating pair (i_1, j_1), (i_2, j_2) of a given sequence $S = s_1 \ldots s_n$ implies $s_{i_1} \ldots s_{j_1} = s_{i_2} \ldots s_{j_2}$ for tandem repeats and sequence motifs [31]. The length of such a repeated pair is $j - i + 1$. Given a sequence AGCC ATCA ATCA ATCA TCGCC, the tandem repeat is ATCA with starting indexes 5, 9, and 13, of length 4 and it is repeated 3 times.

Tandem repeats of DNA may be classified as *perfect* (exact) repeats or *imperfect* (approximate) repeats in which the structure of tandem repeats may have been altered due to mutations. The distance between the repeating sequences can be determined either by the *Hamming distance* which is the number of mismatches between the repeats, or the *edit distance* which is the minimum number of indels and substitutions to transform one sequence to another, or by a scoring matrix which provides weighted edit distances [4]. Tandem repeats can be classified as *microsatellites* (<10 bp), *minisatellites* (between 10 and 100 bp), and *satellites* (>100 bp). Microsatellites tend to be perfect, whereas minisatellites and satellites in general are approximate repeats [13]. A short tandem repeat (STR) is a DNA sequence repeat of length between 5 and 30 base pairs. The STRs in the Y chromosome of an individual are passed from father to son almost unchanged. STRs are commonly used in forensics and genealogical DNA testing to discover identity of individuals. Tandem repeats can be classified as follows [15].

1. *Exact tandem repeats*: These are in the form $SS \ldots S$, for example, GCAT GCAT GCAT.
2. *k-approximate tandem repeats*: These consist of repeats as $S_1 \ldots S_n$ where each S_i has a distance (edit, hamming, or other) of at most k to a consensus repeat S. For example, CAGT CAGG GAGC is a 1-approximate tandem repeat of three sequences with the consensus sequence CAGT.
3. *Multiple length tandem repeats*: A multiple length tandem repeat is shown as $(Ss^n)^m$ where edit distance between S and s is greater than a threshold value t. As an example, (CATTTG GACT GACT) (CATTTG GACT GACT) is a repeat of this kind with $n = 2$ and $m = 2$. These repeats can also be approximate.
4. *Nested tandem repeats*: Considering two sequences S_1 and S_2 which have an edit distance larger than t, these repeats are of the form $S_2^i S_1 S_2^j S_1 \ldots S_1 S_2^z$ where $i, j, \ldots, z > 1$. S_1 is the tandem repeat and S_2 is termed the *interspersed repeat*. The approximate values of these repeats are also possible.

Discovery of tandem repeats can be performed either by combinatorial methods which carry out exhaustive searches of the sequence, or statistical methods that search for perfect short tandem repeats in a small sample of the target sequence first and then attempt to enlarge these short repeats. We have an exact solution using the combinatorial methods in many cases at the expense of high complexity which quickly becomes unacceptable for large sequence sizes. On the other hand, statistical approaches provide suboptimal solutions with favorable performances. If we can determine the repeat sequence to be searched in a DNA or a protein amino acid sequence beforehand, we can search this sequence in a number of other nucleotide or amino acid sequences and in this case, we can implement a pattern-matching algorithm, for example, Knuth–Morris–Pratt algorithm we have analyzed in Sect. 5.3 can be used for this purpose. However, the time complexity of this algorithm being $O(n^2)$ makes it unsuitable even for moderate-sized sequences. An early algorithm that finds primitive repeats in $O(n \log n)$ time was proposed by Crochemore [5] which uses suffix trees. Stoye and Gusfield presented an algorithm that finds primitive and non-primitive tandem repeats as described next.

8.2.1 Stoye and Gusfield Algorithm

Stoye and Gusfield presented simple time and space optimal algorithms using suffix trees to locate tandem repeats (and arrays) in a sequence [32]. The algorithms are based on the property of suffix trees that allow the efficient location of the branching occurrences of tandem repeats. We will briefly review only the basic algorithm here as in [32]. Given a sequence S, a tandem array shown as $w = \alpha^k$ of S can be specified

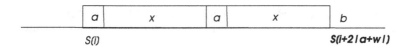

Fig. 8.1 Branching and non-branching repeats. When $b = a$, the tandem repeat is non-branching, otherwise it is branching

as (i, α, k), meaning it starts at location i and is repeated k times. A tandem array is *right-maximal* if there is no other repeat of α immediately after w and *left-maximal* if there is no preceding α preceding w. A tandem array with two repeats is termed a tandem repeat in this study and a tandem repeat is called *branching* when the adjacent character proceeding the repeat is different than the first character of the repeat, otherwise it is called a *non-branching* repeat as shown in Fig. 8.1.

Every tandem repeat is either branching or is included in a chain of tandem repeats obtained by successive rotations starting from a branching tandem repeat. The algorithm presented by the authors makes use of these concepts by first detecting tandem repeats and then obtaining the required non-branching repeats from these branching repeats. Suffix trees can be used to simplify this process; let $T(S)$ be the suffix tree of the sequence S and $L(v)$ the *path-label* of a node v in T which is the concatenation of the edge labels in the path from the root to the node v. The *string-depth* of a node v is $D(v) = |L(v)|$ and a leaf v is labeled with index i if and only if $L(v) = S[i..n]$. For an internal node v, the leaf list $LL(v)$ is defined as the list of the leaf labels below v. Algorithm 8.1 shows the pesudocode of this algorithm.

The leaf list of v can be found by a linear traversal of the subtree rooted at v in a time proportional to the size of $LL(v)$. A depth-first-search traversal of T from the root is performed and successive numbers called DFS numbers are assigned to the leaves in their order of occurrences during traversal. These DFS numbers are stored in an array indexed by the original leaf numbers. The next number to be given to a leaf a is also stored at an internal node v during the first visit of v in a DFS traversal.

Algorithm 8.1 *SG_Alg*

1: **Input** : Suffix tree T
2: $V \leftarrow$ all internal nodes of T
3: **for all** $v \in V$ **do**
4:     **collect** leaf list $LL(v)$ of v
5:     **for all** leaf $i \in LL(v)$ **do**
6:         $j \leftarrow i + D(v)$
7:         **if** $j \in LL(v)$ **then**   ▷ check if $L(v)^2$ is a tandem repeat)
8:             **if** $S[i] \neq S[2D(v)]$ **then**   ▷ check if it is branching
9:                 **report** tandem repeat at position i with length $2D(v)$
10:             **end if**
11:         **end if**
12:     **end for**
13: **end for**

On return from the traversal through v, the most recent DFS number b assigned is also stored at v. Using this approach, leaves of $LL(v)$ all have numbers between a and b.

The time for this algorithm is proportional to the total size of all the leaf lists which is $O(n^2)$ and a simple modification provides $O(n \log n)$ time bound [32]. The idea of this algorithm was used to find tandem repeats using suffix arrays in [1] resulting in less memory usage.

8.2.2 Distributed Tandem Repeat Search

A natural way to find tandem repeats in a sequence S is to partition the sequence to a number of processes. We can use coarse-grain parallelism and assume the supervisor/worker with each process p_i $(0 \leq i < k)$ residing in the processor i for a total of k processors. The supervisor process p_0 distributes the part of the sequence $S_{(i-1)(n/k)+1}, ..., S_{(in/k)}$ to each process p_i which then searches for tandem repeats in its partition. The boarder regions between the parts of the sequence assigned to processes should be carefully considered not to miss a motif that may be partially existent in one partition and partly in the consecutive one. The results from each process p_i can then be gathered at the supervisor p_0 which can merge the results, asses the statistical significance of the tandem repeats found, and output the result. When tandem repeats are searched in a set S of sequences, subsets of S can be distributed to processes and the results can again be collected in the supervisor for post-processing. When the pattern to be searched is determined beforehand, we can use any of the parallel/distributed pattern-matching algorithms. There is hardly any results reported in literature about parallel or distributed implementation of tandem repeat discovery in DNA or protein amino acid sequences.

8.3 Sequence Motifs

Sequence motifs are recurring DNA/RNA nucleotide or protein amino acid patterns that have some biological significance. These motifs may represent binding sites for DNA and binding domains for proteins. Discovering motifs in biological sequences therefore helps to understand the biological functions such as transcriptional regulation, mRNA splicing, and protein complex formation. It also provides an alternative method to find the phylogenetic relationships between organisms by discovering conserved motifs in them.

Finding binding sites in DNA is crucial in understanding the mechanisms that regulate gene expression. Based on their functional importance, there are many research studies in finding these sequence motifs resulting in numerous algorithms. Frequently, the motif sequences undergo changes resulting in variations of the original motif due to mutations. Hence, as in the case of tandem repeats, we can have exact or approximate motif search. Furthermore, the DNA motifs may be *simple* or *structured*. A structured motif consists of a central pattern and one or two motifs called *satellites* placed at some distance to the right and left of the central pattern [16]. The central motifs and satellites are searched using different methods. The motif finding problem has few variants described below:

1. *Planted(l, d) Motif Search* (PMS): Given are a set of *n* sequences $S = \{S_1, \ldots, S_n\}$ over an alphabet Σ, with *m* being the approximate length of all strings; and two integers *l* and *d*, $0 \leq d < l < m$; a substring $M \in S_i$ with length *l* is searched with variants M_i that are at most a Hamming distance of *d* from *M* which is called the *planted motif* and M_i are called the *planted variants* of *M*.
2. *Edited Motif Search* (EMS): We are again given a set of *n* strings $S = \{S_1, \ldots, S_n\}$ over an alphabet Σ, and integers *l*, *d*, and *q*. Our aim is to search for motifs of length *l* that have at most an edit distance of *d* between them and are incident in at least *q* of the sequences $S_1, \ldots, S_n$.
3. *Simple Motif Search* (SMS): In this problem, a simple motif consists of symbols (residues) and ?'s which mean wild characters. The alphabet for DNA in this case is {A, C, G, T, ?}. For example, GCC??T?A?A is a pattern of length 10 with four wild characters. The "?" symbol can be replaced by any character from the original alphabet. A motif cannot begin or end with a wild character. In SMS, we are asked to find all simple motifs of maximum length *l* and the number of their occurrences in an input set of sequences. The (u, v)-class in SMS is defined as the class of simple motifs with length *u* having exactly *v* wild characters.

In an orthogonal setting, the motif finding algorithms are classified as *pattern-based* algorithms and *profile-based* algorithms. In pattern-based algorithms, the aim is to discover the motif subsequence whereas the positions of the motif are identified in the latter [20]. Furthermore, based on the algorithmic techniques employed, we can further classify the motif search algorithms as probabilistic and combinatorial.

The EMS problem has been addressed in some studies employing the pattern-based algorithms mainly [19]. The *Speller* [28], Edit Distance Motif Search 1 and 2 (EDMS) [33], and Deterministic Motif Search (DMS) [26] are some of the example studies. The *Speller* algorithm first constructs a generalized suffix tree for the input set of sequences. It then searches this tree for subsequences of length *l* which have a maximum edit distance *d* from each other and if there are at least *q* of such sequences, they are reported as motifs.

The SMS problem is again studied in the context of pattern-based methods. In the SMS algorithm presented by Rajasekaran et al. to find simple motifs [26], a (u, v)-class is defined as a group of motifs where each motif has a length of *u* and has exactly *v* wild characters. For example, ACC??TC?G is (9, 3) class. The algorithm performs $(u - 2v)$ sorts to find the patterns in a (u, v) class. More specifically, it consists of the following steps. First, all subsequences of length less than *l* of the sequence *S* are formed for each (u, v) class. Then, these subsequences are sorted considering the wildcard positions only. The last step of the algorithm involves scanning through the sorted list and counting the number of occurrences of each pattern. The time complexity of the algorithms is reported as $O((u - 2v)Mu/w)$ for a (u, v)-class where *M* is the number of residues and *w* is the word length of the computer.

Teiresias is proposed by Floratos and Rigoutos to find rigid motifs in a set of biological sequences [7]. Given an input set S of sequences, it aims to find all maximal motifs that exist in at least k sequences of S. A pattern P is called a $\langle L, W \rangle$ pattern if every subpattern of P with length W or more contains at least L residues, with $L \leq W$. The algorithm consists of two steps; the input is scanned to find elementary patterns which are $\langle L, W \rangle$ patterns in the first step. The convolution step inputs these which are recursively combined to find the maximal patterns.

The *consensus sequence* of a set of sequences consists of the most prevalent nucleotide or the amino acid in a biological sequence. We simply check each sequence and record the frequency of characters in each column. The highest frequency character then becomes part of the consensus sequence. The consensus motif is assumed to be the subsequence without any mutations and that all of the remaining motifs have been derived from this motif by mutations. For example, given the below four DNA sequences of lengths 11, the consensus sequence M is formed from the highest occurring characters in each column and hamming distance of each sequence to M is then formed as follows:

```
                                   Hamming distances to M
S1:  A  C  C  G  A  A  C  T  T  A  G              5
S2:  A  G  C  G  C  T  G  T  G  A  C              3
S3:  C  C  C  T  A  T  G  A  G  A  A              4
S4:  A  T  C  G  A  T  G  C  G  T  C              2
M :  A  C  C  G  A  T  G  T  G  T  C
```

In order to find conserved motifs in a set of sequences, we can use multiple alignment methods described in Sects. 6.4 and 6.6. Two commonly used methods for multiple sequence alignments are the BLAST family of algorithms and CLUSTALW as we have seen. A sequence alignment algorithm will provide similar subsequences in these sequences, and assuming motifs are preserved in homologous sequences, we can detect them using this method. This process is referred to as *profile analysis*. However, we need to do post-processing to discover motifs. We can build a *profile matrix* which basically shows the frequency of characters at each conserved region. Then, a consensus sequence as described above can be formed to discover the motif in that region.

For the rest of the chapter, we will review the motif search algorithms in view of the techniques used with focus on the planted motif search problem. The planted motif search (PMS) problem is NP-hard and therefore various heuristic algorithms are usually employed. We can coarsely classify the PMS algorithms based on whether they provide optimal results in which case they are called exact algorithms, or heuristic algorithms which are also called approximate algorithm, not to be confused with the approximation algorithms which always have an approximation ratio, that provide suboptimal results. Viewed from another angle, the PMS algorithms can be categorized as probabilistic or combinatorial ones. Probabilistic algorithms provide suboptimal results and the combinatorial algorithms can provide optimal or suboptimal results. These algorithms can also be classified as profile-based in which the starting position of the motif is searched or the pattern-based algorithms which predict the motif sequence. The PMS series of algorithms are exact and use exhaustive enumeration, whereas MEME and Gibbs samplings are probabilistic methods as we will investigate.

8.3.1 Probabilistic Approaches

Two probabilistic methods called MEME and Gibbs sampling are more frequently used than others as described below.

8.3.1.1 MEME

Multiple expectation maximization for motif elicitation (MEME) is a software tool for motif discovery in a set of DNA or protein sequences [2,3]. It uses the expectation maximization (EM) algorithm to discover motifs in a number of sequences by representing motifs as position weighted matrices (PWMs). The EM algorithm is used in various applications to find the maximum-likelihood estimate of the parameters of a dataset distribution in general. The MEME algorithm does not need the distribution of motifs in the input set of sequences before running the algorithm. MEME constructs statistical models of motifs in which each model is represented by a discrete probability distribution matrix M. An entry m_{ij} of M shows the probability that character i will be present in the jth position of the motif. A high-level pseudocode of MEME is shown in Algorithm 8.2 as adapted from [10].

Algorithm 8.2 $MEME$

1: **Input** : Set of sequences $S = S_1, ..., S_n$
2: **Output** : Suffix tree $T(S)$ representing S
3: **for** $pass = 1$ to n_m **do**   ▷ Loop 1
4:     **for** $w = w_{min}$ to w_{max} **do**   ▷ Loop 2
5:         **for** $sp = 1$ to n_{sp} **do**   ▷ Loop 3
6:             **estimate** score of the model with the initial
7:         **end for**   ▷ end Loop 3
8:         **select** the initial model with the maximum score
9:         **for** $\lambda = \lambda_{min}$ to λ_{max} **do**   ▷ Loop 4
10:            **run** EM for convergence
11:        **end for**   ▷ end Loop 4
12:        **select** model with maximal $G(.)$
13:     **end for**   ▷ end Loop 2
14:     **select** the model with maximal $G(.)$ and report
15:     **delete** the model from the previous probabilities
16: **end for**   ▷ end Loop 1

In the external *for* loop, multiple motifs are searched in the input set $S_1, \ldots, S_n$. All possible motif widths are searched between the lines 4 and 11 by generating a model for each width. Then, a heuristic criterion function $G(.)$ which is based on maximum-likelihood ratio test is used to choose the best model. The search at a specific motif width consists of two steps: in the first step between lines 5 and 7 of the algorithm, the score of a model is estimated; and the expectation maximization is executed until a convergence is reached in the second step between lines 9 and 11 after the selection of the model with the best score. The parameter λ is proportional to the number of incidences of the motif in the sequences. After a motif is discovered, the model is erased from the prior probabilities to prevent finding the same motif again.

Parallel MEME

A parallelization of MEME (ParaMEME) is reported in [10]. The algorithm shown in Algorithm 8.2 contains four loops as labeled and two of these *for* loops can be parallelized as described. Estimating the score of the models is performed in Loop 2 between the lines 5 and 7 and this process can be parallelized easily as the operations are independent. In this loop, a set of initial models is formed and these models can be distributed to a number of processors and each process can analyze a subset of models it is assigned to find the best scoring one. The best of the locally best initial models can then be found by gathering the results in a special process or by other reduction techniques. The determination of the initial models can be parallelized in this manner conveniently.

The running of expectation minimization between lines 9 and 11 involves transfer of larger amount of data among processes than finding the initial models. The authors did not attempt to parallelize the loop between lines 4 and 13 (Loop 2) as the two inside *for* loops between this larger *for* loop (loops 3 and 4) have been parallelized. A load balancing mechanism is also employed especially for the third *for* loop where the initial models are estimated. It was shown that the overall ParaMEME algorithm is scalable with efficiencies greater than 72 % using 64 processors.

8.3.1.2   Gibbs Sampling

Gibbs sampling method proposed by Lawrence et al. is a probabilistic scheme to find motifs [12]. It is a Markov Chain Monte Carlo (MCMC) method in which results at each step depend only on the results of the preceding step [17]. It assumes that the length k of the motif to be searched is known and each sequence contains exactly one instance of the motif. If the starting position of the motif in each input sequence is known, a sequence profile for the motif can be constructed. Gibbs sampling method starts by random positions in each sequence and builds profiles using these positions. It then changes the position of the motif to the place where it best matches the profile, shown as high-level description in Algorithm 8.3.

Algorithm 8.3 *Gibbs_Sampling*

1: **Input** : Set of n sequences $S = \{S_1, ..., S_n\}$
2: **Output** : Set of t motifs $M = \{m_1, ..., m_t\}$
3: **for all** $S_i \in S$ **do**
4:     set $x_i \in S_i$ to a random position
5: **end for**
6: **repeat**
7:     **for** $i = 1$ to n **do**
8:         **build** a profile P using sequences at positions $x_j \in S_j, \forall j \neq i$
9:         **update** x_i to a position in S_i where P best matches in S_i
10:     **end for**
11: **until** the positions $x_1, ..., x_n$ do not change

The randomness can be increased by selecting the new x_i values randomly according to the probability distribution of P in S_i in line 8 of the algorithm. The Gibbs sampling algorithm excludes sequence S_i at each iteration and attempts to optimize the motif location in S_i using the profile values from all other sequences. In practice, it often finds motifs in a set of sequences.

8.3.2 Combinatorial Methods

These methods typically provide exact solutions by exhaustive enumeration but they may also employ heuristics to find approximate solutions. We will describe a basic exact algorithm first, followed by graph-based algorithms which represent the similarity among sequences by the edges of a graph and then attempt to discover dense regions in this graph using heuristics.

8.3.2.1 Exact Algorithms

An exact motif search algorithm finds all occurrences of a motif in a biological sequence. These algorithms may be pattern-based or profile-based algorithms. As examples of exact pattern-based algorithms, Rajasekaran et al. proposed a number of algorithms for the planted motif search problem [27]. These algorithms use the concept of neighbor search of possible patterns as various other algorithms but they also employ techniques such as sampling, local search, and probabilistic approaches. We will review the first algorithm named PMS1 which inputs a set of $S = S_1, \ldots, S_t$ sequences of length t each. It forms all possible l-mers for each sequence in the first step. Then, all l-mers that are at a distance of d from the l-mers found in the first step are generated in the second step. These l-mers in each sequence are sorted, merged, and l-mers occurring in each sorted list are reported as motifs as shown in Algorithm 8.4. The time complexity of this algorithm is reported as $O(tn(ld)|\Sigma|^d l/w)$ where w is the word length of the computer and n is the length of the input sequences.

Algorithm 8.4 $PMS1\_Alg$

1: **Input** : Set of t sequences $\mathcal{S} = \{S_1, ..., S_t\}$
2: **Output** : Set of k motifs $\mathcal{M} = \{m_1, ..., m_k\}$
3: **for all** $S_i \in \mathcal{S}$ **do**
4:      $C_i \leftarrow$ all l-mers of S_i
5: **end for**
6: **for all** C_i **do**
7:    **for all** $u \in C_i$ **do**
8:       **build** all l-mers v such that u and v are at a hamming distance of d.
9:       $C_i' \leftarrow$ all such l-mers
10:   **end for**
11: **end for**
12: **for all** C_i' **do**
13:    $L_i \leftarrow$ lexicographically sorted list of l-mers in C_i'
14: **end for**
15: **merge** all L_is
16: **output** l-mers that exist in all L_is

8.3.2.2 Graph-Based Algorithms

In graph-based algorithms, subsequences are represented by the vertices of a graph G and an edge is formed between two similar subsequences. The motif discovery problem is therefore reduced to finding cliques or maximum density subgraphs of G. The WINNOWER algorithm is an earlier attempt using this approach as described below.

WINNOWER

Pevzner and Sze proposed a graph-based motif detection algorithm for the Planted(l, d) motif search problem in a set of sequences $\mathcal{S} = \{S_1, \ldots, S_n\}$ called WINNOVER [25]. The similarity graph $G(V, E)$ is first constructed with nodes representing all l-mer subsequences of the sequences. Two nodes u and v of G are connected if the Hamming distance between them is at most d and they are from two different sequences. The expected distance between two motifs in fact is $2d - \frac{4d^2}{3l}$. The graph G built in this manner is n-partite as there will not be any edges between l-mers of the same sequence. Since we are looking for a motif in all of the n sequences, such a motif will be represented by a clique of size n in G. Let us illustrate this idea by a simple example with three DNA segment inputs S_1, S_2, and S_3 as shown below:

```
S1 = A C T T T C
S2 = T A C T T C
S3 = G C A G T T
```

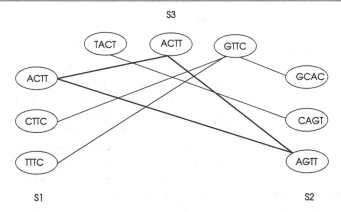

Fig. 8.2 The 3-partite graph representing the three DNA segment 4-mers. The motifs ACTT, ACTT, and AGTT have a maximum of 2 distance between them and form a clique shown in *bold*. The consensus sequence for this motif is ACTT

We have $n = 3$ and $m = 6$ in this case and let us assume that we are searching for motifs of length 4 and hence $l = 4$. We form consecutive 4-mers of these sequences and construct a graph with 4-mers as nodes and connect the nodes from different sequences that have a maximum hamming distance of 2 between them for $d = 1$. We are therefore searching for the solution of Planted$(4, 1)$ problem in these sequences. The graph with these values is shown in Fig. 8.2 where the clique shown in bold represents the motif.

Finding cliques in a graph is NP-hard, and a heuristic is used in WINNOWER to find cliques. The authors, observing G is n-partite and almost random, proposed the concept of *extendable clique*. A clique is extendable if there are neighbors of this clique in each partition of the n-partite graph such that if $C_1 = \{V_1, \ldots, V_k\}$ is a clique, the neighbor u of C_1 ensures that $C_1 = \{V_1, \ldots, V_k, u\}$ is also a clique. Based on the observation that every edge in a maximal n-clique is part of a minimum of $\binom{n-2}{k-2}$ extendable k-cliques, spurious edges are eliminated in each iteration of the algorithm to yield cliques. The time complexity of this algorithm is $O(N^{2d+1})$ where $N = nm$.

The SP-STAR algorithm [25] proposed by the same authors is also a graph-based motif search algorithm that finds cliques, but different than WINNOWER, it associates weights with edges and l-mers and can eliminate more spurious edges than WINNOWER at each iteration. This algorithm is faster than WINNOWER and uses less memory.

Other Graph-Based Algorithms

PRUNER is another graph-based algorithm for motif detection [30]. It also constructs a graph G based on pairwise similarity as WINNOWER and eliminates the vertices that cannot be part of a clique. However, the approach in the second step of the algorithm is different, in which it groups the edges of G as l-grams that are different in more than d positions into Group1 and less than or equal to d positions into Group2. The edges in these groups are removed iteratively to discover the cliques. This algorithm has two versions as PRUNER-I and PRUNER-II; the time and space complexities of the first one are reported as $O(n^2 t^2 l^{d/2} |\Sigma|^{d/2})$ and $O(ntl^{d/2} |\Sigma|^{d/2})$, respectively, where d is the maximum number of allowed mismatches, t is the number of sequences, and l is the length of the motif. The second algorithm has $O(n^3 t^3 l^{d/2} |\Sigma|^{d/2})$ and $O(l^{d/2} |\Sigma|^{d/2})$ time and space complexities.

Structured motifs inference, localization and evaluation (SMILE) uses suffix trees to discover structured motifs [14]. It constructs a generalized suffix tree T for the set of sequences S first and traverses T to discover motifs. Its time complexity is an exponential function of the number of gaps. Mismatch tree algorithm (MITRA) uses efficient depth-first-search tree traversal to find the leaves which correspond to motifs [6]. Its time complexity is $O(l^{d+1} |\Sigma| nN)$ and space complexity $O(nNl)$ where n is the number of input strings and N is the size of the strings with d denoting maximum allowed number of mismatches for l-size motifs.

8.3.3 Parallel and Distributed Motif Search

We have already described how the probabilistic MEME algorithm can be parallelized using the SPMD model. We will now first describe a simple generic distributed algorithm that can be used for any motif search method followed by a distributed graph-based motif search algorithm proposal and then review some of the recent parallel/distributed methods.

8.3.3.1 A Generic Distributed Algorithm Proposal

Our main approach for distributed motif discovery in general would be again to partition the data space among a number of processes, $p_0, \ldots, p_{k-1}$. Each process p_i can then search for the motifs in its local partition. If search is for motifs in a single sequence S, it can be partitioned among processes and each p_i can search for the motifs in its subsequence $s_i \in S$. Otherwise, when the search is made with a set of n sequences $S = \{S_1, \ldots, S_n\}$, then we can assign sequences of S to k processes which can search motifs in their allocated sequences. The supervisor process then performs post-processing of the discovered motifs by each process. It can gather the motifs from each worker process and check whether they exist in all sequences. Algorithm 8.5 shows one way of performing this distributed procedure at the end of which a decision can be made whether the discovered motifs are planted or edited. This way is convenient as we do not need to specify the type of motif searched beforehand.

Algorithm 8.5 $Dist\_Motif\_Alg$

1: **Input** : Set of n sequences $S = \{S_1, ..., S_n\}$
2:          Set of k processes $P = \{p_0, ..., p_{k-1}\}$
3: **Output** : Set of t motifs $M = \{m_1, ..., m_t\}$
4: $received \leftarrow \emptyset$
5: **if** $p_i = p_0$ **then**   ▷ I am the supervisor process
6:     **for** $i = 1$ to $k - 1$ **do**   ▷ send a subset of sequences to each process
7:         **send** $S_i \in S$ to p_i
8:     **end for**
9:     **while** $received \neq P$ **do**   ▷ get motifs from processes
10:         **receive** M_i from p_i
11:         $M \leftarrow M \cup M_i$
12:         $received \leftarrow received \cup p_i$
13:     **end while**
14:     **merge**   ▷ check existence of motifs in all sequences
15:     **for all** $m \in M$ **do**
16:         **if** $m \in S_i, for\ all\ 1 \leq i \leq n$ **then**
17:             **output** m as a *planted motif*
18:         **else if** m exists in at least q sequences **then**
19:             **output** m as an *edited motif*
20:         **end if**
21:     **end for**
22: **else**   ▷ I am a worker process
23:     **receive** S_i from p_0 ▷ receive my set of sequences
24:     $M_i \leftarrow$ motifs in S_i ▷ search approximate motifs in my sequences
25:     **send** M_i to p_0 ▷ send found motifs to supervisor
26: **end if**

8.3.3.2   Distributed Graph-Based Algorithm Proposal

Graph-based motif finding algorithms required construction of an n-partite similarity graph G for n sequences with nodes representing l-mers first, and then detection of cliques in G. The time-consuming parts of these approaches are the discovery of cliques which is NP-hard. There are various approaches for detecting cliques in parallel in general graphs; however, our interest in the context of motif search is finding k-cliques which are the cliques with k nodes, with each node of the clique from a partite group in a k-partite graph.

As a simple attempt to parallelize the graph-based motif search, we propose the following distributed algorithm based on the WINNOVER algorithm using the supervisor–worker model. The supervisor broadcasts the sequence set to all workers. Each worker process then finds the l-mers set L_i of its sequence S_i, computes the distances of l-mers in L_i to all other l-mers and forms the adjacency matrix rows for the l-mers it owns. The adjacency matrix A will have no elements along the blocks around its diagonal since it is k-partite. Each process p_i then first checks which of the l-mers it owns has at least one edge connected to all other l-mers held by other processes. These nodes are the candidate members of possible cliques. Let us call such nodes as L_i^1 which are then distributed to all other processes using a broadcast operation. Each process then uses L_i^1 as seed to

Fig. 8.3 The 4-partite graph representing the three DNA segment 3-mers from four sequences. There are two cliques in this graph represented by the nodes s_1, s_5, s_8, s_{12} and s_2, s_6, s_8, s_{10} shown in *bold*. The node s_8 is a member of both cliques

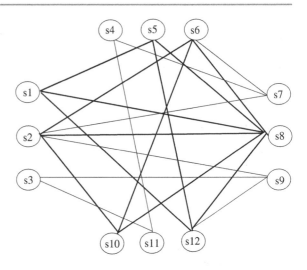

implement the WINNOVER algorithm for the nodes in its partition as shown in Algorithm 8.6. The detected cliques with this size are the motifs.

Let us illustrate the operation of this algorithm using a simple example. A graph is first built from the 3-mers of four input sequences S_1, S_2, S_3, and S_4 where an edge in the graph depicts a maximum distance of $2d$ between the nodes it connects as shown in Fig. 8.3. The 3-mers from these sequences are $s_1, \ldots, s_{12}$ with $s_1, s_2, s_3 \in S_1$, and others similarly from S_2, S_3, and S_4 in sequence.

Algorithm 8.6 *Dist_Graph_Alg*

1: **Input** : Set of n sequences $\mathcal{S} = \{S_1, ..., S_n\}$
2:          Set of k processes $P = \{p_0, ..., p_{k-1}\}$
3: **Output** : Set of t motifs $\mathcal{M} = \{m_1, ..., m_t\}$
4: **if** $p_i = p_0$ **then**   ▷ I am the supervisor process
5:     **broadcast** $\mathcal{S}$ to all workers
6:     **receive** cliques from all processes
7:     **output** cliques as motif set $\mathcal{M}$
8: **else**   ▷ I am a worker process
9:     **receive** $\mathcal{S}$ from p_0 ▷ receive my set of sequences
10:    **find** l-mers L_i of S_i
11:    $L_i^1 \leftarrow l$-mers that have at least 1 edge to l-mers of other sequences
12:    **broadcast** L_i^1 with the connections
13:    **receive** full set $\mathcal{L}^1$ from all other processes
14:    **compute** cliques of size k for my l-mers using WINNOVER
15:    **send** cliques to p_0
16: **end if**

Let us assume that we have four processes $p_0, \ldots, p_3$. Each process p_i with the rows it is responsible are shown below.

	s_1	s_2	s_3	s_4	s_5	s_6	s_7	s_8	s_9	s_{10}	s_{11}	s_{12}	
s_1	0	0	0	0	1	0	0	1	0	0	0	1	p_0
s_2	0	0	0	0	0	1	0	1	0	1	0	0	
s_3	0	0	0	0	0	0	0	0	1	0	1	0	
s_4	0	0	0	0	0	0	1	0	0	0	1	0	p_1
s_5	1	0	0	0	0	0	0	1	0	0	0	1	
s_6	0	1	0	0	0	0	0	1	0	1	0	0	
s_7	0	1	0	1	0	1	0	0	0	0	0	0	p_2
s_8	1	1	0	0	1	1	0	0	0	1	0	1	
s_9	0	1	1	0	0	0	0	0	0	0	0	1	
s_{10}	0	1	0	0	0	1	0	1	0	0	0	0	p_3
s_{11}	0	0	1	1	0	0	0	0	0	0	0	0	
s_{12}	1	0	0	0	1	0	0	1	1	0	0	0	

Each process now checks the nodes it has with the nodes in other processes. It then forms a vector for each row that has an element with entries shown as below. These elements represent the nodes that have at least one connection to an l-mer of each partition of the k-partite graph and are broadcast with their connected neighbors.

	s_1	s_2	s_3	s_4	s_5	s_6	s_7	s_8	s_9	s_{10}	s_{11}	s_{12}	
s_1	1	0	0	0	1	0	0	1	0	0	0	1	p_0
s_2	0	1	0	0	0	1	0	1	0	1	0	0	
s_5	1	0	0	0	1	0	0	1	0	0	0	1	p_1
s_6	0	1	0	0	0	1	0	1	0	1	0	0	
s_8	1	1	0	0	1	1	0	1	0	1	0	1	p_2
s_{10}	0	1	0	0	0	1	0	1	0	1	0	0	p_3
s_{12}	1	0	0	0	1	0	0	1	1	0	0	1	

Each process then performs the WINNOVER algorithm for the nodes in its partition using the global connection data. The final cliques in this graph represented by the nodes s_1, s_5, s_8, s_{12} and s_2, s_6, s_8, s_{10} are shown in bold. The node s_8 is a member of both cliques.

8.3.4   A Survey of Recent Distributed Algorithms

The distributed algorithms for motif search are scarce; however, there has been few research efforts in this direction recently. One such study is ACME by Sahli et al. [29] which uses parallelization of suffix trees. It supports *supermaximal* motifs which are motifs not included in any others. The model adapted is the supervisor–worker model with thousands of worker processes. The claimed novelty of this algorithm is the traversal order of the search space and the order of accessing data in the suffix tree. The supervisor inputs the sequence data from the disk, decomposes the search space, and generates tasks. Each worker receiving the input sequence builds the annotated suffix tree and requests a task from the supervisor. The search space is horizontally partitioned into a large number of sub-tries for each process. The algorithm is implemented in MPI and also using multi-threading, and experimented with the human genome of 2.6 GB size and protein sequence of 6 GB. It was assessed as scalable based on the obtained results.

Nicolae and Rajasekaran presented a parallel exact algorithm for planted motif search called PMS8 which can work on large values of l and d [21]. Exact PMS algorithms frequently employ finding the common neighbors of three l-mers. The PMS8 has two main steps as sample-driven and pattern-driven steps. In the first step, tuples of l-mers from different sequences are generated, and in the pattern-driven step of the algorithm, common d-neighborhoods of these tuples are formed. The parallel implementation considers $m - l + 1$ independent subtasks. Equal number of such subtasks can be assigned to each process. The only interprocess communication needed is the broadcasting of the input to processors. In order to manage load balancing efficiently, process 0 is assigned as the scheduler and the other processes as the workers. The scheduler inputs the sequences, broadcasts it to the workers, and waits for requests from workers. Workers continue requesting subtasks from the scheduler and solving it until no more subtasks remain. They send the motifs found to the scheduler which outputs them in the end. This method was tested in a cluster of 64 nodes using OpenMPI and was found to be scalable.

The exact algorithm PMSP which is based on PMS1 has been parallelized recently in [18]. A bit vector mapping method which allows detection of large motif lengths with greater Hamming distances is used in this study. The tasks which find the motifs are distributed evenly to the cluster nodes, and both coarse gain and fine grain parallelism methods are employed. The authors report that the evaluation on two-node SMP system shows that the algorithm is scalable.

The simple motif search (SMS) problem can also be parallelized. There are various approaches for this task; in a simple and basic approach, the sequences can be partitioned to a number of processes and the results found can be merged. Alternatively, the sorting operations needed can be considered as tasks and each process can handle one task under the arbitration of a central scheduling process which provides load balancing.

A very recent study by Ikebata and Yoshida is called repulsive parallel Markov Chain Monte Carlo (RPMCMC) algorithm and aims to find motifs by parallelizing Gibbs sampling method in a distributed memory system [11]. The authors criticize the inability of the classical Gibbs sampling method to discover latent diverse motifs even for a small number of inputs. This is claimed to be the result of posterior distribution containing many locally high probability regions. The presented algorithm runs the Gibbs motifs sampler in parallel by all-at-once sampling which discovers the diverse motifs as shown by the experiments.

8.4   Chapter Notes

Repeating subsequences occur both in DNA nucleotide and protein amino acid sequences. These patterns are speculated to have some biological significance. Some of the repeats in DNA indicate transcription factor binding sites and therefore discovering them helps to find the location of genes. An increased number of such repeats in DNA may signal existence of some diseases and finding them is helpful to determine the disease state of an organism. In forensics, discovery of repeats is frequently used to identify an individual. Last but not least, search of repeats in a number of biological sequences is used to compare them and find conserved regions in the organisms represented by these sequences which may be used to infer phylogenetic relationships along them.

A tandem repeat consists of repeating patterns, whereas a sequence motif is dispersed in various locations in DNA or protein. The search of these repeats may be viewed from several different angles. First, the pattern may be known before the search or not, and in the former case, the problem is reduced to pattern matching, whereas the latter is the general case. We may be looking for approximate repeats both in tandem repeats or motifs, rather than exact ones, to allow mutations and this is what is commonly encountered in reality. We may want to discover the pattern itself only as in pattern-based search, or its locations in the sequence as in profile-based search or both. Our aim may be to extract repeats in a single sequence or to find repeats in a number of sequences to find their similarities. Yet, another distinction is based on the method used, and we can employ an exact algorithm which finds every occurrence of a repeat or a heuristic algorithm that only provides an approximate solution.

Statistical algorithms provide approximate solutions and combinatorial algorithms present exact results in general, but combinatorial methods may also employ heuristic approaches to result in approximate results. We reviewed few sequential representative algorithms: an exact combinatorial algorithm, suffix tree-based tandem repeat, and motif extraction algorithms, and the probabilistic algorithms MEME and Gibbs sampler. We then investigated ways of parallelizing repeat search process on distributed memory processors. A simple method is based on the supervisor–worker model of parallel processing and relies on partitioning dataset of sequences to a number of processes, and letting each worker process search for pattern in its subset. In another approach, we discussed how the suffix trees can be constructed and analyzed for repeating patterns in parallel. Using graphs to detect repeats opens another door for parallelism as bringing the problem to this domain provides some already existing parallel methods such as searching for cliques in parallel in a graph to our immediate use. All l-mers of a sequence are formed as a partition of a k-partite graph for k sequences and edges are connected between close l-mers. A clique in such a k-partite graph represents a motif. We proposed a distributed algorithm that finds the cliques in a k-partite graph in k rounds. The repeat search in DNA sequences and also in proteins will continue to be a fundamental area of research in bioinformatics due to the imperative information gained about molecular biology by the analysis of these repeats.

Exercises

1. Implement Stoye and Gusfield algorithm to find the tandem repeats in the follow-ing DNA sequence. Draw the suffix tree for the sequence and mark the locations of the repeats in this tree.

$$A\ A\ T\ T\ A\ T\ T\ A\ T\ T\ C\ C\ G\ A$$

2. Find the Planted$(3, 1)$ motif in the following DNA sequence using the naive algorithm by checking every subsequence.

$$T\ G\ G\ A\ A\ G\ G\ C\ T\ G\ G\ A$$

3. Work out the consensus sequence of the following protein sequences. Work out also the distance of each sequence to the consensus sequence. What should be the value of the maximum distance d for these four sequences to be considered as a planted$(11, d)$ motif?

```
S1:  R  C  Q  E  P  D  X  F  D  H  N
S2:  H  C  B  M  I  V  E  Q  D  H  M
S3:  R  C  B  P  L  V  X  S  D  H  M
S4:  A  W  K  M  P  V  X  Q  Z  H  M
```

4. Draw the similarity graph for the following DNA sequences for distance $d = 1$ and 4-mers. Check this graph visually for the existence clique of size 4 and if it exists, test whether the vertices of this clique are Planted$(7, 1)$ motifs.

```
S1:  C  C  C  G  A  G  C
S2:  G  G  A  G  C  T  G
S3:  C  C  C  T  A  T  G
S4:  T  A  G  C  A  T  G
```

5. Sketch the high-level pseudocode for the parallel MEME algorithm as described in Sect. 8.3.1.

6. Describe how multiple sequence alignment can be used to discover motifs in a set of sequences. Would you use global or local alignment and why?

7. Show the operation of the distributed graph-based motif finding algorithm for the following DNA sequences to find Planted$(3, 1)$ motifs. There are four processes $p_0, \ldots, p_3$ and the adjacency matrix is row-wise distributed to three processes p_1, p_2, and p_3 by p_0 assuming that process p_0 handles only data flow and synchronization.

```
S1:  G A C G A A C G T
S2:  A G A C C T G A C
S3:  C G A T A T G C A
S4:  A T C C A C G C T
```

References

1. Abouelhoda MI, Kurtz S, Ohlebusch E (2002) The enhanced suffix array and its applications to genome analysis. In: Proceedings of WABI 2002, LNCS, vol 2452, pp 449–463. Springer
2. Bailey TL, Elkan C (1995b) The value of prior knowledge in discovering motifs with MEME. In: Proceedings of thethird international conference on intelligent systems for molecular biology, pp 21–29. AAAI Press
3. Bailey TL, Elkan C (1995a) Unsupervised leaning of multiple motifs in biopolymers using EM. Mach Learn 21:51–80
4. Chun-Hsi H, Sanguthevar R (2003) Parallel pattern identification in biological sequences on clusters. IEEE Trans Nanobiosci 2(1):29–34
5. Crochemore M (1981) An optimal algorithm for computing the repetitions in a word. Inf Process Lett 12(5):244–250
6. Eskin E, Pevzner PA (2002) Finding composite regulatory patterns in DNA sequences. Bioinformatics 18:354–363
7. Floratos A, Rigoutsos I (1998) On the time complexity of the TEIRESIAS algorithm. In: Research report RC 21161 (94582), IBM T.J. Watson Research Center
8. Gatchel JR, Zoghbi HY (2005) Diseases of unstable repeat expansion: mechanisms and common principles. Nat Rev Genet 6:743–755
9. Goldstein DB, Schlotterer C (1999) Microsatellites: evolution and applications, 1st edn. Oxford University Press, ISBN-10: 0198504071, ISBN-13: 978-0198504078
10. Grundy WN, Bailey TL, Elkan CP (1996) ParaMEME: a parallel implementation and a web interface for a DNA and protein motif discovery tool. Comput Appl Biosci 12(4):303–310
11. Ikebata H, Yoshida R (2015) Repulsive parallel MCMC algorithm for discovering diverse motifs from large sequence sets. Bioinformatics 1–8: doi:10.1093/bioinformatics/btv017
12. Lawrence CE, Altschul SF, Boguski MS, Liu JS, Neuwald AF, Wootton JC (1993) Detecting subtle sequence signals: a Gibbs sampling strategy for multiple alignment. Science 262:208–214
13. Lim KG, Kwoh CK, Hsu LY, Wirawan A (2012) Review of tandem repeat search tools: a systematic approach to evaluating algorithmic performance. Brief Bioinform. doi:10.1093/bib/bbs023
14. Marsan L, Sagot MF (2000) Algorithms for extracting structured motifs using a suffix tree with application to promoter and regulatory site consensus identification. J Comput Biol 7(3/4):345–360

15. Matroud A (2013) Nested tandem repeat computation and analysis. Ph.D. Thesis, Massey University
16. Mejia YP, Olmos I, Gonzalez JA (2010) Structured motifs identification in DNA sequences. In: Proceedings of the twenty-third international florida artificial intelligence research society conference (FLAIRS 2010), pp 44–49
17. Modan K, Das MK, Dai H-K (2007) A survey of DNA motif finding algorithms. BMC Bioinform 8(Suppl 7):S21
18. Mohantyr S, Sahu B, Acharya AK (2013) Parallel implementation of exact algorithm for planted (l,d) motif search. In: Proceedings of the international conference on advances in computer science, AETACS
19. Mourad E, Albert YZ (eds) (2011) Algorithms in computational molecular biology: techniques, approaches and applications. Wiley series in bioinformatics, Chap. 18
20. Mourad E, Albert YZ (eds) (2011) Algorithms in computational molecular biology: techniques, approaches and applications, pp 386–387. Wiley
21. Nicolae M, Rajasekaran S (2014) Efficient sequential and parallel algorithms for planted motif search. BMC Bioinform 15:34. doi:10.1186/1471-2105-15-34
22. Pardalos PM, Rappe J, Resende MGC (1998) An exact parallel algorithm for the maximum clique problem. In: De Leone et al (eds) High performance algorithms and software in nonlinear optimization, vol 24. Kluwer, Dordrecht, pp 279–300
23. Parson W, Kirchebner R, Muhlmann R, Renner K, Kofler A, Schmidt S, Kofler R (2005) Cancer cell line identification by short tandem repeat profiling: power and limitations. FASEB J 19(3):434–436
24. Pelotti S, Ceccardi S, Alu M, Lugaresi F, Trane R, Falconi M, Bini C, Cicognani A (2008) Cancerous tissues in forensic genetic analysis. Genet Test 11(4):397–400
25. Pevzner P, Sze S (2000) Combinatorial approaches to finding subtle signals in DNA sequences. In: Proceedings of the eighth international conference on intelligent systems on molecular biology. San Diego, CA, pp 269–278
26. Rajasekaran S, Balla S, Huang C-H, Thapar V, Gryk M, Maciejewski M, Schiller M (2005) High-performance exact algorithms for motif search. J Clin Monitor Comput 19:319–328
27. Rajasekaran S, Balla S, Huang C-H (2005) Exact algorithms for planted motif problems. J Comput Biol 12(8):1117–1128
28. Sagot MF (1998) Spelling approximate repeated or common motifs using a suffix tree. In: Proceedings of the theoretical informatics conference (Latin98), pp 111–127
29. Sahli M, Mansour E, Kalnis P (2014) ACME: Efficient parallel motif extraction from very long sequences. VLDB J 23:871–893
30. Satya RV, Mukherjee A (2004) New algorithms for Finding monad patterns in DNA sequences. In: Proceedings of SPIRE 2004, LNCS, vol 3246, pp 273–285. Springer
31. Srinivas A (ed) (2005) Handbook of computational molecular biology. Computer and information science series, Chap. 5, December 21, 2005. Chapman & Hall/CRC, Boca Raton
32. Stoye J, Gusfield D (2002) Simple and flexible detection of contiguous repeats using a suffix tree. Theor Comput Sci 1–2:843–856
33. Thota S, Balla S, Rajasekaran S (2007) Algorithms for motif discovery based on edit distance. In: Technical report, BECAT/CSE-TR-07-3
34. Zambelli F, Pesole G, Pavesi G (2012) Motif discovery and transcription factor binding sites before and after the next-generation sequencing era. Brief Bioinform 14:225–237
35. Zhang S, Li S, Niu M, Pham PT, Su Z (2011) MotifClick: prediction of cis-regulatory binding sites via merging cliques. BMC Bioinf 12:238

Genome Analysis

9

9.1 Introduction

Basic sequence analysis we have analyzed until now involved pattern matching, comparing two or more sequences by alignment, clustering of sequences, and searching for repeats. Our aim in this chapter is to analyze genome and proteins at a higher, more macroscopic level of subsequences rather than at nucleotide and amino acid level. We will investigate the genome at three coarser and more observable levels: the genes, genome rearrangements, and haplotyping.

A gene is basically a subsequence of DNA which encodes for a protein. Transfer of genetic information from a gene to a protein is called the *gene expression*. Genes are the coding parts of DNA and constitute about 3 % of DNA. *Gene finding* or *gene prediction* is the process of discovering the location and contents of genes in a DNA strand where the latter term in the general sense implies that some statistical processing is involved during search. Finding genes helps to understand their function better and we will see there are a number of algorithms for this purpose.

Point mutations in which a nucleotide changes by substitution or indels are not the only evolutionary changes in DNA. Various changes at subsequence level, such as reversal or exchange of subsequences, are some of the dynamic events which may cause generation of new organisms and also may be one of the reasons behind various diseases. This process in which parts of a DNA are modified at subsequence level is termed *genome rearrangement* and discovering these macroscopic mutations is one of the fundamental research areas in bioinformatics as it gives us insight into the evolutionary process in a larger time frame and understanding of the disease states.

A haplotype is part of a single chromosome of a genome. The biotechnical methods usually produce data from both chromosomes of an organism for practical and cost-effective reasons, however, the sequence information from a haplotype only is needed as it is meaningful in the study of genetic information for a number of applications including disease analysis. Obtaining data of a haplotype from genetic data is called *haplotype inference* and there are many methods to accomplish this task. We start this chapter by first describing fundamental sequential and distributed methods for

© Springer International Publishing Switzerland 2015
K. Erciyes, *Distributed and Sequential Algorithms for Bioinformatics*,
Computational Biology 23, DOI 10.1007/978-3-319-24966-7_9

gene finding followed by analysis of genome rearrangements and conclude by the investigation of haplotype inference methods. These are the three distinct problems encountered while analyzing the genome as a whole.

9.2 Gene Finding

Genes are the basic units of hereditary information distributed over the chromosomes. There are over twenty thousand genes in human genome and functioning of about 50% of them is not fully understood. In humans, chromosome 1 is the largest and contains the highest number of genes in excess of 3,000 whereas the Y chromosome of males is the smallest and contains only about few hundred genes.

Genes are placed between long sequences of DNA that are believed to be nonfunctional. The genome of an organism is the whole DNA including functional (genes) and nonfunctional subsequences. The size of the human genome is about three billion base pairs (bps) of nucleotides. A complementary copy of the nucleotide sequence in a gene is made in the *transcription* process. The nonfunctional parts of the transcribed DNA sequence is thrown out in the *splicing* process as commonly observed in eukaryotes, to result in the messenger RNA (mRNA) sequence which is forwarded to the protein synthesis process. Each codon of the mRNA is translated into an amino acid sequence which makes up a protein in the final translation process. These consecutive steps of transcription, splicing, and translation are termed as the *central dogma of life*.

The structure of a gene is depicted in Fig. 9.1. The functional, coding regions in a gene called *exons* are placed between the nonfunctioning sequences called *introns* both of which have varying lengths, introns commonly being much longer in length than exons in humans. The splicing process involves excising out the introns from the transcribed sequence and the exons are combined to form the mRNA. The mRNA coding region always starts with the same codon of ATG to generate the amino acid *methionine* and ends with one of the three stop codons of TAA, TAG, or TGA which are not translated into amino acids. There are 20 amino acids and 64 codons which means few codons correspond to the same amino acid as shown in the genetic code. The amino acid sequence that is translated is called a *polypeptide* and a protein consists of one or more polypeptides. The gene sequence between the start codon and the stop codon of a gene is called the *open reading frame* (ORF).

As a gene has start and stop codons specifying its beginning and ending locations, gene finding can be performed by simply searching for the start codon ATG over the genome and once this is found, we can search for one of the stop codons to locate the whole gene. In this case, the problem is reduced to that of pattern matching, however, this time consuming approach is hardly used on its own due to the huge size of genome. There are various problems associated with gene finding. First, measurements may be erronous resulting in false signals. Genes may overlap, that is, a part of a gene may be a part of another gene. Moreover, a gene may be nested in another gene. There are occasions where several genes may encode for a single

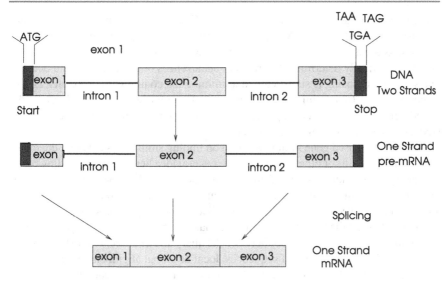

Fig. 9.1 The structure of a gene. Splicing forms the mRNA which is used for translation process

protein, for example, hemoglobin is made of several polypeptide chains each of which is the result of a single gene expression [2].

Finding genes helps to analyze them and understand their functionality better. There are numerous algorithms and methods for gene finding problem in prokaryotes and eukaryotes, we will review only the fundamental ones in the next sections.

9.2.1   Fundamental Methods

Gene finding in prokaryotes is relatively easier as they have smaller genomes and their genes occupy a much higher percentage of the genome than eukaryotes. On the contrary, finding genes in eukaryotes is more difficult as genes are more sparsely distributed, separated by long noncoding DNA segments. Moreover, the gene in an eukaryote consists of exons and introns and therefore the coding region is not contiguous as we have seen. Also, detecting the gene signals such as promoters are more difficult due to their more complex structure than in prokaryotes. Our main emphasis in this section is gene finding in eukaryotes.

Algorithms to find genes can be broadly classified as *ab initio* ("from the beginning" in Latin) methods and comparison-based methods. *Ab initio* methods typically employ statistical algorithms that search for certain signals associated with genes. Two such commonly used sequence information are the *signal sensors* and *content sensors*. Signal sensors are the sequence repeating patterns such as the ones found in transcription binding sites; signals related to splice sites, and start and stop codons. We have already seen methods to detect sequence repeats in Chap. 8 and such algorithms can be used to find the repeats in transcription binding sites to detect genes.

The first exon of a gene starts with the codon ATG and ends with one of the three stop codons: TAA, TAG, or TGA. Moreover, the boundary between an exon and an intron of a gene is always signaled by the sequence AG. Using this information, it is possible to statistically determine the frequent occurrences of these signals and deduce genes. Content sensing is statistically determining the coding sequences such as the usage of codons to determine exon sequences. There are several *ab initio* methods to find genes using these concepts such as dynamic programming, Hidden Markov Models (HMMs), and neural networks (NNs).

Comparison-based gene finding is based on finding similarities between the input genome and proteins or other genomes. The general idea here is that the coding parts of a genome are more conserved through the evolution than the other noncoding parts. Therefore, the similar parts of two or more genomes or proteins may indicate shared genes between two or more species. Two major approaches for comparison are either between the amino acid sequence that may be produced by a genome and some known protein sequences; or between the query and a number of known genome sequences. To find similarity of sequences, local or global alignment algorithms we have seen in Chap. 6 can be used. BLAST family of algorithms can be adopted for this purpose [1], and as another example, the GeneWise program performs global alignment of the translated genomic sequence of the query to a homologous protein sequence [8].

9.2.2 Hidden Markov Models

A *Markov chain* is a state machine where the future state is only dependent on the current state and the probability of moving from the current state to the future state. The sum of the probabilities of transitions from a state is equal to unity. Formally, a Markov chain is a triplet (Q, p, A) such that

- Q is a finite set of states from an alphabet Σ.
- p is the initial state probabilities.
- A is the state transition probabilities with elements a_{st} for transition from state s to t such that $a_{st} = P(x_i = t | x_{i-1} = s)$.

The probability of each variable x_i is dependent only on the value of the preceding variable x_{i-1}. The probability of observing a given sequence of events in a Markov chain is the product of all observed transition probabilities as shown below.

$$Pr(x_1) = \prod_{i=2}^{L} Pr(x_i | x_{i-1}) \tag{9.1}$$

A *Hidden Markov Model* (HMM) is a nondeterministic state machine that emits variable-length sequences of discrete symbols. It is a 5-tuple (Q, V, p, A, E) with Q, p, and A as defined above and additionally, V and E as below:

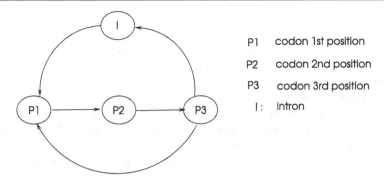

P1	codon 1st position
P2	codon 2nd position
P3	codon 3rd position
I:	intron

Fig. 9.2 An HMM to model genes

- V is a finite set of observation variables per state.
- E is a probability emission matrix $e_{sk} = P(v_k$ at time $t|q_t = s)$

A HMM hides the internal factors represented by states and describes the observable events represented by symbols that are dependent on states.The hidden states form a Markov chain with the probability distribution of the observed symbol depending on the internal state. A HMM is basically a Markov Chain in which the output of the current state is observable but the states are not directly observable. The output from a state is randomly selected based on some probability distribution. In the generalized Hidden Markov Model (GHMM), the output from a state may consist of a sequence, rather than a single character, based on some probability distribution.

HMMs provide reliable results and hence are widely used in a number of applications including speech recognition, sequence alignment, and gene finding. Figure 9.2 displays a simple HMM for gene recognition. Each codon of a gene will have three symbols which are dependent on states P_1, P_2, and P_3. A noncoding intron part of a gene results in transfer of state to I. Different emission probabilities are assigned to states P_i to show the symbol statistics at ith position of a codon. Once an HMM is constructed with known genes, it can be used to find undiscovered genes. HMMs are used for gene detection in [30] and various tools adopt the HMM method for gene discovery including GENSCAN [11,25], HMMGene [34,35], Genie [26], and GeneZilla [27].

9.2.3 Nature Inspired Methods

Nature inspired methods mimic the processes in nature, in this case, the biochemical processes in organisms.These methods are considered as *ab initio* procedures as they predict the location of a gene with some probability. Two such methods are the use of artificial neural networks (ANNs) and genetic algorithms (GAs) described next.

9.2.3.1 Artificial Neural Networks

An ANN has a similar structure to the neural network of the brain as we have briefly reviewed in Chap. 2. A neuron has a number of input channels called *dendrites* and its output is transmitted to its terminal buttons via its *axon* using synapses. A similar structure is adopted in ANNs with the connections between the neurons having weights. Each neuron *fires* or not, based on whether the sum of the input signals it receives is above a threshold or not. Adjusting the weights of the ANN based on a pattern or some vector to obtain the desired output from the ANN is termed as *training* of the ANN and this process is also referred to as *learning*. An ANN consists of a number of layers with the first layer being the input layer and the last layer is the output layer. A typical ANN will have three layers with a hidden layer between the input and output layers as shown in Fig. 9.3.

In the *supervised learning* model of ANNs, an external entity to the network is needed which describes what the output response to the input signal should be whereas the *unsupervised learning* does not need external intervention and uses the available information in the network.

ANNs can be classified as *feedforward*, *feedback*, and *self-organizing* networks [32]. The output of each layer in a simple feedforward ANN feeds the input of the next layer, providing a unidirectional information flow. A feedback ANN implies that some of the output is processed as input to the preceding layers providing loops in the network. The self-organizing ANNs do not require any external involvement and perform learning themselves and hence employ unsupervised learning.

An ANN can be used as a signal sensor to detect genes in a genome. The use of an ANN for gene prediction aims to train it with known genes to detect genes in other organisms with high probability. Training the ANN to be sensitive to the length of exons is difficult as this length varies significantly. However, training to detect splice sites, and the translation start and stop sites is possible. Brunak et al. provided one of the first studies to predict splice sites in DNA using ANNs [10].

Fig. 9.3 A three-layer ANN

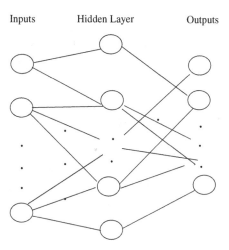

Gene recognition and analysis internet link-I (GRAIL-I) is an early study employing ANNs for gene prediction. It uses a multilayer feedforward ANN and in the later improved version called GRAIL-II, the problem of missing short exons in GRAIL-I was solved using variable-length windows. Gene identification using neural nets and homology information (GIN) [12] and GENSCAN [11] are the tools that also utilize ANNs for gene prediction. More recent work using ANNs to detect genes are in [39,42]. A detailed survey of these methods can be found in [23].

9.2.3.2 Genetic Algorithms

A *genetic algorithm* (GA) is a heuristic method that simulates the natural selection process. These algorithms are used to find solutions to optimization problems employing processes such as *mutation, inheritance,* and *selection* as in evolution. A candidate population is randomly selected and evolved by forming new generations in a GA. As each generation is constructed, the fitness function value of each individual is computed and the individuals with better fitness values are selected to form the next generation. The algorithm stops when a stopping condition such as the number of generations constructed or some criteria is reached, or a combination of various attributes determines the end of the algorithm. These algorithms can be used to detect genes by forming fitness functions based on content statistics of genes or may be used to classify biological phenotypes based on gene expression.

There are only few studies on the use of GAs for gene prediction. Earlier attempts of using GAs for protein structure prediction was reported in 2000 in [43] and in 2003 in [18]. An exclusive GA-based gene prediction method is described in [41] and a GA for gene regulatory activity prediction is proposed in [21].

9.2.4 Distributed Gene Finding

As the general approach we have adopted in distributed sequence processing until now, we can employ data partitioning and split the genome and distribute genome segments to a number of processes $p_0, \ldots, p_{k-1}$. The root process p_0 will be responsible for this initial step. Each process p_i then performs the gene finding procedure in its local data. The results are transferred to the root process which then merges them and submits as output to the user. There are not any studies using this approach reported in literature although the implementation is trivial.

HMMs can be constructed in parallel to ease the burden on the single processor. One such study is reported in [38] where the forward and Viterbi algorithms are parallelized and shown to be scalable for models with small address space.

9.3 Genome Rearrangement

Genomes of organisms undergo rearrangements during evolution in which the order of their subsequences change radically. These rearrangements occur at much lower frequencies than point nucleotide mutations and hence, we can trace evolution in a larger time frame with the analysis of these arrangements than can be obtained by point mutations. Also, the size of data that can be inspected as a whole is larger, providing a statistically better represented evolutionary process. Furthermore, genome rearrangements are believed to be one of the causes of complex diseases such as cancer and finding them helps to understand the mechanisms of these diseases better to provide improved drug therapies.

In the search of relatedness of one species to another via genome rearrangements, the main idea is to assume that evolution applies minimum number of steps when one species is transformed to another. Hence, if we can find this minimum distance, we can determine the closeness of the two species in real life. For these reasons, discovering genome rearrangements by computational methods is needed and is one of the active research areas in bioinformatics.

Genes are the main units of rearrangements in the genome; two genomes may have many common genes arranged in different orders resulting in different organisms. Assuming a genome consists of segments ABCDEF each of which may consist of a number of genes or single genes, the observed mutations at sequence level in a genome can be classified as follows:

1. *Insertion*: A segment is inserted to the genome, ABCDEF → ABCDGEF
2. *Deletion*: A segment is deleted from the genome, ABCDEF → ABDEF
3. *Duplication*: A segment is either as repeated tandem or as distributed as we analyzed these repeated patterns in Chap. 8. ABCDEF → ABBCDEF or ABCDEF → ABCDEBF. The whole genome duplication in which a copy of the full genome or some of its chromosomes is also possible.
4. *Reversals/Inversions*: The order of a number of segments is reversed, ABCDEF → ABEDCF
5. *Transpositions*: A segment of a genome is moved to another location, ABCDEFG-HIJ → ABFGHCDEIJ. *Inverted transpositions* are also possible, ABCDEFGHIJ → ABHGFCDEIJ

Moreover, two chromosomes may be involved in the following mutations:

1. *Fusion*: Two chromosomes combine to form a new chromosome.
2. *Fission*: A chromosome is divided into two chromosomes.
3. *Translocation*: The transfer of one segment from a chromosome to another chromosome.

In unichromosomal genomes, reversals are the most frequently encountered rearrangements whereas in multichromosomal genomes in which genes are distributed over a number of chromosomes; reversals, translocations, fissions, and fusions are commonly found.

9.3.1 Sorting by Reversals

A permutation of n numbers such as (1 2 3 4 5 6) is a reordering of these numbers, such as (3 1 2 4 6 5); the first sequence is called the *identity permutation* (I). If two such sequences represent two genomes, our interest is to find the minimum number of rearrangement steps from the above-described mutation activities, to transform one genome to another in order to compare these two genomes and find their evolutionary distance. The basic assumption here is that the evolutionary process has taken a similar path of minimum distance.

One of the first studies in genome rearrangement analysis was that of Dobzhansky and Sturtevant who examined *Drosophila* fruit fly to find that it has inversions [20] and proposed that the degree of disorganization of genes in two genomes is an indicator of the evolutionary distance between them. In unsigned permutations, the order of nucleotides within a segment remains unchanged during a rearrangement operation and in signed permutation algorithms, this order is reversed in addition to the rearrangement. As an example, assume an unsigned genome permutation $\pi_A = 1\ 2\ 3\ \underline{4\ 5}\ 6$. Then, we can convert this to another genome $\pi_B = 1\ 2\ 5\ 4\ 3\ 6$ by just one unsigned reversal. We can obtain a signed genome permutation $\pi_C = 1\ 2\ -5\ -4\ -3\ 6$ of π_A again with one step. Studying the mitochondrial genomes (mtDNAs) of cabbage and turnip, Palmer and Herbon found these are 99 % similar, however, their order of genes were different in the genomes [40]. Three reversals between the genomes would provide transforming of cabbage to turnip as below where a negative sign indicates the order reversal of the elements of a segment.

```
Cabbage   1   -5    4   -3    2
          1   -5    4   -3   -2   (inversion of 2)
          1   -5   -4   -3   -2   (inversion of 4)
Turnip    1    2    3    4    5   (reversal and inversion of -5,...,-2)
```

As we have seen in this example, *sorting by reversals* may result in forming new species and is more frequently observed in unichromosomal genome rearrangements than all other rearrangements in a single chromosome.

Definition 9.1 (*reversal*) Let us assume a genome having n subsequences is represented by a permutation $\pi = (\pi_1, \ldots, \pi_n)$. A *reversal* $\rho(i, j)$ transforms π into a permutation $\pi \times \rho(i, j)$ which performs a reversal starting from location i of π with length j.

For example, for $\pi = 3\,\underline{1\,4}\,\overline{6\,7}\,2\,5$, $\pi \times \rho(2, 3) = 3\,6\,4\,1\,7\,2\,5$. The *reversal distance* $d(\pi_A, \pi_B)$ between two permutations π_A and π_B is the minimum number of reversals that transforms π_A into π_B which can be expressed as $\pi_A \times \rho_1, \dots, \rho_{d(\pi_A, \pi_B)}$. Given two permutations π and σ, the *reversal distance problem* searches for a minimum number of k reversals $\rho_1, \dots, \rho_k$ such that $\pi \times \rho_1 \times \rho_2, \dots, \rho_k = \sigma$. If one of the two permutations is assumed to be the identity permutation I, $d(\pi)$ is the reversal distance between π and I. Let us try to find the reversal distance of the permutation $\pi = 2\ 3\ 4\ 1$ using unsigned and signed permutations. We can see that unsigned permutations result in one less step.

```
unsigned permutations     signed permutations
      2 3 4 1                   2 3 4 1
      4 3 2 1                  -4 -3 -2  1
      1 2 3 4                  -4 -3 -2 -1
                                1 2 3 4

      2 steps                   3 steps
```

Definition 9.2 (*breakpoint*) Given a genome with permutation π, a *breakpoint* $b(\pi)$ of π is the position i in π where $|\pi_{i+1} - \pi_i| \neq 1$. The breakpoint $b(\pi) \geq 1$ if and only if $\pi \neq I$ [37].

For example, with $\pi = 1\ 2\ |\ 5\ 4\ 3\ |\ 6\ 7$, there are two breakpoints at $(2, 5)$ and $(3, 6)$ locations. Approximation algorithms to find the minimum number of reversals to convert one permutation to another generally rely on removing breakpoints from a permutation to sort it. Every reversal removes at most two breakpoints, and hence:

$$\lceil b(\pi)/2 \rceil \leq d(\pi) \leq n - 1 \tag{9.2}$$

A strip of a permutation can be defined as follows:

Definition 9.3 (*strip*) A strip is an interval $[i, j]$ where $(i - 1, i)$ and $(j, j + 1)$ are breakpoints with no other breakpoints between them. In other words, a strip is the maximal increasing or decreasing subsequence that does not contain any breakpoints.

A permutation can be divided into increasing and decreasing strips shown by overlined and underlined sequences, respectively, as in $\pi = \overline{1}\,\underline{2}\,\overline{435}\,\overline{67}\,\underline{9}\,\overline{8}$. A single element strip can be labeled as an increasing or a decreasing strip, commonly as a decreasing strip. However, π_0 and π_{n+1} are always labeled as increasing when they are single element strips. Sorting by reversals process attempts to transform a permutation sequence to identity sequence using a minimum number of reversals. Since the identity permutation has no breakpoints, sorting by reversal is equivalent to eliminating all breakpoints from the permutation.

9.3.2 Unsigned Reversals

Sorting by reversals for unsigned permutations is known to be NP-hard [13] which necessitates the use of approximation algorithm to solve this problem. The approximation algorithms in chronological order to find unsigned reversal distance are 2-approximation by Kececioglu and Sankoff [33], 1.75 approximation by Bafna and Pevzner [4], 1.5-approximation by Christie [14], and 1.375-approximation by Berman et al. [7]. We will start the analysis by first the naive algorithm and gradually search algorithms with better time bounds.

9.3.2.1 The Naive Algorithm
As a first and naive attempt, we can greedily implement a reversal that transfers i into the ith position in the permutation. We need to perform the reversals for all positions in the given permutation π as follows:

1. **for** $i = 1$ **to** $n - 1$
2.   **if** $\pi_i \neq i$ **then**
3.   **do** a reversal to put i in ith place in π

Given a permutation $\pi = 3\ 4\ 1\ 5\ 2$, the steps of this naive algorithm would be as below.

$$
\begin{array}{ccccc}
3 & 4 & 1 & 5 & 2 \\
1 & 4 & 3 & 5 & 2 \\
1 & 2 & 5 & 3 & 4 \\
1 & 2 & 3 & 5 & 4 \\
1 & 2 & 3 & 4 & 5
\end{array}
$$

As another example, if we need to find the reversal distance of the permutation $\pi = 6\ 1\ 2\ 3\ 4\ 5$, there will be six iterations of the algorithm although we could simply have two reversals that transforms $\pi = 6\ \underline{1\ 2\ 3\ 4\ 5}$ to $\pi = \underline{6\ 5\ 4\ 3\ 2\ 1}$ and then to $I = 1\ 2\ 3\ 4\ 5\ 6$. The maximum number of reversals in this algorithm is $n - 1$ as the for loop is executed a $n - 1$ times which is unacceptable for very long sequences.

9.3.2.2 A 4-Approximation Algorithm
Our aim in sorting by reversal is to decrease the number of breakpoints. An observation of a permutation π shows that if there is at least one decreasing strip, we can find a reversal that reduces the number of breakpoints. We can therefore sketch an algorithm based on this idea as shown in Algorithm 9.1. Whenever the permutation contains a decreasing strip, we can reverse it to decrease the number of breakpoints; and if there are still breakpoints left with only increasing strips, we convert one of the increasing strips to a decreasing one so that we have a decreasing strip to reverse in the next step. During each iteration, the decreasing strip with the minimum element

k among all decreasing strips of π is selected and the strip between k and the element $k - 1$ of π is reversed.

Algorithm 9.1 *4-approx_Alg*

1: **Input** : permutation sequence $\pi = \{x_1, \ldots, x_n\}$
2: **Output** : reversals $\rho_1 \rho_2 \ldots \rho_m$
3: **while** $b(\pi) > 0$ **do**   ▷ if there is a remaining breakpoint in π
4:    **if** π has a decreasing strip **then**
5:        **find** the decreasing strip S_k with the smallest element k
6:        **reverse** the substring between k and $k - 1$
7:    **else**
8:        **do** a reversal to convert an increasing strip into a decreasing strip
9:    **end if**
10: **end while**
11: **return** i, $\rho_1 \rho_2 \ldots \rho_i$

Let us consider the permutation $\pi = 0\ 3\ 4\ 6\ 5\ 1\ 2\ 7$ with 0 added at the beginning and 7 at the end for convenience, and implement the algorithm for this permutation. The beginning and ending values are considered as increasing strips and any single element that is not part of any strip other than these two are considered as decreasing strips. Looking at π, there is one decreasing strip, (6,5) with minimum element 5. Hence, the subsequence between 5 and 4, including 5, is reversed in the first iteration. This results in a permutation with no decreasing strips, therefore, we reverse one of the increasing strips, (3 4 5 6). Continuing in a similar manner, the identity permutation is obtained as shown below.

```
0 3 4 6 5 1 2 7        k = 5, k-1 = 4
0 3 4 5 6 1 2 7        reverse an increasing strip
0 6 5 4 3 1 2 7        k = 3, k-1 = 2
0 6 5 4 3 2 1 7        k = 1, k-1 = 0
0 1 2 3 4 5 6 7
```

If there is a decreasing strip in π, every reversal in this algorithm reduces the number of breakpoints $b(\pi)$ by at least 1. If there are no decreasing strips, a reversal produces at least one decreasing strip. Therefore, there will be at most $2b(\pi)$ reversals in this algorithm. Since the optimum number of reversals is at least $\lceil b(\pi)/2 \rceil$, the approximation ratio is $2b(\pi)/\lceil b(\pi)/2 \rceil = 4$.

9.3.2.3   A 2-Approximation Algorithm

Let us assume that π represents the permutation sequence, in order to consider the boundaries, $\pi_0 = 0$ and $\pi_{n+1} = n+1$ are added to the original definition. Our aim is to transform π to the identity permutation I by iteratively removing the breakpoints using sorting by reversals. A general observation with the 4-approximation algorithm shows that as long as we have a decreasing strip in π, we can reduce the number of breakpoints by performing a reversal on this strip. When there are no decreasing strips, we can no longer reduce the number of breakpoints. In the 2-approximation algorithm provided by Kececioglu and Sankoff for unsigned reversals using breakpoints [33], the reversal that removes the highest number of breakpoints is chosen in each iteration until the identity permutation is obtained as shown in Algorithm 9.2 [33].

Algorithm 9.2 *2-approx_Alg*

1: **Input** : permutation sequence $\pi = \{x_1, \ldots, x_n\}$
2: **Output** : reversals $\rho_1\rho_2...\rho_m$ and distance $d(\pi)$
3: $i \leftarrow 0$
4: **while** π contains a breakpoint **do**
5:     $i \leftarrow i+1$
6:     $\rho_i \leftarrow$ reversal that removes the breakpoints of π by largest amount
7:     ties are resolved preferring reversals which leave a decreasing strip
8:     $\pi \leftarrow \pi \times \rho_i$
9: **end while**
10: **return** i, $\rho_1\rho_2...\rho_i$ ▷ i is the distance $d(\pi)$

Choice of such reversals is not trivial and we will show the detailed operation of this algorithm as in [45]. Let us assume that the permutation π contains decreasing strips with S_m being the string with the maximum element of all elements in decreasing strips, and S_k as the minimum element containing strip as before. Let S_{k-1} be the strip containing the element $k-1$ and S_{m-1} containing $m+1$. Let also ρ_k be the reversal that merges S_k and S_{k-1}; and ρ_m be the reversal that merges S_m and S_{m-1}. If neither $\rho_m \times \pi$ nor $\rho_k \times \pi$ contains any decreasing strip, then the reversal $\rho_k = \rho_m$ removes two breakpoints. This test is done in line 6 of Algorithm 9.2 shown in detail below. In the first two cases, number of breakpoints is reduced by at least one and the last case reduces it by two.

1. **if** $\pi \times \rho_k$ contains a decreasing strip **then**
2. $\pi \leftarrow \pi \times \rho_k$
3. **else if** $\pi \times \rho_m$ contains a decreasing strip **then**
4. $\pi \leftarrow \pi \times \rho_k$
5. **else**
6. $\pi \leftarrow \times\pi(\rho_k = \rho_m)$
7.   reverse any increasing strip in π to get a decreasing strip

If a step of the algorithm forms a π without a decreasing strip, this step has reduced $b(\pi)$ by two. This algorithm will provide a two decrease in the number of breakpoints for every two reversals. Therefore, there will be at most $b(\pi)$ reversals using this algorithm and since the optimum number of reversals is at least $\lceil b(\pi)/2 \rceil$, the approximation ratio is $b(\pi)/\lceil b(\pi)/2 \rceil = 2$.

9.3.3 Signed Reversals

In signed permutations, each element of the genome can have a positive or a negative sign as described to reflect the orientation in the order of genome elements such as genes. Our problem again is to sort a signed permutation with minimum number of traversals. A 1.5-approximation algorithm to find signed reversal distance was provided by Bafna and Pevzner [4] using breakpoint graphs. Hannenhalli and Pevzner showed in 1995 that this problem can be solved exactly in polynomial time with $O(n^4)$ time complexity [28]. The running time of the polynomial algorithm was improved to $O(n^2)$ by Kaplan et al. [31] and the computation time of the reversal distance was obtained as $O(n)$ by Bader et al. [3]. The polynomial time algorithms used the concept of the *breakpoint graph* as described next.

9.3.3.1 Breakpoint Graph

Bafna and Pevzner proposed the use of the *breakpoint graph* model to solve the sorting by reversals problem [4,5]. The permutation π is expanded with the 0 and $n+1$ permutations as before and an edge-colored graph $B(\pi)$ with each node representing an element of π is formed. A *black* edge in $B(\pi)$ connects two consecutive elements of π and the *gray* edges connect the vertices in the order of the identity permutation in this graph. Figure 9.4a displays the breakpoint graph for the permutation $\pi = 3\ 1\ 6\ 5\ 4\ 2$ enlarged by elements 0 and 7. The identity permutation can be displayed in the horizontal axis with still dotted edges between the consecutive nodes and black edges between the nodes in the permutation, to result in an equivalent graph shown in (b).

A cycle in an edge-colored graph is called alternating if the colors of every two consecutive edges of this cycle are different, black and gray in this case. Each node in $B(\pi)$ has equal number of black and gray edges incident to it. Therefore, edges of $B(\pi)$ can be decomposed into a set of edge-disjoint alternating cycles, edges of which change colors in each vertex. Let us form cycle decompositions of $B(\pi)$, and select the decomposition with the maximal number of cycles in it. Let $c(\pi)$ be the number of cycles in this decomposition. Bafna and Pevzner showed that [4]

$$d(\pi) \geq n + 1 - c(\pi) \tag{9.3}$$

where $d(\pi)$ is the reversal distance and $c(\pi)$ the number of cycles, since every reversal changes $c(\pi)$ by at most one. This equation shows that increasing the number of alternating cycles in $B(\pi)$ results in π getting closer to the identity permutation

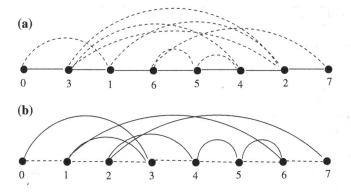

Fig. 9.4 The breakpoint graphs for the permutation $\pi = 0\ 3\ 1\ 6\ 5\ 4\ 2\ 7$. These two representations are equivalent

and therefore to the solution. The reversal distance problem is hence reduced to *maximal cycle decomposition* problem. The reader is referred to [4,5] for a detailed analysis of this algorithm. A modified breakpoint graph can also be used for sorting by transpositions [15].

9.3.3.2   Sorting by Oriented Pairs

An alternative simple and effective algorithm was proposed by Bergeron [6] for sorting signed reversals. An *oriented pair* (π_i, π_j), $i < j$ of a permutation π is defined as a pair of consecutive integers in π with opposite signs. For example, in the below-signed permutation, the oriented pairs are $(2,-1)$ and $(2,-3)$.

$$(\ 0\ \ 2\ \ 4\ \ -3\ \ 6\ \ 5\ \ -1\ \ 7)$$

Reversals of elements of π between oriented pairs result in consecutive integers which provides a method for sorting signed permutations. The reversals to be considered by an oriented pair (i, j) will be

$$\rho(i, j-1), \ \text{if } \pi_i + \pi_j = 1 \tag{9.4}$$

and

$$\rho(i+1, j), \ \text{if } \pi_i + \pi_j = -1 \tag{9.5}$$

The *score* of an oriented reversal is the count of the oriented pairs in the formed permutation. The algorithm presented in [6] simply iterates by always selecting the oriented pair which gives the highest score until there are no oriented pairs left as shown in Algorithm 9.3

Algorithm 9.3 *Orient_Alg*

1: **Input** : permutation sequence $\pi = \{x_1, \ldots, x_n\}$
2: **Output** : reversals $\rho_1 \rho_2 \ldots \rho_m$
3: $i \leftarrow 0$
4: **while** $\pi \neq I$ **do**
5:    $i \leftarrow i + 1$
6:    $P \leftarrow$ oriented pairs in π
7:    **find** the oriented pair $(a, b) \in P$ with the highest score when strip between is reversed
8:    $\rho_i \leftarrow$ the reversal between a and b according to Eqs. 9.4 and 9.5
9:    $\pi \leftarrow \pi \times \rho_i$
10: **end while**
11: **return** i, $\rho_1 \rho_2 \ldots \rho_i$ $\triangleright$ i is the distance $d(\pi)$

Let us apply this algorithm to the permutation $\pi = 0\ 2\ 1\ 4\ 6\ -3\ 5\ 7$. The iterations which always pick the oriented pair that gives the maximum pairs in the resulting permutation are shown below with the reversals underlined.

```
                        oriented pairs

0  4  5  1  2 -3  6  7     (2,-3) and (4,-3); pick (4,-3)
                          as (2,-3) does not yield any pairs.

0 -2 -1 -5 -4 -3  6   7    (0,-1) and (-5,6); pick (0,-1)
                          arbitrarily as they both yield 2 pairs.

0  1  2 -5 -4 -3  6  7     (2,-3) and (-5,6); picking any results
                          in identity permutation.

0  1  2  3  4  5  6  7     identity permutation

0 -2 -1  3  4  5  6  7     Had we picked (-5,6) in the second step;
                          (0,-1) and (-2,3) are the oriented pairs now,
                          picking any results in identity permutation.

0  1  2  3  4  5  6  7     identity permutation
```

The difficult part of this algorithm is the determination of which oriented pair to apply the reversal with. We need to check each pair and count the pairs in the resulting permutation had we applied this reversal. We then need to find the one that gives the maximum score. For sequences with many genes, this would be time consuming. Bergeron showed that if applying Algorithm 9.3 to a permutation π using k reversals yields a permutation π', then $d(\pi) = d(\pi')+k$. A second algorithm based on *hurdles* was also presented in [6]. A framed interval of a permutation π is defined as follows:

$$i\ \pi_{j+1}\ \pi_{j+2} \ldots \pi_{j+k-1}\ i + k \qquad\qquad (9.6)$$

where all values between i and $i + k$ are in the interval $[i...i + k]$. For example, in the permutation π = 2 4 3 1 5 7 6; [2 4 3 1 5], [3 1 5 7], and [1 5 7] are examples of framed intervals. A hurdle is defined as a framed interval that does not contain any shorter framed intervals. The last sequence in our example is a hurdle. *Hurdle cutting* is the process of reversing the segment inside the hurdle and *hurdle merging* is defined as reversing the segment between the end point of a hurdle and the beginning point of another hurdle. Furthermore, a *simple hurdle* is defined as a hurdle, cutting of which decreases the number of hurdles. Using these concepts, a second algorithm defined in [6] is as follows. This algorithm together with the first one can be used to sort signed permutations.

1. **if** a permutation π contains $2k$ hurdles ($k \geq 2$) **then**
2.   merge any two nonadjacent hurdles
3. **else if** π has $2k+1$ hurdles ($k \geq 1$) **then**
4.   **if** it has one simple hurdle **then**
5.   cut the hurdle
6.   **else if** it has no hurdles **then**
7.   merge two nonadjacent hurdles or adjacent ones if $k = 1$

9.3.4 Distributed Genome Rearrangement Algorithms

Genome rearrangement algorithms are difficult to parallelize due to the dependencies in global data. We will attempt to sketch a distributed implementation of the oriented pair-based algorithm as follows. Examining of Algorithm 9.3, it can be seen the line where the oriented pair with the maximum score is determined is the most time-consuming task and this can be parallelized. The permutation π is distributed to k processes $p_0, \ldots, p_{k-1}$ of the distributed system so that each process p_i has a copy. Each p_i is responsible for the element of lower index in an oriented pair (a, b) in its segment from $(in/k) + 1$ to $(i+1)n/k$ where n is the length of π. In each round, each process p_i searches for the scores of oriented pairs in its segment and these scores are broadcasted to determine the highest one. Each p_i then updates its copy of the permutation π and continues with the next round until π is converted to the identity permutation I. This distributed algorithm is shown in pseudocode in Algorithm 9.4.

We assume there are no supervisors this time and each process is provided with a copy of the permutation π initially. There is an extra burden of sorting the maximum score values received from k processes which requires $O(k \log k)$ time. In a tightly coupled system where interprocess communication costs are usually negligible, the supervisor–worker model we have used can still be adopted where the supervisor collects maximum values together with the pairs, sorts the values, and sends the maximum value to workers in each round. In this approach of using a central coordinator, synchronization at the end of each round is more straight forward, mainly by the exchange of messages between the workers and the supervisor. This is the main reason the supervisor–worker model is relatively easier to implement. Otherwise as in this particular implementation, we need to have a mechanism to ensure that each process p_i has finished its task in the current round before the next round starts, as in line 23 of the algorithm.

A reported work about a distributed implementation of the breakpoint graph method is in [17]. The authors propose first the construction of the breakpoint graph $B(\pi)$ in parallel. The optimum distance in $B(\pi)$ is also computed in parallel in the second step. The authors state that this second step of the algorithm is the parallel version of the algorithms in [9,44]. Further algorithms are used to find the possible next reversals sequence and a parallel algorithm in the last step inputs the results from the first three stages and outputs the final sorting of the signed permutation. There are no implementation details and no analysis information given.

9.4 Haplotype Inference

Our DNA consists of 23 pairs of chromosomes with each pair consisting of a chromosome inherited from father and a chromosome from mother. Humans are *diploid*, meaning they have pairs of chromosomes. A *locus* or a *site* is a location in a pair of chromosomes. The observable physical characteristics of an organism such as its appearance and behavior is called its *phenotype*. For example, the color of the hair and the blood group type of an individual belong to her phenotype.

Algorithm 9.4 *Distributed_Oriented, code for process p_i*

1: **Input** : permutation sequence $\pi = \{\pi_1, ..., \pi_n\}$
2: **Output** : reversals $\rho_1 \rho_2 ... \rho_m$ and distance $d(\pi)$
3: $S_i \leftarrow \{\pi_{(in/k)+1}, ..., \pi_{(i+1)n/k}\}$ ▷ initialize my segment
4: $r \leftarrow 0$
5: **while** $\pi \neq I$ **do**
6:     $r \leftarrow r + 1$ ▷ start of a round
7:     $max_i \leftarrow 0$
8:     **for all** oriented pair (a, b) such that $a \in S_i$ **do** ▷ find the highest scoring pair
9:         $score \leftarrow$ number of oriented pairs in π if reversal is applied to (a, b) strip
10:        **if** $score > max_i$ **then**
11:            $max_i \leftarrow score$
12:        **end if**
13:    **end for**
14:    **broadcast** max_i to all processes
15:    **for all** $p_i \in P$ **do** ▷ receive maximum score resulting by pair (a, b) from each process
16:        **receive** $max_i, (a, b)$ from p_i
17:        $M \leftarrow M \cup max_i$
18:    **end for**
19:    **sort** M
20:    $max\_r \leftarrow$ maximum element in M; (a, b) is the oriented pair of max_r
21:    $\rho_i \leftarrow$ the reversal between a and b according to Eqs. 9.4 and 9.5
22:    $\pi \leftarrow \pi \times \rho_i$
23:    **synchronize** with all processes ▷ end of a round
24: **end while**

An *allele* is a member of a pair of genes in a particular locus on a chromosome. For example, given two chromosomes $X_1 = $ CAGGTA and $X_2 = $ ACTCGA, the alleles at locations 2 and 5 are A,C and T,G, respectively. An allele for eye color may code for blue and another one dark which may result in a dark eye color in phenotype. *Genotype* refers to the two alleles at a locus. A genotype is *homozygous* if there are two identical alleles at a particular site and is called *heterozygous* if the alleles are different in a locus. Some alleles can have *dominant* or *recessive* characteristics. For the characteristic regulated by a recessive allele to be present in the phenotype, the allele should exist in both chromosome loci. Let us denote the dominant allele by A and a recessive allele by a; we will find AA or aa in a homozygous locus and Aa in a heterozygous locus. The phenotype affected by the heterozygous locus will exhibit the dominant allele A characteristics.

Three persons P1, P2, and P3 with genotypes are shown below. For example, the genotypes at locus 3 for these individuals are CG. The haplotypes between the locus 3 and 5 is CCT for P1 and CGA for P2.

```
position    1 2 3 4 5 6 7

P1          A G C C T G A
            A T G G C C T

P2          A G C G A G A
            G C G A T C T
```

A single nucleotide polymorphism (SNP) is a mutation at a single position in the DNA sequence among individuals. Some SNPs may be associated with certain diseases but most SNPs do not lead any observable differences between individuals. SNPs are inherited from parents to children, and individuals develop their own SNPs in their lifetime which are passed to their children.Therefore, we can find genetic relations between individuals by comparing their SNPs. For example, a child will inherit approximately half of her SNPs from each parent.

A *haplotype* is a group of genes or more commonly, a group of SNPs inherited together from a single parent. Study of haplotypes and discovering them is crucial in understanding the health and disease states of an organism. For example, if a certain haplotype can be related to a disease, then the genes in that region of the chromosome can be identified and investigated further. Also, since SNPs are passed through generations from parents to children, haplotype data can be used to find evolutionary history of an organism.

9.4.1 Problem Statement

The process of determining the genotype of an organism is called *genotyping*. In order to infer haplotypes of an organism from its genotype, a model can be provided as follows. At each locus, each chromosome has either state 0 or 1; and for each locus of an individual, if both haplotypes have sates 0, the genotype has state 0 and the same is valid for 1 state. If the haplotypes at a particular locus have states 0 and 1, or 1 and 0, the genotype is assigned a state 2. For example, if we are given the genotype $G =$ (2102), we can have the haplotype combination sets $\mathcal{H}_1 = \{(0100), (1101)\}$, $\mathcal{H}_2 = \{(0101), (1100)\}$, $\mathcal{H}_3 = \{(1100), (0101)\}$, $\mathcal{H}_4 = \{(1101), (0100)\}$ to result in this genotype. The total set of haplotypes will be $\mathcal{H} = \{(0100), (0101), (1100), (1101)\}$.

Although DNA sequencing techniques have advanced significantly in the last decade, it is still difficult and costly to obtain haplotype data directly from DNA. The results of the DNA extraction and sequencing processes commonly provide genotype data at a relatively cheap price. Haplotype allele rather than genotype information is needed as it can be related to health and disease states and also to determine phylogenetic relationships. Therefore, there is a need for efficient computational methods to discover haplotype data from genotype data. This process of resolving haplotypes is known as *haplotype inference* or *genotype phasing* and sometimes, however less frequently, called *haplotype reconstruction*. This problem is known to be NP-hard and there are many heuristic methods and algorithms to solve it. We will describe the main methods as Clark's algorithm and expectation maximization in the next sections together with a brief survey of recent available tools.

9.4.2 Clark's Algorithm

Clark's algorithm is an early algorithm proposed to discover haplotypes from genotype data in 1990 [16]. It is based on the maximum parsimony concept which states that the best hypothesis to explain a complex process is the one that requires the least number of assumptions. The main idea of this algorithm is to search for the easy to find haplotypes in the genotype data first and use these resolved haplotypes to discover the remaining haplotypes. Algorithm 9.5 shows the high-level description of the steps of Clark's algorithm. It first checks the genotype set $\mathcal{G}$ for homozygote or heterozygote genotype with one heterozygote alleles as the haplotypes for these can be determined precisely. These discovered haplotypes are included in the set $\mathcal{H}$ to aid the solution of more complicated genotypes that have more than one heterozygote. The algorithm iterates until there are no more unresolved genotypes remain.

Algorithm 9.5 *Clark_Alg*

1: **Input**: genotype set $\mathcal{G} = \{G_1, ..., G_k\}$
2: **Output**: haplotype set $\mathcal{H} = \{H_1, ..., H_n\}$
3: $\mathcal{H} \leftarrow \emptyset$
4: **repeat**
5:     $G_i \leftarrow$ homozygote or a heterozygote genotype with one heterozygote allele
6:     $\mathcal{H}_i \leftarrow$ haplotypes for G_i
7:     $\mathcal{H} \leftarrow \mathcal{H} \cup \mathcal{H}_i$
8:     $\mathcal{G} \leftarrow \mathcal{G} \setminus G_i$
9: **until** all homozygote and heterozygote genotypes with one heterozygote allele are processed
10: **repeat**
11:     **for all** $G_i \in \mathcal{G}$ **do**
12:         $\mathcal{H}_i \leftarrow$ haplotypes for G_i using $\mathcal{H}$
13:         $\mathcal{H} \leftarrow \mathcal{H} \cup \mathcal{H}_i$
14:         $\mathcal{G} \leftarrow \mathcal{G} \setminus G_i$
15:     **end for**
16: **until** $\mathcal{G} = \emptyset$

The implementation steps of this algorithm for a sample genotype set is shown below.

```
                    G                        Gi      H

Step 1 (1101),(1100),(0102),(2102)   (1101) (1101)

Step 2 (1100),(0102),(2102)          (1100) (1101),(1100)

Step 3 (0102),(2102)                 (0102) (1101),(1100),(0101),(0100)

Step 4 (2102)                        (2102) (0100),(1101),(1100),(0101)
```

Clark's algorithm found many applications when first proposed and is still used for inferring haplotypes. It can be used for large data but requires haplotypes to be not very diversified as this would result in prolonged processing. One immediate problem with this algorithm is that it requires the existence of homozygote or a heterozygote genotype with one heterozygote allele to start. Also, it may not give unique solutions due to the order of processing.

9.4.3 EM Algorithm

The expectation maximization (EM) algorithm first proposed by Dempster et al. in 1977 [19] is implemented for haplotype inference from unphased genotype data in various studies [22,29,36]. We will give a brief description of the EM algorithm as applied to haplotype inference here, referring the reader to the mentioned references for detailed analysis. The EM algorithm aims to find the maximum-likelihood estimate of the parameters of a data set distribution in general. The implementation of the EM algorithm for genotype phasing consists of the following main steps:

1. Guess the initial haplotype frequencies
2. **repeat**
3.   Use the current guess to compute the expected number of occurrences.
4.   Compute the new estimates of haplotype frequencies using the expected number of occurrences.
5. **until** the haplotype frequencies converge

The complexity for one step of the EM algorithm can be stated as $O(nk)$, where n is the number of genotype data and k is the maximum number of heterozygous loci in the genotypes to be analyzed.

9.4.4 Distributed Haplotype Inference Algorithms

The method to be employed would again be the partitioning of the data space to perform distributed haplotype inference. We will describe an implementation of distributed Clark's algorithm and show how EM algorithm can be designed to run in parallel on a distributed computer system.

9.4.4.1 Distributed Clark's Algorithm Proposal

Clark's algorithm searched the genotype set for homozygote or heterozygote alleles. It then used the found haplotypes to construct other haplotypes for the remaining genotype data. Intuitively, we can attempt to perform distributed implementation of this algorithm as follows. Given a genotype set $\mathcal{G} = \{G_1, \ldots, G_m\}$, we partition it to a set $P = \{p_0, \ldots, p_{k-1}\}$ of k processes. Each process in each round attempts to resolve the genotype data in its partition and broadcasts the haplotypes found in all others. The procesess make use of this discovered haplotypes for the next round using the ones found up to that point. Algorithm 9.6 shows one way of implementing the described algorithm with a supervisor process p_0. The role of the supervisor is confined to initial distribution of genotype data, gather and broadcast discovered haplotypes and the final collection of haplotype data to present to user. A slightly different parallel implementation of Clark's algorithm is reported in [46] where the processes pick the genotypes randomly to find the haplotypes for them.

Algorithm 9.6 *Distributed_Clark*

1: **Input** : $\mathcal{G} = \{G_1, G_2, ..., G_n\}$ ▷ set of n genotypes
2:        $P = \{p_0, ..., p_{k-1}\}$ ▷ set of k processes
3: **Output** : $\mathcal{H} = \{H_1, H_2, ..., H_m\}$ ▷ set of m haplotypes
4: **if** $p_i = p_0$ **then**   ▷ I am the supervisor
5:     **for** $i = 1$ to $k - 1$ **do**   ▷ send partial genotype list to workers
6:         **send** $((i - 1)n/k) + 1$ to (in/k) elements of $\mathcal{G}$ to p_i
7:     **end for**
8:     **for all** $p_i \in P$ **do**   ▷ receive initial discovered haplotypes from workers
9:         **receive** partial H_i from p_i
10:         $\mathcal{H} \leftarrow \mathcal{H} \cup H_i$
11:         $\mathcal{G} \leftarrow \mathcal{G} \setminus G_i$
12:     **end for**
13:     **broadcast** $\mathcal{H}$ to all workers
14:     **repeat**
15:         **for all** $p_i \in P$ **do**   ▷ receive discovered haplotypes from workers
16:             **receive** partial H_i from p_i
17:             $\mathcal{H} \leftarrow \mathcal{H} \cup H_i$
18:             $\mathcal{G} \leftarrow \mathcal{G} \setminus G_i$
19:         **end for**
20:         **broadcast** $\mathcal{H}$ to all workers
21:     **until** $\mathcal{G} = \emptyset$
22:     **broadcast** *stop* to all processes
23: **else**   ▷ I am a worker
24:     **receive** my genotype values in $\mathcal{G}_i$
25:     $\mathcal{H}_i \leftarrow$ haplotypes for homozygotes and heterozygotes with one site in G_i
26:     **send** $\mathcal{H}_i$ to p_0
27:     **repeat**
28:         **receive** newly discovered haplotypes in $\mathcal{H}_i$
29:         **resolve** new haplotypes $\mathcal{H}'_i$ in G_i using $\mathcal{H}_i$
30:         $\mathcal{G} \leftarrow \mathcal{G} \setminus G_i$
31:         **send** $\mathcal{H}'_i$ to p_0
32:     **until** *stop* received from p_0
33: **end if**

9.4.4.2 Distributed EM Algorithm Proposal

The EM algorithm consists of two main steps which are executed until convergence is reached. We will attempt to parallelize this procedure by partitioning the data to a set of processes $pr_0, \ldots, pr_{k-1}$ with pr_0 as the supervisor again to control the overall operation. It initializes the estimate vector and broadcasts it along with partitions of data to the processes initially which compute their current partial estimates as shown in Algorithm 9.7. The results are gathered at the supervisor which then computes the new estimates and then broadcasts it again. This process continues when the supervisor determines that convergence is reached and sends a *stop* message to all workers. It is possible to manage without a supervisor so that every process pr_i broadcasts its results.

Algorithm 9.7 *Distributed_EM*

1: **Input:** $\mathcal{G} = \{G_1, G_2, ..., G_n\}$ ▷ set of n genotypes
2:     $PR = \{pr_0, pr_1, ..., pr_{k-1}\}$ ▷ set of k processes
3: **Output:** $\mathcal{H} = \{H_1, H_2, ..., H_m\}$ ▷ set of m haplotypes
4: **if** $pr_i = pr_0$ **then**   ▷ I am the supervisor
5:     $\mathcal{P} \leftarrow$ initial haplotype frequency estimates
6:     **for** $i = 1$ to $k - 1$ **do**   ▷ send split of genotype data to workers
7:         **send** $((i - 1)n/k) + 1$ to (in/k) elements of $\mathcal{G}$ to p_i
8:     **end for**
9:     **repeat**
10:         **broadcast** $\mathcal{P}$ to all processes
11:         $\mathcal{P} \leftarrow \emptyset$
12:         **for all** $pr_i \in PR$ **do**   ▷ receive partial clusters from workers
13:             **receive** partial current estimate P_i from pr_i
14:             $\mathcal{P} \leftarrow \mathcal{P} \cup P_i$
15:         **end for**
16:         **compute** new estimate $\mathcal{P}'$ using $\mathcal{P}$ and $\mathcal{G}$
17:         $\mathcal{P} \leftarrow \mathcal{P}'$
18:     **until** haplotypes converge
19:     **broadcast** *stop* to all processes
20: **else**   ▷ I am a worker
21:     **receive** my data in G_i
22:     **repeat**
23:         **receive** previous estimates in $\mathcal{P}$
24:         **compute** new estimates P_i using $\mathcal{P}$ and G_i
25:         **send** P_i to pr_0
26:     **until** *stop* received from p_0
27: **end if**

9.5 Chapter Notes

We have briefly reviewed three distinct genome analysis problems. The first problem was finding the location and contents of genes and this is needed as the first step of gene analysis. We saw the two main methods for this purpose as *ab initio* and comparison-based approaches. *Ab initio* methods attempt to discover statistically more frequently appearing subsequences in or around gene regions. HMMs are a widely used *ab initio* scheme, and ANNs and GAs provide alternative techniques for gene prediction.

Comparative methods assume that the evolutionary changes in the genes are slower when compared to the mutations in the noncoding parts of the genome. Therefore, comparing two similar genomes, we can detect regions of high similarity using sequence alignment methods, and these are potentially the coding sequences of the genome. We need to do post-processing of these similarity regions to detect the starting and stopping locations of the gene. There are hardly any algorithms for parallel/distributed gene finding, however, we can conveniently employ any parallel/distributed methods specified in Sect. 7.3 such as distributed BLAST and CLUSTALW-MPI for distributed alignment.

Genome rearrangements are mutations of the genome at macroscopic scale and are believed to be the cause of forming new organisms as well as being one of the major causes for various complex diseases. The main rearrangements are in the form of reversals as observed in single chromosomes. Assuming evolution follows the path of minimal changes when a new species is formed, we would be interested in following a similar path to find the real distance between the species to compare them. There are various methods and algorithms to find this distance between species and we briefly reviewed approximation algorithms to find unsigned and signed sorting by reversals including the breakpoint graph method. The parallel/distributed algorithms are very few and we described how to parallelize the oriented pair method in a distributed computing system.

Data obtained from the genome of an organism comes from two chromosomes and we need data from a single chromosome for various genome analysis tasks including disease analysis. Haplotype inference is the process of obtaining single chromosome data from the data of two chromosomes and there are various methods such as Clark's algorithm and the expectation maximization procedure for this task. Perfect phylogeny haplotyping is another method for genotype phasing as described in [24]. We proposed two algorithms to describe how Clark's algorithm and the expectation maximization method can be modified to execute in a distributed computer system.

In summary, we can deduce that although there are many sequential algorithms and tools for these three difficult tasks, parallel/distributed algorithms to solve them are very scarce. Since we are dealing with the whole genome, data partitioning of the genome to a number of processes is a convenient way of distributed processing. However, there is a need for more sophisticated distributed processing schemes and this is probably another potential research area in bioinformatics.

Exercises

1. Discuss how and why sequence alignment methods can be used to find genes in a number of closely related organisms.
2. Given the permutation $\pi = 4\ 2\ 1\ 3\ 5\ 6$, convert π to the identity permutation using the naive algorithm. Show each step of the algorithm.
3. For the permutation $\pi = 3\ 2\ 4\ 5\ 6\ 1\ 8\ 7$, mark the increasing and decreasing strips. Then, implement the 4-approximation algorithm to reduce π to the identity permutation.
4. For the permutation $\pi = 4\ 3\ 1\ 2\ 5\ 7\ 8\ 6$, mark the increasing and decreasing strips. Then, implement the 2-approximation algorithm to reduce π to the identity permutation.
5. Draw the breakpoint graph for the permutation $\pi = 5\ 4\ 1\ 2\ 6\ 8\ 3\ 7\ 9$ by showing the black and gray edges.
6. Sketch a distributed version of the oriented pair-based algorithm using the supervisor–worker model with the supervisor involved in computation. Show

the implementation steps of this algorithm for the permutation $\pi = -5\ 3\ 1\ 4\ -2$ $-7\ 8\ 6$ for three processes p_0, p_1, and p_2.

7. Show the implementation of Clark's algorithm for the genotype set $\mathcal{G} = \{(1012),$ $(1101), (1202)\}$.

8. Provide the distributed implementation of Clark's algorithm for the genotype set $\mathcal{G} = \{(1012), (1101), (1202), (1000), (1001), (2110)\}$ using two processes p_0 and p_1. The supervisor is also involved in finding the haplotypes. Show all iterations of the algorithm.

References

1. Altschul SF, Gish W, Miller W, Myers EW et al (1990) Basic local alignment search tool. J Mol Biol 215(3):403–410
2. Axelson-Fisk M (2010) Comparative gene finding: models, algorithms and implementation: Chap. 2, Computational Biology Series, Springer
3. Bader DA, Moret BME, Yan M (2001) A linear-time algorithm for computing inversion distance between signed permutations with an experimental study. In: FKHA Dehne, J-R Sack, R Tamassia (eds) WADS, LNCS, vol 2125. Springer, pp 365–376
4. Bafna V, Pevzner PA (1993) Genome rearrangements and sorting by reversals. In: Proceedings of the 34th annual symposium on foundations of computer science, pp 148–157
5. Bafna V, Pevzner PA (1996) Genome rearrangements and sorting by reversals. SIAM J Comput 25(2):272–289
6. Bergeron A (2005) A very elementary presentation of the Hannenhalli-Pevzner theory. Discrete Appl Math 146(2):134–145
7. Berman P, Hannenhalli S, Karpinski M (2002) 1.375-approximation algorithm for sorting by reversals. In: Proceedings of the 10th annual european symposium on algorithms, series ESA 02, Springer, London, UK, pp 200–210
8. Birney E, Durbin R (2000) Using GeneWise in the Drosophila annotation experiment. Genome Res 10:547–548
9. Braga MDV, Sagot M, Scornavacca C, Tannier E (2007) The solution space of sorting by reversals. In: Mandoiu II, Zelikovsky A (eds) ISBRA 2007, vol 4463, LNCS (LNBI)Springer, Heidelberg, pp 293–304
10. Brunak S, Engelbrecht J, Knudsen S (1991) Prediction of human mRNA donor and acceptor sites from the DNA sequence. J Mol Biol 220(1):49–65
11. Burge C, Karlin S (1997) Prediction of complete gene structures in human genomic DNA. J Mol Biol 268(1):78–94
12. Cai Y, Bork P (1998) Homology-based gene prediction using neural nets. Anal Biochem 265(2):269–274
13. Caprara A (1997) Sorting by reversals is difficult. In: Proceedings of the 1st ACM conference on research in computational molecular biology (RECOMB'97), pp 75–83
14. Christie DA (1998) A 3/2-approximation algorithm for sorting by reversals. Proceedings the ninth annual ACM-SIAM symposium on Discrete algorithms, series SODA 98. Society for Industrial and Applied Mathematics, Philadelphia, PA, USA, pp 244–252
15. Christie DA (1999) Genome Rearrangement Problems. Ph.D. thesis, The University of Glasgow
16. Clark AG (1990) Inference of haplotypes from PCR-amplified samples of diploid populations. Mol Biol Evol 7:111–122

17. Das AK, Amritanjali (2011) Parallel algorithm to enumerate sorting reversals for signed permutation. Int J Comp Tech Appl 2(3):579–589

18. Day RO, Lamont GB, Pachter R (2003) Protein structure prediction by applying an evolutionary algorithm. In: Proceedings of the international parallel and distributed processing symposium

19. Dempster AP, Laird NM, Rubin DB (1977) Maximum likelihood from incomplete data via the EM algorithm. J R Stat Soc 39(1):1–38

20. Dobzhansky T, Sturtevant A (1938) Inversions in the chromosomes of drosophila pseudoobscura. Genetics 23:28–64

21. Duc DD, Le T-T, Vu T-N, Dinh HQ, Huan HX (2012) GA_SVM: a genetic algorithm for improving gene regulatory activity prediction. In: IEEE RIVF international conference on computing and communication technologies, research, innovation, and vision for the future (RIVF)

22. Excoffier L, Slatkin M (1995) Maximum-likelihood estimation of molecular haplotype frequencies in a diploid population. Mol Biol Evol 12(5):921–927

23. Goel N, Singh S, Aseri TC (2013) A review of soft computing techniques for gene prediction. Hindawi Publishing Corporation ISRN Genomics, vol 2013, Article ID 191206. http://dx.doi.org/10.1155/2013/191206

24. Gusfield D (2002) Haplotyping as perfect phylogeny: conceptual framework and efficient solutions. In: Proceedings of the 6th annual international conference computational biology, pp 166–175

25. http://genes.mit.edu/GENSCAN.html. The GENSCAN Web Server at MIT

26. http://www.fruitfly.org/seq_tools/genie.html. The Genie web server

27. http://www.genezilla.org/. The GeneZilla web server

28. Hannenhalli S, Pevzner PA (1999) Transforming cabbage into turnip: polynomial algorithm for sorting signed permutations by reversals. J ACM 46(1):1–27

29. Hawley ME, Kidd KK (1995) HAPLO: a program using the EM algorithm to estimate the frequencies of multi-site haplotypes. J Heredity 86(5):409411

30. Henderson H, Salzberg S, Fasman KH (1997) Finding genes in DNA with a Hidden Markov Model. J Comput Biol 4(2):127–141

31. Kaplan H, Shamir R, Tarjan RE (2000) A faster and simpler algorithm for sorting signed permutations by reversals. SIAM J Comput 29(3):880–892

32. Karayiannis NB, Venetsanopoulos AN (1993) Artificial neural networks, learning algorithms, performance evaluation, and applications. Springer Science+Business Media, New York

33. Kececioglu J, Sankoff D (1993) Exact and approximation algorithms for the inversion distance between two permutations. In: Proceedings of the 4th annual symposium on combinatorial pattern matching, volume 684 of Lecture Notes in Computer Science, Springer, New York, pp 87–105

34. Krogh A (1997) Two methods for improving performance of a HMM and their application for gene finding. In: Gaasterland T, Karp P, Karplus K, Ouzounis C, Sander C, Valencia A (eds) Proceedings of the fifth international conference on intelligent systems for molecular biology. AAAI Press, Menlo park, CA, pp 179–186

35. Krogh A (1998) An introduction to hidden Markov models for biological sequences. In: Salzberg SL, Searls DB, Kasif S (eds) Computational methods in molecular biology Chapter 4 . Elsevier, Amsterdam, The Netherlands, pp 45–63

36. Long JC, Williams RC, Urbanek M (1995) An E-M algorithm and testing strategy for multiple-locus haplotypes. Am J Hum Genet 56(3):799–810

37. Mourad E, Albert YZ (eds) (2011) Algorithms in computational molecular biology: techniques, approaches and applications. Wiley Series in Bioinformatics, Chap 33

38. Nielsen J, Andreas Sand A (2011) Algorithms for a parallel implementation of Hidden Markov Models with a small state space. IPDPS Workshops 2011:452–459

39. Palaniappan K, Mukherjee S (2011) Predicting essential genes across microbial genomes: a machine learning approach. In: Proceedings of the IEEE international conference on machine learning and applications, pp 189–194

40. Palmer JD, Herbon LA (1988) Plant mitochondrial DNA evolves rapidly in structure, but slowly in sequence. J Mol Evol 28(1–2):87–97

41. Perez-Rodriguez J, Garcia-Pedrajas N (2011) An evolutionary algorithm for gene structure prediction. Industrial engineering and other applications of applied intelligent systems II 6704:386–395

42. Rebello S, Maheshwari U, Safreena Dsouza RV (2011) Back propagation neural network method for predicting lac gene structure in streptococcus pyogenes M group A streptococcus strains. Int J Biotechnol Mol Biol Res 2:61–72

43. Schulze-Kremer S (2000) Genetic algorithms and protein folding. Protein Struct Prediction Methods Mol Biol 143:175–222

44. Siepel AC (2002) An algorithm to find all sorting reversals. Proceedings of the 6th annual international conference computational molecular biology (RECOMB 2002). ACM Press, New York, pp 281–290

45. Sung W-K (2009) Algorithms in bioinformatics: a practical introduction. CRC Press, Taylor and Francis Group, pp 230–231

46. Trinca D, Rajasekaran S (2007) Self-optimizing parallel algorithms for haplotype reconstruction and their evaluation on the JPT and CHB genotype data. In: Proceedings of 7th IEEE international conference on bioinformatics and bioengineering

Part III
Biological Networks

Analysis of Biological Networks

10

10.1 Introduction

Biological processes are commonly represented as networks where nodes of a network serve as the biological entities and the edges connecting them show the interactions between them. The size of such a network is huge and the data obtained is often noisy making the analysis difficult. The modeling and analysis of biological networks is needed as it provides important clues about the underlying biological processes which help to evaluate health and disease states of an organism and possibly provide drug treatment. Understanding the structures of biological networks is essential to develop therapeutic treatment procedures for complex diseases, such as cancer [30], and Parkinson's disease and schizophrenia [15,26].

The biological networks can be broadly classified as the networks in the cell at molecular level and the networks outside the cell at a much coarser level. The main molecular networks in the cell are the metabolic networks, gene regulatory networks and protein–protein interaction networks. Neural networks and the brain functional networks are the two networks of the brain at different hierarchies, and the phylogenetic networks show the evolutionary relations between organisms whereas the food web is based on the predator-prey relationships of organisms.

The main challenges in biological networks analysis are their error-prone structures and dynamicity, and the problems associated with sampling. Biological systems at intracellular level are dynamic and the classical graph representations of these networks may not be adequate in many cases. Network models which include details about the kinetics and quantities of the molecules are needed as a solution to this problem. Analyzing data of the whole network is not feasible in many cases due to its size and evaluation of samples to represent the network is commonly preferred. This method however provides only approximate solutions as a sample will not precisely reflect the actual structure of the whole network.

In this chapter, we first describe the networks in the cell and outside the cell and then describe properties of biological networks from the view of graph theory. We continue with the modeling of these networks stated as representatives of general

© Springer International Publishing Switzerland 2015

K. Erciyes, *Distributed and Sequential Algorithms for Bioinformatics*,
Computational Biology 23, DOI 10.1007/978-3-319-24966-7_10

complex networks which are networks with very large number of nodes and edges. We then define the fundamental problems in biological networks which are module detection, network motif search, and network alignment which constitute the main topics in this part of the book along with phylogeny.

10.2 Networks in the Cell

A cell contains many different types of chemical compounds with DNA located at its nucleus. We have already seen in Chap. 2 the central paradigm of molecular biology in which genetic code in DNA is read by the RNA polymerase complex and is *transcribed* into the corresponding RNA. In the following *translation* phase, amino acid sequences which make up the proteins are produced by the ribosomes that bind to messenger RNA. The proteins formed interact with each other forming protein–protein interaction (PPI) networks. Important biological networks at cell level are related to DNA, RNA, genes, proteins, and metabolites. We will briefly describe the main networks in the cell which are metabolic networks, gene regulatory networks, and PPI networks in the next sections.

10.2.1 Metabolic Networks

The *metabolism* of an organism is its basic chemical system that produces essential cell ingredients, such as sugars, lipids, and amino acids. *Metabolic networks* model all possible biochemical reactions in the cell to generate metabolisms. The nodes in these networks are the biochemical metabolites and the edges represent either the reactions which convert one metabolite to another or the enzymes that catalyze these reactions [9,25,28]. The graphs representing the metabolic networks can be directed or undirected depending on the reversibility of the reaction representing an edge. A *metabolic pathway* is a sequence of biochemical reactions to perform a specific metabolic function. For example, glycolysis is one such metabolic process where energy supply is generated in which glucose molecules are broken into two sugars and these sugars are used to generate adenosine triphosphates (ATPs). Metabolic networks have small diameters and hence exhibit *small-world* characteristics in which the average path length between any two nodes is small. They also have few high-degree and many low-degree nodes as found in *scale-free* networks. Figure 10.1 displays the simplified metabolic reaction network of *E. coli* which shows these properties.

The study of metabolic networks and pathways within these networks is an important area of research in biomedicine as understanding the metabolic mechanisms in the cell helps to find cures for diseases. The infections can be controlled for example, by discovering the differences between the metabolic networks of humans and pathogens causing the infections [12].

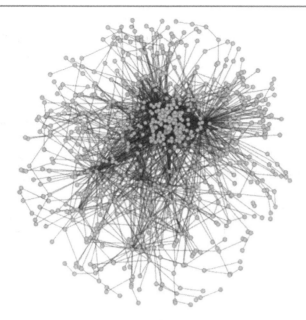

Fig. 10.1 The simplified metabolic reaction network of *E. coli* where endogenous metabolites are the nodes and the edges represent the reactions (taken from [16])

10.2.2   Gene Regulation Networks

Genes code for proteins via transcription and translation, which are essential for the functioning of an organism. This process is known as gene expression. The regulation of gene expression can be done at transcriptional, translational or post-translational stages of gene expression. Gene expression is controlled by proteins produced by other genes resulting in *regulatory interactions*. In simple terms, gene *A* regulates gene *B* if a change in expression of gene *A* induces a change in the expression of gene *B*. This regulation can take two forms; it can be an *up-regulation* which activates the gene expression process in the other gene, or *down-regulation* in which case the expression is inhibited. A gene regulation network (GRN) consists of genes, proteins, and other small molecules which all make up the nodes of the network, and their interactions form the edges. The regulators are usually proteins and referred to as *transcription factors*. A GRN is commonly represented by a directed graph where arrows show the direction of regulation. Figure 10.2 shows a GRN with three genes.

   The GRNs are sparse, that is, they have low density which means genes are regulated by few other genes. The out-degrees of nodes in a GRN follows *power-law* and *scale-free* properties meaning there are only few high out-degree nodes in these networks and the rest of the nodes have low out-degrees. The maximum distance between the nodes, the diameter of the gene networks, is low which is a property of the small-world networks. The GRNs of the lysis/lysogeny cycle regulation of

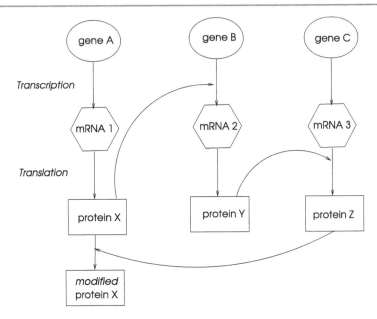

Fig. 10.2 A gene regulatory network with three genes A, B, C; three mRNAs 1, 2, 3; and three proteins X, Y, and Z. Gene A regulates gene B by protein X at transcription, gene B regulates gene C at translation by protein Y, and gene C regulates gene A at post translation by protein Z to modify protein X

bacteriophage-e [20], the endomesoderm development network in Sea Urchin [4], and the segment polarity network in Drosophila development [1] have been determined [10].

10.2.3   Protein Interaction Networks

Proteins consist of sequences of amino acids and they carry out vital functions such as acting as enzymes for catalysis of metabolic processes, signaling compounds, or serving as transporters for various substances like oxygen in the cell. A PPI network is built from interacting proteins. The interaction is needed in the cell for the activation of a protein by another one or to build protein complexes which are a group of proteins that form clusters. An onset of a disease is usually related to alterations in certain protein–protein interactions. This observation necessitates the study of protein interactions from a system level to understand the disease process better.

Detecting protein–protein interactions have been traditionally studied using physical methods such as affinity chromatography or co-immunoprecipitation [12,19,29]. Two techniques which are widely used currently are the yeast two-hybrid (Y2H) system analysis of protein complexes and affinity purification coupled to mass spectrometry (AP-MS) method. The Y2H system uses the structures of transcriptional factors and AP-MS method is based on the purification of protein complexes used with mass spectrometry [12]. A PPI network can be modeled conveniently by an undirected graph $G(V, E)$ where V is the set of proteins and E is the set of edges showing the interactions. The graph is undirected as it is usually not possible to determine which protein binds the other, in other words, which protein influences the other.

As can be seen in Fig. 1.2 of Chap. 1, this sample PPI network and PPI networks in general are not homogenous. There are central nodes with many number of connections and the majority of nodes have few connections. The central nodes provide the connections between the nodes with few connections. Although the number of nodes in a PPI network as in any molecular biological network is very large, any node can reach any other node with only few number of hops. We can therefore state that these networks have small diameters. Indeed, this physical property is very useful in fast transmission of signals between proteins. On one hand, we need these central nodes function properly for the healthy state of an organism. On the other hand, if the PPI network is formed or modified by the disease state of an organism, the therapy, and the drug design method should target these central nodes as their failure will help stop the functioning of the disease process. The PPI networks have this small-world property and the scale-free property as commonly found in other molecular biological networks [6,24]. We can also detect groups of nodes with high density which are called *clusters*. These clusters may have some important functionality attributed to them as there is a strong interaction in these groups. Another area of interest is the search of repeating patterns of subgraphs in these networks which are called *motifs*. These motifs may have certain attributed functionality and they may be the building blocks of these networks. Alignment of two or more networks is the network analogy of the sequence alignment we have reviewed in Chap. 6, our aim in this case is to compare and find how similar two or more networks are. Aligning PPI networks of different organisms helps us to discover the evolutionary relationships between them as well as detecting their functional similarities. We can therefore specify main topological problems in PPI and molecular biological networks as detecting the central nodes and the clusters, searching motifs, and aligning these networks.

10.3 Networks Outside the Cell

In this section, we review the main biological networks outside the cell which are the networks of the brain, phylogenetic networks, and the food web.

10.3.1 Networks of the Brain

Networks in the brain can be investigated at cellular level or at a coarser functional level. There are two main networks of the brain: neural networks at cell level and the brain networks at a coarser level. The structures and implications of these two networks will be described next.

10.3.1.1   Biological Neural Networks

The neural system of human body consists of three parts as receptors, neural networks, and effectors. The input stimuli to receptors may come externally or internally which are transferred to *neurons* as electrochemical signals. The neurons, which are the basic nerve cells in the brain, process the inputs and provide signals to activate effectors which provide the output to the external world.

There are about 10 million neurons in the human brain. Information between neurons is transmitted using biochemical reactions. The neuron cell body is called *soma* and a neuron receives electrochemical inputs from other neurons via *dendrites* which act as inputs to it. An impulse received by a dendrite from another neuron may be *excitatory* meaning it helps firing, or *inhibitory* meaning it tries to prevent firing. If the sum of the input electrochemical signals is greater than a threshold value, the neuron fires and transmits a signal of constant magnitude along its output channel called *axon*. The excitatory signals may be assumed to have positive weights and the inhibitory signals have negative weights while modeling these networks. The end points of an axon are connected to the dendrites of other neurons across small gaps called *synapses*. Each neuron of the brain is connected to 10,000 other neurons on the average. The sum of the input signals may be slightly or very much higher than the threshold but the output signal strength is the same in both cases. Figure 10.3 displays the basic model of a neuron with described parts.

A neuron basically has a binary output, it either sends a signal (fires) or not. These simple cells in the brain perform complex tasks such as cognition and storage of information. Artificial neural networks attempt to mimic the operation of biological neurons and they have been successfully implemented for various tasks such as image processing and learning where the digital computers have not performed well. However, understanding of most of the cognitive processes in the brain has not been

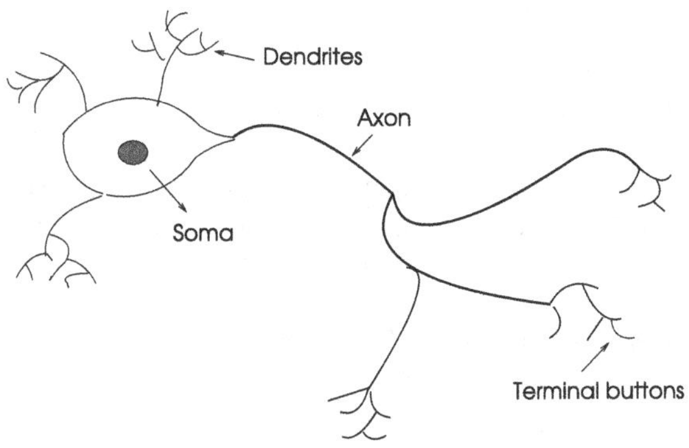

Fig. 10.3 The parts of a neuron

achieved by using the neural network models alone and models at much coarser level are needed as described next.

10.3.1.2   Brain Networks

The processing in the brain can be better understood at a higher conceptual level than the neural networks. There are two ways of investigating the modeling of processes in the brain. We can either look at the *structural networks* or *functional networks* and search for models based on these properties. The structural connectivity of the brain can be obtained by examining the physical properties, such as cortical thickness, gray matter volume, white matter fiber connections using the MRI data. The functional connectivity can be obtained by evaluating the temporal correlations between the activities using fMRI and EEG/MEG techniques [7,22]. When the connectivity information is gathered, brain networks can be modeled. We can form two types of networks based on the connectivity information gathered as the *brain structural networks* or the *brain functional networks* (BFNs). In a brain structural network, the connectivity information basically represents white matter integrity whereas the connections in a BFN shows the association between the functional regions of the brain. Based on the data obtained, we can form the adjacency matrix of the connections in a BFN and carry out the network analysis. The network analysis of BFNs show they exhibit common properties of complex networks such as the social networks and the Web. They are small-world networks with small diameters [5,23] which provides flow of electrochemical signals between different parts of the brain in short time. They also follow power law with few very high-degree hubs and many low-degree nodes [14]. These network properties were found to change during normal development, aging, and various neurological and neuropsychiatric diseases such as schizophrenia [3] and Alzheimer's disease [7,8].

10.3.2   Phylogenetic Networks

Phylogenetics is the study of reproducing the evolutionary relationships from data of living organisms. This relationship is usually represented by a rooted *phylogenetic tree* also called *evolutionary tree*. Leaves of this tree are the living organisms and the internal nodes represent common ancestors. Time increases from root to the leaves in phylogenetic trees and the edges of the tree are usually drawn in proportion to the time passed between an ancestor and a descendant. Distance between two organisms is computed using the variation in their DNA and protein sequences.

The main methods of phylogenetic tree construction are the distance-based methods, maximum parsimony, maximum likelihood, and Bayesian methods. Distance-based methods compute the evolutionary distances between organisms and predict their common ancestors. In maximum parsimony methods, trees are constructed by evaluating possible mutations between the sequences representing the organisms and maximum likelihood uses the probability of events. Bayesian methods are related to maximum likelihood approaches.

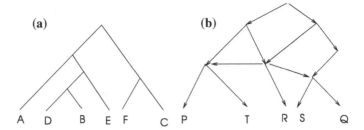

Fig. 10.4 **a** A rooted phylogenetic tree of organisms A,..., F. **b** A rooted phylogenetic network of organisms P,..., T

A phylogenetic tree may not be adequate to represent the evolution of organisms due to events such as horizontal gene transfer and hybridization in which two lineages recombine to create new species. A *phylogenetic network* is a generalization of a phylogenetic tree in which a node may have more than one parent. In essence, a phylogenetic network is an acyclic directed graph, however, the induced undirected graph may have cycles. Phylogenetic network construction methods are relatively more recent and are still at the stage of development. A rooted phylogenetic tree and a rooted phylogenetic network are depicted in Fig. 10.4.

10.3.3 The Food Web

All living organisms need to consume food for survival. Plants make their own food using a process called *photosynthesis*. Many organisms however, are not able to produce their own food and they rely on other organisms such as plants or animals for survival. A *food chain* displays the relationships between the producers and the consumers. An arrow in a food chain diagram indicates the predator-prey relationship with the arrow pointing to the predator. The food chain in sea is shown in Fig. 10.5 in simplified form. At the bottom, the *plankton* which is a microscopic plant is

Fig. 10.5 The food chain in the sea

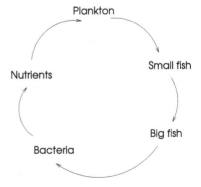

consumed by small fish and the small fish is a prey to the big fish. When big fish dies, it is consumed by *bacteria* which provide nutrition to the environment to be consumed by planktons.

In general, many consumers intake more than one type of food such as humans who eat plants and animals and therefore many food chains are interconnected. A *food web* is a complex network of many interconnected food chains. Cycles are common in a food web as the dead organisms are consumed by other organisms which provide nutrition to some other organisms. A food web is considered as a static networks as this relation is hardly altered over time. One fundamental research topic in food webs as biological networks is on investigation of the effect of removal of an organism from a food web, which does happen due to major climate changes.

10.4   Properties of Biological Networks

We can investigate the properties of biological networks from the view of graph theory as global or local properties. Global properties reflect characteristics of a biological network when considered as a whole. Local properties on the other hand, describe the node properties specific to the surroundings of a node. It is however possible to deduce global network properties from local network properties in many cases.

10.4.1   Distance

The distance $d(u, v)$ between two nodes u and v of a graph is defined as the shortest path between them. For undirected graphs, this distance is specified as the minimum number of edges (hops) and for weighted graphs, it is the sum of the weights of the minimum-weight path between the two vertices. There may be more than two such paths between the vertices u and v which is more common for unweighted and undirected graphs. The distance between the two vertices in two directions may not be equal in a directed graph and hence the relation is not symmetric in general. There will be $n(n - 1)$ number of such distances, counting in both directions. The distance from a node to all other nodes can be computed using Dijkstra's shortest path algorithm we saw in Sect. 3.6 in $O(m + n \log n)$ time. Computation for all nodes can be done by running this algorithm for all nodes. The *average distance* of a graph G, $d_G(av)$, is the average of all distances between each pair of nodes in G. This parameter provides us with the information on how easy it is to reach from one vertex to another.

10.4.2 Vertex Degrees

The degree of a vertex v, $\delta(v)$ is defined as the number of edges incident to v. For directed graphs, the *in-degree* of a vertex v is the number of edges that end at v and similarly, the *out-degree* of v is the number of edges that start at v. The average degree, $\delta(av)$, of a graph is the average of all degrees. The sum of degrees of vertices of an undirected graph is equal to twice the number of edges ($2m$) as we count them in both directions. The *density* of a graph, $\rho(G)$, is defined as the number of its edges to the maximum possible number of edges as $2m/(n(n-1))$. We can therefore state that $\rho(G) = \delta(av)/(n-1)$. In *dense graphs*, ρ does not change significantly as n gets large, otherwise the graph is called *sparse*.

The degree sequence of a graph specifies the list of its vertices in decreasing or increasing order, usually in decreasing order. Isomorphic graphs have the same degree sequence but graphs with the same degree sequence may not be isomorphic. We will need to generate similar graphs randomly to test our algorithms so that we can evaluate the significance of the results obtained in biological networks. The degree sequence is one parameter we try to conserve while generating these random test graphs as we will see in Chap. 13.

The *degree distribution* is another important global property of a graph. This parameter specifies the percentage of the vertices with the same degree to the total number of vertices as follows:

$$P(k) = \frac{m_k}{n}, \tag{10.1}$$

where the number of vertices with degree k is shown by m_k. The degree distribution is Binomial in random networks. Figure 10.6 displays these concepts. In *assortative networks*, the high-degree nodes tend to connect to high-degree nodes like themselves as in social networks. Conversely, in *disassortative networks*, the high-degree nodes have the tendency to form connections with low-degree nodes. In molecular interaction networks, the disassortative property is prevalent, for example, high-degree nodes called *hubs* in PPI networks are usually linked to nodes with few connections.

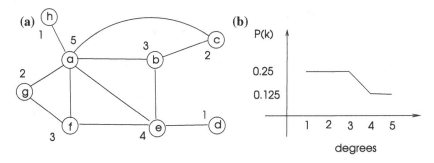

Fig. 10.6 a A sample graph with node degrees is shown. The average degree for this graph is 2.63, the degree sequence is 5, 4, 3, 3, 2, 2, 1, 1 and its density is $(2 \times 11)/(8 \times 7) = 0.39$. **b** The degree distribution of this graph

10.4.3 Clustering Coefficient

The *clustering coefficient* of a graph provides more information about the connectivity of a graph than its average degree. The clustering coefficient of a vertex v, CC (v), is the ratio of the connections between its neighbors to the maximum possible connections between these neighbors, as follows.

$$CC(v) = \frac{n_v}{k_v(k_v - 1)}, \tag{10.2}$$

where k_v is the number of neighbors vertex v and n_v is the number of existing edges between them. The clustering coefficient shows the density of connections around a vertex and it reflects the connectedness of the neighborhood of a vertex. Clustering coefficient is 0 for a vertex with no connected neighbors such as the central vertex of a star network and it is 1 for any vertex of a complete network.

The average clustering coefficient, or the clustering coefficient of a graph as more commonly used, is the average value of the node clustering coefficients and shows the tendency of the network to form clusters. It is the probability that there is an edge between the two neighbors of a randomly selected vertex. This parameter is defined as follows:

$$CC(G) = \frac{1}{n} \sum_{v \in V} CC(v) \tag{10.3}$$

Figure 10.7 shows the clustering coefficients of the vertices of a sample graph. The average clustering coefficient of the nodes with degree k is the clustering function $C(k)$. In [21], the clustering coefficients for the metabolic networks of 43 organisms were calculated and all were found to be about an order of magnitude higher than expected for a scale-free network of similar size.

10.4.4 Matching Index

The matching index shows the similarity of two nodes in a network. The matching index between two vertices u and v of a graph G is defined as the ratio of the

Fig. 10.7 A sample graph with vertex clustering coefficients shown next to nodes. The average degree for this graph is 2.625 and the average clustering coefficient for this graph is the arithmetic average of all the vertex values which is 0.46

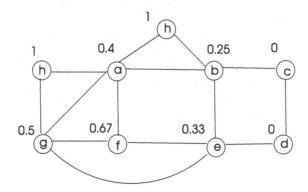

total number of common neighbors of these two vertices to total number of their distinct neighbors. The vertices b and f of Fig. 10.7 have a and e as their common neighbors and they have five vertices a, c, e, g, h as distinct neighbors, therefore, their matching index M_{bf} is $2/5 = 0.4$. We can extend this concept to neighborhoods of vertices to more than one hop and define *k-hop matching index* as the ratio of common neighbors of two vertices that are within k hops to all distinct neighbors that are within k hops. We can then consider two vertices as similar if their matching indices to all other vertices in the network are approximately equal. The matching index can be used to relate different parts of a network based on some property. It has been used to analyze spatial growth in BFNs and to predict the connections of private cortical networks [17].

10.5   Centrality

Centrality measures reflect the relative importance of a node or an edge in a biological network. Centrality is a function that assigns a real value to a node or an edge based on its rank in the network. The main centrality metrics we will consider are the *degree centrality, closeness centrality, betweenness centrality* and the *eigenvalue centrality*.

10.5.1   Degree Centrality

The degree centrality of a node is simply its degree in the network. The degree of a vertex provides us with information about its importance in the network. For example, a high-degree protein in a PPI network is involved in many interactions with its neighbors and removal of such a node from this network may have lethal effects than removing a protein with a lower degree [12,13]. The degree centrality of the nodes of a graph $G(V, E)$ can be computed using its adjacency matrix A as the sum of the rows of this matrix yields degree centrality values of the nodes. In matrix notation, the centrality vector $C = A \times [1]$, that is, it can be formed by multiplying the adjacency matrix with a vector of all 1's.

The degree centrality shows the local importance of a node. We can calculate the average degree of a network which will give us some idea about the structure of the nodes, however, the graph structure as a whole will be difficult to predict from this parameter. For example, a PPI network has few protein nodes with very high degrees and the rest of the nodes have only few connections to their neighbors. Evaluating the average degree centrality for such a network will not reveal this scale-free structure. In general, this metric alone is not sufficient to evaluate the functional importance of a node.

10.5.2 Closeness Centrality

Closeness centrality is an improvement over degree centrality as it considers the easiness to reach from one node to all other nodes in the network. If a node is located near the center of a network, it will be close to all nodes and hence we can deduce it is an important node. The formal definition of this parameter for a graph $G(V, E)$ representing a network is as follows.

$$C_C(v) = \frac{1}{\sum_{u \in V} d(u, v)}, \tag{10.4}$$

where $d(u, v)$ is the distance between the nodes u and v. We find the sum of distances between a node v to all other nodes in the network and take the reciprocal of this value to obtain the closeness centrality of the node v. Finding shortest distances between a source node and all other nodes in an unweighted graph can be done by the BFS algorithm of Sect. 3.3.5 in $O(n + m)$ time and for all nodes the time required would be $O(n^2 + nm)$. However, we need to provide a simple modification to the BFS algorithm to store the sum of distances for each node. Computation of closeness values for an unweighted and undirected sample graph is shown in Fig. 10.8.

For weighted graphs, we can use the Dijkstra's shortest path algorithm of Sect. 3.3.5 and need to run it for all nodes or use Floyd-Warshall all-pairs-shortest-paths (APSP) algorithm which requires $O(n^3)$ time [11].

10.5.3 Betweenness Centrality

A different approach is taken in betweenness centrality in which we analyze the importance of a vertex or an edge in terms of shortest paths running through them. If there are significantly more shortest paths passing through a vertex or an edge, it is assumed that vertex or the edge is more fundamental for the functioning of the network than another vertex or an edge that has less paths.

10.5.3.1 Vertex Betweenness
The vertex betweenness of a vertex v is defined as follows.

$$C_C(v) = \sum_{s \neq t \neq v} \frac{\sigma_{st}(v)}{\sigma_{st}}, \tag{10.5}$$

where σ_{st} is total number of shortest paths between vertices s and t, and $\sigma_{st}(v)$ is the total number of shortest paths between vertices s and t that pass through vertex v. For an unweighted graph, we could run a modified version of the BFS algorithm in which we keep records of multiple shortest paths between the source and destination node pairs (s, t). Afterwards, we can find the ratio of the shortest paths that pass through a vertex to find its betweenness centrality. The multiple shortest paths to each node is found in a sample graph in Fig. 10.9. We then find the ratio of shortest paths through each vertex and the sum of these values yield the betweenness value

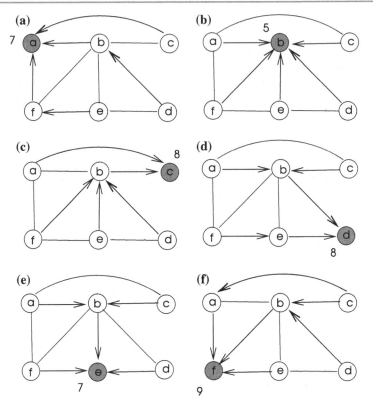

Fig. 10.8 Closeness centrality example. The sum of the shortest path values are shown next to the source nodes. The closeness centrality values for vertices $a, \ldots, f$ are 1/7,1/5,1/8,1/8,1/7, and 1/9 respectively. The vertex b has the largest closeness centrality which can also be seen visually as it is closer to all vertices than all others

for vertices. For example, vertex c in (d) has the shortest paths (a, d), (f, d), (b, d), and (e, d) running through it. There is one shortest path between a and d so the contribution for this path is 1. Similarly, (b, d) and (e, d) paths are unique raising the betweenness of c to 3. However, the (f, d) path has two alternatives both of which pass through c and each contribute 0.5 resulting in a betweenness value of 4 for node c. We need to sum all values for shortest paths to all source nodes which gives 8 for node c.

10.5.3.2   Edge Betweenness
The edge betweenness centrality is an evaluation of the percentages of shortest paths that run through and edge instead of a vertex. This parameter has been found to

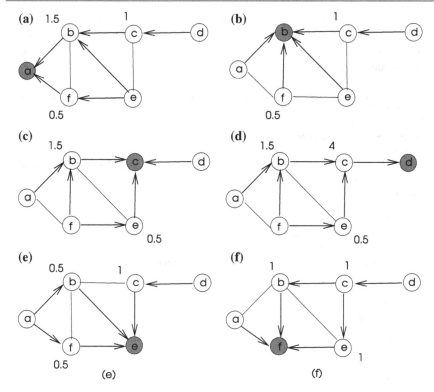

Fig. 10.9 Vertex betweenness centrality example. The betweenness values computed for vertices $a, \ldots, f$ are 0, 6, 8, 0, 2, 1.5 with vertex c having the highest value and vertex b with a close value to c which can also be seen visually as these vertices have several shortest paths running through them

provide favorable clustering when used in biological networks as we will see in Chap. 11. Given a graph $G(V, E)$, it is defined as follows for an edge $e \in E$.

$$C_C(v) = \sum_{s \neq t \neq v} \frac{\sigma_{st}(e)}{\sigma_{st}} \tag{10.6}$$

Newman and Girvan proposed an algorithm (NG algorithm) to find edge betweenness values in a graph in two phases. In the first phase, each vertex is labeled with a weight representing the number of shortest paths that pass through it. In the second phase of the algorithm, weights are assigned to edges using the vertex weights assigned in the first phase. The vertex labeling method is shown in Algorithm 10.1 where each source vertex s is assigned a distance 0 and a weight of unity initially. Each neighbor u of s are then labeled with distance 1 and weight 1 as we know u is one hop from source and it is at the shortest distance to s. Then a check is made for each neighbor v of these neighbors; if v is not assigned a distance yet, meaning it is not visited before, distance of v is made equal to 1 plus the distance of the neighbor on the route to s. Otherwise, if the distance of v is equal to the distance of u plus 1, it means v has an extra shortest path to s other than through u. In this case the weight of v is made equal to the sum of the weights of u and v as it has multiple shortest paths running through it. The total weight of a vertex is the sum of weights found for all source vertices.

Algorithm 10.1 *Label_Vertex*

1: **Input** : $G(V, E)$
2: **Output** : $\forall v \in V : w_v$ weight labels of vertices
3: $d_s \leftarrow 0; w_s \leftarrow 1$ ▷ source vertex s is initialized
4: **for all** $v \in N(s)$ **do**   ▷ neighbors of s are initialized
5:      $d_v \leftarrow d_s + 1, w_v \leftarrow 1$
6: **end for**
7: **repeat**
8:      **for all** $u \in N(v)$ **do**
9:          **if** u is not assigned a distance **then**
10:             $d_u \leftarrow d_v + 1; w_u \leftarrow w_v$
11:         **else if** u is assigned a distance **and** $d_u = d_v + 1$ **then**   ▷ check multiple shortest paths
12:             $w_u \leftarrow w_v + w_u$
13:         **end if**
14:     **end for**
15: **until** all vertices have assigned distances

Having found vertex weights, the next step is to determine the edge weights representing edge betweenness values. In the algorithm proposed by the authors, the leaf vertices from the vertex labeling algorithm are first identified, and for each vertex v that is a neighbor to such a leaf vertex u , the edge (u, v) is assigned a weight w_u/w_v. Then, moving upwards towards the source vertex s, each edge (u, v) with u being farther to s than v, is assigned a weight that is 1 plus the sum of the weights of edges below it multiplied by w_v/w_u. This procedure continues until s is reached. Algorithm 10.2 displays the pseudocode for this algorithm.

Algorithm 10.2 *Edge_Betweenness*

1: **Input** : $G(V, E)$
2: **Output** : $\forall v \in V : w_v$ weight labels of vertices
3: **for all** $v \in V$ that is a leaf vertex **do**
4:      **for all** $u \in N(v)$ **do** $w_{uv} \leftarrow w_u/w_v$
5:      **end for**
6: **end for**
7: **repeat**
8:      moving upwards to the source vertex s, with u being closer to s:
9:      $w_{uv} \leftarrow (1 + \text{sum of the scores below } v) \times w_u/w_v$
10: **until** vertex s is reached

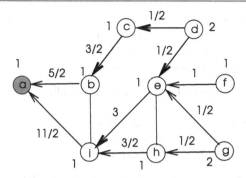

Fig. 10.10 Edge betweenness centrality example. For the source vertex a, the vertex weights are computed using Algorithm 10.1 and then the edge weights are computed using Algorithm 10.2. This process has to be repeated for all vertices and the edge betweenness value for an edge is the sum of all the values obtained

   This process is repeated for all source vertices and the edge betweenness value of an edge is the sum of all edge betweenness values found for each source vertex which can be performed in $O(mn)$ time. In order to find clusters of a network, the final step involves removing the edge with the highest edge betweenness value from the graph at each step, until the clusters that meet a required criteria are discovered giving a total time complexity of $O(m^2 n)$ or $O(n^3)$ on sparse graphs. Figure 10.10 displays the vertex weights and edge betweenness values obtained in a sample graph for the source vertex a using this algorithm.

10.5.4   Eigenvalue Centrality

Eigenvalue centrality is based on the idea that a node connected to an important node is also important. Therefore, this centrality value is high for nodes connected to important nodes. In a social network for example, let us assume a person X who may not be famous but is a friend of a famous person Y. The person X is important as reaching Y through X is possible by her. Google's PageRank algorithm uses this concept in determining the importance of Web pages. Therefore, we can assign an importance parameter to a node based on the importance of its neighbors. The relative importance (score) of a node i can then be defined in terms of the scores of its neighbors as follows:

$$x_i = \frac{1}{\lambda} \sum_{j \in N(i)} x_j = \frac{1}{\lambda} \sum_{j \in V} a_{ij} x_u \qquad (10.7)$$

where $N(i)$ is the set of neighbors of node i and a_{ij} is the ijth entry of the adjacency matrix A of the graph $G(V, E)$ which shows neighborhood of the nodes. This equation can be rewritten in matrix notation as:

$$Ax = \lambda x \qquad (10.8)$$

The adjacency matrix A has a number of eigenvalues and eigenvectors associated with these eigenvalues. By Perron-Probenius theorem, only the largest eigenvalue provides the centrality score of the vertices [18]. The procedure to find eigenvalue centralities of vertices then consists of the following steps:

1. Form the adjacency matrix A of the graph $G(V, E)$ that represents the biological network.
2. Find eigenvalues $\lambda_1, \ldots, \lambda_n$ of G using the equation $\det(A - \lambda I) = 0$ and select the largest eigenvalue λ_m from these eigenvalues.
3. Compute the eigenvector X_m associated with λ_m.

Each vertex v_i has $x_i \in X_m$ as it's eigenvalue centrality. Eigenvalue centrality has been used to discover genetic interactions and gene-disease associations [17].

10.6 Network Models

Studies to investigate the topological properties of biological networks showed that these networks have the following common characteristics:

- Biological networks have small average path length which provides easy access from any node to any other node.
- There are many low-degree nodes and few very high-degree nodes.
- The average clustering coefficient of these networks is high.

Our aim in this section is to investigate network models that best describe these properties.

10.6.1 Random Networks

The *random graph model* was proposed by Erdos and Renyi in 1950s. This model assumes that there are n vertices, $\{v_1, \ldots, v_n\}$, and m edges are to be formed in the network. An edge (v_i, v_j) is placed between vertices v_i and v_j with probability $p = 2m/(n(n-1)$ and this process is repeated for each pair of vertices. It has been shown that the degree distribution in these networks is Binomial which can be approximated by a Poission distribution for large networks as shown in Fig. 10.11. Also, the average clustering coefficient, or simply the clustering coefficient, is inversely proportional to the size of the network in random networks and the average path length is small, proportional to the logarithm of the network size. However, many real-world complex networks do not exhibit the homogenous degree distribution and small clustering coefficient observed in these networks.

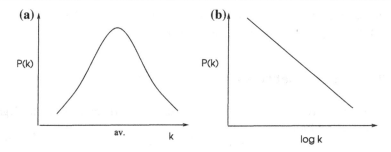

Fig. 10.11 Degree distribution of **a** An ER random network. **b** Scale-free network

10.6.2   Small World Networks

Networks with small average path length usually with co-existing large cluster-ing coefficients are called *small-world networks*. The average shortest path length between the nodes of such networks is small compared with the size of the network and usually is proportional to $\log n$. PPI networks, metabolic networks, and gene regulation networks have small average path lengths and large average clustering coefficients. Watts and Strogatz proposed a model (WS model) which accounts for small average path length and high clustering coefficient as observed in many real-life complex networks including biological networks [31]. A simple way to generate a small world network based on this model is to start from a regular lattice type of network where each node is placed on a one dimensional ring connected to its $n/2$ neighbors, providing high clustering coefficient. Then, rewiring of a node to one of the distant vertices with a probability p_n is provided. The network becomes closer to an ER random network with low clustering coefficient and short average path distance proportional to the logarithm of the network size when p_n increases, and it becomes an ER random network when $p_n \to 1$ as shown in Fig. 10.12. However, the WS model exhibits only these two properties of biological networks, namely, high

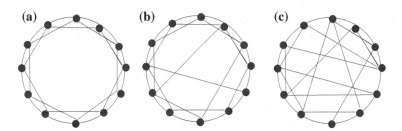

Fig. 10.12 WA model. **a** A regular network which has high average clustering coefficient but also large average path length. **b** A rewired scale-free network with high average clustering coefficient and small average path length. **c** A network approaching ER random network as rewiring is increased

clustering coefficient and short average path length but fails to exhibit another very important property of these networks as described next.

10.6.3   Scale-Free Networks

Many biological networks have the power-law degree distribution where the degree distribution has the form:

$$P(k) \sim k^{-\gamma}, \ \gamma > 1 \tag{10.9}$$

where γ is termed the *power-law exponent*. The plot of degree distribution displays a heavy-tailed curve in these networks which means there are few nodes with high degrees and most of the nodes have low degrees. The networks with power-law degree distribution are called *scale-free networks*. The majority of the nodes in these networks have only few neighbors and a small fraction of them have hundreds and sometimes thousands of neighbors. Many biological networks are scale-free, for example, degree distributions of the central metabolic networks of 43 organisms were shown to have heavy tails in agreement with Eq. 10.9 with $2 < \gamma < 3$ [9]. The PPI networks of *E. coli, D. melanogaster, C. elegans* and *H. pylori* were also shown to be scale-free [6].

The high-degree nodes in scale-free networks are called *hubs*. In a PPI network, the removal of a hub protein will presumably have an important effect for the survival of the network than the removal of a protein with low degree. In simplest terms, the PPI network will be disconnected and the transfer of signals between many of the nodes will cease. A hub in a PPI network may be formed as a result of a disease and detecting of such disease-related hubs is important in a PPI network as drug therapies can be targeted to them to stop their functioning.

Barabasi and Albert was first to propose a method to form scale-free networks [2]. The so-called Barabasi-Albert model (BA model) assumes that these networks are dynamic in nature and given an initial network G_0, the dynamics of the network is governed by two rules:

1. *Growth*: A new node u is added to the network at each discrete time t.
2. *Preferential attachment*: The new node u is attached to any node v in network with a probability proportional to the degree of v. This rule can be stated as follows:

$$P(v, t) = \frac{\delta_v}{\sum_{w \in V(G_{t-1})} \delta_w} \tag{10.10}$$

This method provided forming of a scale-free network with power law $P(k) \sim k^{-\gamma}$ with $\gamma = 3$ [2]. The BA model mainly shows the distribution of dynamically evolving complex networks including biological networks. However, some issues have been raised about the validity of this model for biological networks [15]. The techniques used to identify interactions in biological networks are error-prone and only samples of these networks can be evaluated. These samples may not reflect the overall network structure and it is possible to have a scale-free sample from an ER random network or vice versa [27].

10.6.4   Hierarchical Networks

The clustering coefficient of a network shows the clustering structure in a biological network. Moreover, we defined the clustering function $C(k)$ in these networks as the average clustering coefficient of the nodes with degree k. Investigation of this function in biological networks revealed surprising results. This function was found to be related to the degree as $C(k) \sim k^{-1}$ which means the clustering coefficient of nodes decreases as their degree gets higher [15,21]. In other words, the neighbors of low-degree nodes in such networks are highly clustered whereas the nodes around the high-degree nodes are sparsely connected. These networks are termed *hierarchical networks*. We are then confronted with the task of providing a model for most of the biological networks which exhibits scale-free, small-world, and hierarchical network properties. It was suggested in [21] that these networks have modules consisting of densely clustered low-degree nodes and different modules are connected by high-degree hubs. Various other studies reported similar findings for $C(k)$ in the PPI networks of *S. cerevisiae*, *H. pylori*, *E. coli*, and *C. elegans*. This model captures all of the afore mentioned properties of biological networks such as being scale-free and small-world and based on these studies, we can assume that the biological networks have hierarchical structures with high-degree nodes mainly connecting low-degree dense clusters [15]. Figure 10.13 displays such a hierarchical network which is regular and the high-degree nodes in the middle of triangles do not have high clustering coefficients as their neighbors are not highly connected, however, the outer other nodes have highly connected neighbors and have clustering coefficients of varying degrees. We can see all of the small-world, scale-free, and diminishing clustering coefficients with increased degrees properties in this sample network which comes closer to modeling a biological network than other described models.

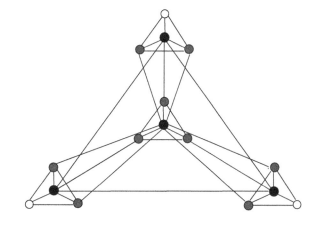

Fig. 10.13 A regular hierarchical network. *Black* nodes have more unconnected neighbors than others and they have 4/15 as clustering coefficient. *Gray* nodes have a clustering coefficient of 2/3 and *white* nodes have 1 as all their neighbors are connected (adapted from [21])

10.7 Module Detection

A cluster in a graph is loosely defined as a group of nodes that have significantly higher number of edges between them than to the rest of the graph. A cluster in a biological network usually has some biological significance, for example it may represent a functional module where density means high interaction therefore significant interactions between its nodes. *Module detection* or *clustering* in a network is the process of finding such subgraphs of dense connections. Clusters in a sample network are depicted in Fig. 10.14.

Detection of clusters in biological networks provides insights to the organization of such systems and may give clues on their functional structures. Clustering in graphs is a thoroughly studied topic and there are numerous algorithms for this purpose. For clustering in biological networks, we may exploit the scale-free and small-world properties of these networks to design effective algorithms. The main challenge in these networks is their huge size which demands high computational power. Clustering algorithms for biological networks can be broadly classified as hierarchical, density-based, flow-based, and spectral algorithms as we will see in Chap. 11. Hierarchical algorithms construct nested clusters at each step and density-based algorithms mainly employ graph-theoretic approaches to cluster. Flow-based algorithms on the other hand, attempt to detect clusters by finding where the flow accumulates in the network. Spectral clustering algorithms take a different approach and make use of the Laplacian matrix of the graph to find clusters. Clustering is an NP-hard problem with no known solution in P and heuristics are commonly employed.

Fig. 10.14 Three clusters in a sample network. The density inside the clusters is higher than the density between them

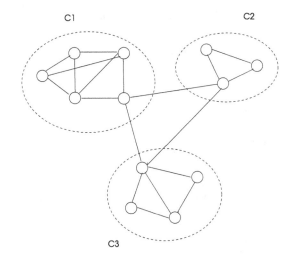

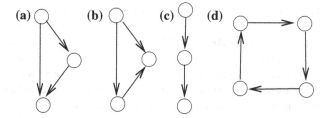

Fig. 10.15 Network motifs **a** Feed-forward loop found in PPI networks and neural networks.
b Three-node feedback found in GRNs. **c** Three chain; found in food webs. **d** Four-node feedback
found in GRNs

10.8 Network Motifs

Network motifs are repeating patterns of subnetworks in a complex biological network. These small networks are usually associated with a special function and their
search is one of the main problems in biological networks. We can analyze and
determine various functions in a biological network by discovering the underlying
motifs, we can compare networks of various species by discovering their common
motifs and can infer evolutionary relationships between them. Finding subgraphs of
a given size in a network is NP-hard, and determining whether a repeating subgraph
pattern is a motif or not is another problem to be addressed. The general approach
is to generate a number of random graphs which are similar to the target graph to
search and check the number of motifs in these random graphs. If the number of a
subgraph in the target graph is significantly higher than its occurrences in randomly
graphs statistically, it can be determined as a motif. Some common motifs found in
biological networks are shown in Fig. 10.15.

   We can proceed in two ways to search for motifs in a biological network. We can
search for all occurrences of connected subgraphs of a given k number of vertices
and select the ones with high frequency, or we can have a predicted input motif
pattern G_1 and search for all subgraphs of the target graph G_2 that are isomorphic to
G_1. In the first case, we need to group isomorphic graphs and we need to search for
isomorphic subgraphs in the second. In both cases, graph isomorphism problem is
encountered which is NP-hard for subgraph isomorphism and in NP but not known
to be NP-complete for graph isomorphism. For this reason, heuristic algorithms are
frequently used to detect motifs in biological networks.

10.9 Network Alignment

Investigation of the evolutionary relationships between species is one of the main
problems in system biology. Network alignment problem attempts to find how similar two or more networks of species are. We can analyze the molecular biological

networks of two species and assess their similarity based on their topological structure resemblance. However, exact topological matches is not possible because of the dynamic nature of these networks which results in various modifications such as addition or removal of edges or nodes. Furthermore, the measurements to detect interactions in them are error-prone resulting in detection of false positive or negative interactions. For this reason, we need to assess their similarity level rather than testing whether they match exactly. This problem is related to subgraph isomorphism problem which is NP-hard and hence, heuristic algorithms are usually preferred. Global network alignment is the comparison of two or more networks as a whole to determine their overall similarity as shown in Fig. 10.16. Local alignment algorithms on the other hand, aim to discover conserved structures across species by comparing their subnetworks. Two species may have a very similar subnetwork discovered by a local alignment algorithm whereas they may not be similar as a whole when searched by a global alignment algorithm. Network alignment is basically the process of finding the level of similarity between two or more networks.

10.10   Chapter Notes

In this chapter, we first described the main networks in the cell which are the metabolic networks, gene regulation networks, the PPI networks and then various other networks outside the cell. We then defined the parameters needed to analyze the global properties of biological networks. The average degree is one such parameter and although this parameter is meaningful in a random network, it does not provide us with much information in other types of networks. Degree distribution on the other hand, gives us a general idea about how the nodes are connected. The average path length or distance shows the easiness of reaching from one node to another. The clustering coefficient of the network is another important parameter that shows the probability of the two neighbors of a randomly selected vertex to be connected. Centrality measures evaluate the importance of a node or an edge in a network and betweenness centrality values provide more realistic results than others in general.

We then reviewed basic network models which are the ER random network, small-world, and scale-free network models. Small-world networks are characterized by small average distance between their nodes and scale-free property of a network means the number of nodes decreases sharply as their degrees increase. We find most of the biological networks exhibit small-world and scale-free properties, however, the clustering coefficients of the nodes in these networks decreases when their degrees increase as experiments show. This fact allowed researchers to look for a new model to explain the aforementioned properties of these networks.

The hierarchical model proposed captures all of the observed properties of biological networks. This model assumes that the low-degree nodes in biological networks constitute dense clusters with high clustering coefficients. A high-degree node or a hub on the other hand has low clustering coefficient and its main function is to act as gateway between the small dense clusters. It is possible to explain all of the

Fig. 10.16 Global network alignment of two networks. The edge (d, b) is deleted and a new edges (u, z) and (y, v) with a new node v are formed in the second network; 6 edges and 6 nodes are preserved between the networks

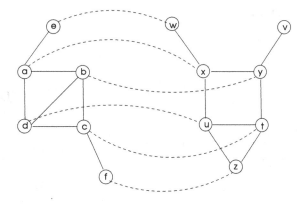

properties observed with this model, for example, a hub in a PPI network serves as a main switch to transfer signals between the small and dense protein clusters. It also provides short paths between these clusters, contributing to the small-world property.

The experiments are performed on samples rather than the whole network for practical reasons. Two major problems associated with evaluating biological network data is that the measurements are frequently error-prone with false positive and false positive interaction detections. Secondly, the sample may not reflect the whole network. As technology advances, we will have less errors and larger sample sizes and this problem will have less effect on the accuracy of the measurements.

In the final section, we reviewed the module detection, network motif search and network alignment problems in biological networks. Module detection involves discovering the cluster structures which usually represent some important activity in that part of the network. Network motifs are repeating patterns of small subgraphs in a biological network which may display basic functions of the network. We can also compare the motifs found in networks of species to investigate their evolutionary relationships. In the network alignment problem, we search for similar subgraphs in two or more species networks to compare them similar to what we did in comparing DNA or protein sequences. We will investigate these problems and sequential and distributed algorithms to solve them in detail in the rest of this part of the book.

Exercises

1. Work out the degree sequence and sketch the degree distribution of the sample graph of Fig. 10.17.
2. Find the clustering coefficents for all of the nodes in the sample graph of Fig. 10.18. Plot the clustering coefficient distribution for this graph. What does it show?
3. Compute the degree centrality values for the nodes of the graph in Fig. 10.19 by first forming the adjacency matrix of this graph and then multiplying this matrix by a vector of all 1s.

Fig. 10.17 Sample graph for
Exercise 1

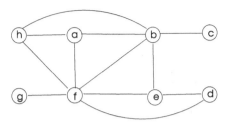

Fig. 10.18 Sample graph for
Exercise 2

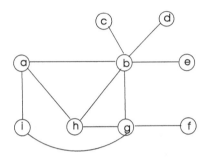

4. *Eccentricity centrality* of a vertex in an undirected, connected graph $G(V, E)$ is
   defined as the reciprocal of its maximum distance to any vertex in the graph as
   follows:

$$C_E(v) = \frac{1}{\max(d(u, v), v \in V)} \qquad (10.11)$$

   For the example graph of Fig. 10.19, work out the eccentricity centrality values
   for all vertices.
5. Given an undirected connected graph, modify the BFS algorithm to compute
   closeness centralities for all nodes in this graph.
6. *Katz Centrality* is a generalization of degree centrality where neighbors within
   k-hops from a node are also taken into account. However, the contributions from
   remote neighbors have less effect and are decreased by the attenuation factor
   $\alpha \in (0, 1)$ such that the effect of a neighbor that is k hops away is α^k. Work out
   the Katz centrality values for the nodes of the graph of Fig. 10.19 with $\alpha = 0.5$.

Fig. 10.19 Sample graph for
Exercises 3 and 4

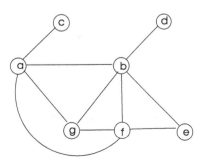

Fig. 10.20 Sample graph for
Exercises 7 and 8

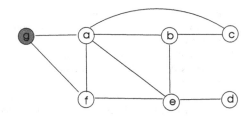

7. Work out the vertex betweenness centrality values for the vertices of the graph of Fig. 10.20.
8. Find the edge betweenness centrality values for the vertices of the graph of Fig. 10.20 for source vertex g.
9. Describe briefly the basic topological properties of biological networks and why hierarchical network model is a better model than the small-world and scale-free models to explain the behaviour of these networks.

References

1. Albert R, Othmer HG (2003) The topology of the regulatory interactions predicts the expression pattern of the drosophila segment polarity genes. J Theor Biol 223:1–18
2. Albert R, Barabasi A (2002) The statistical mechanics of complex networks. Rev Mod Phys 74:47–97
3. Bassett DS, Bullmore E, Verchinski BA et al (2008) Hierarchical organization of human cortical networks in health and schizophrenia. J Neurosci 28:9239–9248
4. Davidson EH, Rast JP, Oliveri P, Ransick A et al (2020) A genomic regulatory network for development. Science, 295:1669–1678
5. Eguiluz VM, Chialvo DR, Cecchi GA, Baliki M, Apkarian AV (2005) Scale-free brain functional networks. Phys Rev Lett 94:018102
6. Floyd RW (1962) Algorithm 97: shortest path. Comm ACM 5(6):345
7. Goh K, Kahng B, Kim D (2005) Graph theoretic analysis of protein interaction networks of eukaryotes. Physica A 357:501–512
8. He Y, Chen Z, Evans A (2010) Graph theoretical modeling of brain connectivity. Curr Opin Neurol 23(4):341–350
9. He Y, Chen Z, Evans A (2008) Structural insights into aberrant topological patterns of large-scale cortical networks in Alzheimers disease. J Neurosci 28:4756–4766
10. Identifying gene regulatory networks from gene expression data
11. Jeong H, Tombor B, Albert R, Oltvai ZN, Barabasi AL (2000) The large-scale organization of metabolic networks. Nature 407:651–654
12. Junker B (2008) Analysis of biological networks, Chap. 9. Wiley ISBN: 978-0-470-04144-4
13. Koschtzki D, Lehmann KA, Tenfelde-Podehl D, Zlotowski O (2005) Advanced centrality concepts. Springer-Verlag LNCS Tutorial 3418:83-111, In: Brandes U, Erlebach T (eds) Network analysis: methodological foundations
14. Kaiser M, Martin R, Andras P, Young MP (2007) Simulation of robustness against lesions of cortical networks. Eur J Neurosci 25:3185–3192
15. Mason O, Verwoerd M (2007) Graph theory and networks in biology. IET Syst Biol 1(2):89–119

16. Pablo Carbonell P, Anne-Galle Planson A-G, Davide Fichera D, Jean-Loup Faulon J-P (2011) A retrosynthetic biology approach to metabolic pathway design for therapeutic production. BMC Syst Biol 5:122

17. Pavlopoulos GA, Secrier M, Moschopoulos CN, Soldatos TG, Kossida S, Aerts J, Schneider R, Bagos Pantelis GPG, (2011) Using graph theory to analyze biological networks. Biodata Mining 4:10. doi:10.1186/1756-0381-4-10

18. Perron O (1907) Zur Theorie der Matrices. Mathematische Annalen 64(2):248–263

19. Phizicky EM, Fields S (1995) Proteinprotein interactionsmethods for detection and analysis. Microbiol Rev 59:94–123

20. Ptashne M (1992) A genetic switch: phage lambda and higher organisms, 2nd edn. Cell Press and Blackwell Scientific

21. Ravasz E, Somera AL, Mongru DA, Oltvai ZN, Barabsi AL (2002) Hierarchical organization of modularity in metabolic networks. Science 297:1551–1555

22. Rubinov M, Sporns O (2010) Complex network measures of brain connectivity: uses and interpretations. NeuroImage 52(3):1059–1069

23. Salvador R, Suckling J, Coleman MR, Pickard JD, Menon D, Bullmore E (2005) Neurophysiological architecture of functional magnetic resonance images of human brain. Cereb Cortex 15:1332–1342

24. Salwinski L, Eisenberg D (2003) Computational methods of analysis of proteinprotein interactions. Curr Opin Struct Biol 13:377–382

25. Schuster S, Fell DA, Dandekar T (2000) A general definition of metabolic pathways useful for systematic organization and analysis of complex metabolic networks. Nat Biotechnol 18:326–332

26. Seidenbecher T, Laxmi TTR, Stork O, Pape HC (2003) Amygdalar and hippocampal theta rhythm synchronization during memory retrieval. Science 301:846–850

27. Stumpf MPH, Wiuf C, May RM (2005) Subnets of scale-free networks are not scale-free: Sampling properties of networks. Proc National Acad Sci 102(12):4221–4224

28. Vidal M, Cusick ME, Barabasi AL (2011) Interactome networks and human disease. Cell 144(6):986–998

29. Vitale A (2002) Physical methos. Plant Mol Biol 50:825–836

30. Vogelstein B, Lane D, Levine A (2000) Surfing the p53 network. Nature 408:307–310

31. Watts DJ, Strogatz SH (1998) Collective dynamics of small-world networks. Nature 393:440–442

Cluster Discovery in Biological Networks

11

11.1 Introduction

Clustering is the process of grouping similar objects based on some similarity measure. The aim of any clustering method is that the objects belonging to a cluster should be more similar to each other than to the rest of the objects under consideration. Clustering is one of the most studied topics in computer science as it has numerous applications in bioinformatics, data mining, image processing, and complex networks such as social networks, biological networks, and the Web.

We will make a distinction between clustering data points commonly distributed in 2D plane which we will call *data clustering* and clustering objects which are represented as vertices of a graph in which case we will use the term *graph clustering*. Furthermore, graph clustering can be investigated as *inter-graph clustering* where a subset from a given set of graphs are clustered based on their similarity or *intra-graph clustering* in which our object is to find clusters in a given graph. We will assume the latter when we investigate clustering in biological networks in this chapter.

Intra-graph clustering or graph clustering in short, considers the neighborhood relationship of the vertices while searching for clusters. In unweighted graphs, we try to cluster nodes that have strong neighborhood connections to each other and this problem can be viewed as finding cliques of a graph in the extreme case. Our aim in edge-weighted graphs, however, is to place neighbors that are close to each other in the same cluster using some metric.

Biological networks are naturally represented as graphs as we have seen in Chap. 10, and any graph clustering algorithm can be used to detect clusters in biological networks such as the gene regulation networks, metabolic networks, and PPI networks. There are, however, important differences between a graph representing a general random network and the graph of a biological network. First of all, the size of a biological network is huge, reaching tens of thousands of vertices and hundreds of thousands of edges, necessitating the use of highly efficient clustering algorithms as well as usage of distributed algorithms for this computation-intensive task. Secondly, biological networks are scale-free with few very high-degree nodes and many low-degree nodes. Third, they exhibit small-world property having small diameters relative to their sizes. These last two observations may be exploited to design efficient clustering algorithms with low time complexities but this alone does not provide the needed performance in many cases and using distributed algorithms is becoming increasingly more attractive to solve this problem.

Our aim in this chapter is to first provide a formal background and a classification of clustering algorithms in biological networks. We then describe and review efficient sample algorithms, most of which are experimented in biological networks and have distributed versions. In cases where there are no distributed algorithms known to date, we propose distributed algorithm templates and point potential areas of further investigation which may lead to efficient algorithms.

11.2   Analysis

We can have overlapping clusters where a node may belong to two or more clusters or a node of the graph becomes a member of exactly one cluster at the end of a clustering algorithm, which is called *graph partitioning*. Also, we may specify the number of clusters k beforehand and the algorithm stops when there are exactly k clusters, or it terminates when a certain criteria is met. Another distinction is whether a node belongs fully to a cluster or with some probability. In *fuzzy clustering*, membership of a node to a cluster is specified using a value between 0 and 1 showing this probability [47].

Formally, a clustering algorithm divides a graph $G(V, E)$ into a number of possibly overlapping clusters $\mathcal{C} = C_1, \ldots, C_k$ where a vertex $v \in C_i$ is closer to all other vertices in C_i than to vertices in other clusters. This similarity can be expressed in a number of ways and a common parameter for graph clustering is based on the average density of the graph and the densities of the clusters. We will now evaluate the quality of a graph clustering method based on these parameters.

11.2.1   Quality Metrics

A basic requirement from any graph clustering algorithm is that the vertices in a cluster output from the algorithm should be connected which means there will be at least one path between every vertex pair (u, v) in a cluster C_i. Furthermore, the path between u and v should be internal to the cluster C_i meaning u is close to v [40], assuming the diameter of the cluster is much smaller than the diameter of the graph. However, a more fundamental criteria is based on evaluating the densities of the graph in a cluster and outside the cluster and comparing them. We will describe two methods to evaluate these densities next.

11.2.1.1 Cluster Density

The quality of a clustering method is closely related to the density of vertices in a cluster which can be evaluated in terms of the density of the unweighted, undirected graph as a whole. Let us first define the density of a graph. The density $\rho(G)$ of an unweighted, undirected simple graph G is the ratio of the size of its existing edges to the size of maximum possible edges in G as follows:

$$\rho(G) = \frac{2m}{n(n-1)} \tag{11.1}$$

Let us now examine the edges incident to a vertex v in a cluster C_i. Some of these edges will be connecting v to other vertices in C_i which are called *internal* edges and the rest of the edges on v that connect it to other clusters are called *external* edges. Clearly, degree of vertex v is the sum of its internal and external edges. The size of internal edges ($\delta_{int}(v)$) and external edges ($\delta_{ext}(v)$) of a vertex v gives a good indication of the appropriateness of v being in C_i. Considering the ratio $\delta_{int}(v)/\delta_{ext}(v)$, if this is small, we can conclude v may have been wrongly placed in C_i, and on the contrary, a large ratio reveals v is properly situated in C_i. We can generalize this concept to the clustering level and define the *intra-cluster density* of a cluster C_i as the ratio of all internal edges in C_i to all possible edges in C_i as follows [40].

$$\delta_{int}(C_i) = \frac{2 \sum_{v \in C_i} \delta_{int}(v)}{|C_i||C_i - 1|} \tag{11.2}$$

We can then define the intra-cluster density of the whole graph as the average of all intra-cluster densities as follows:

$$\delta_{int}(G) = \frac{1}{k} \sum_{i=1}^{k} \delta_{int}(C_i) \tag{11.3}$$

where k is the number of clusters obtained. For example, the intra-cluster densities for clusters C_1, C_2, and C_3 in Fig. 11.1a are 0.66, 0.33, and 0.5 respectively and the average intra-cluster density is 0.50. We divide the same graph into different clusters in (b) with intra-cluster densities of 0.7, 0.5, and 0.6 for these clusters and the average density becomes 0.6. We can say that the clusters obtained in (b) are better as we have a higher average intra-cluster density, as can be observed visually. The *cut size* of a cluster is the size of the edges between C_i to all other clusters it is connected. The *inter-cluster density* $\delta_{ext}(G)$ is defined as the ratio of the size of inter-cluster edges to the maximum possible size of edges between all clusters as shown below [40]. In other words, we subtract the size of maximum total possible intra-cluster edges from the size of the maximum possible edges between all nodes in the graph to find the size of the maximum possible inter-cluster edges, and the inter-cluster density should be as low as possible when compared with this parameter.

$$\delta_{ext}(G) = \frac{2 \times \text{sum of inter-cluster edges}}{n(n-1) - \sum_{i=1}^{k}(|C_i||C_i - 1|)} \tag{11.4}$$

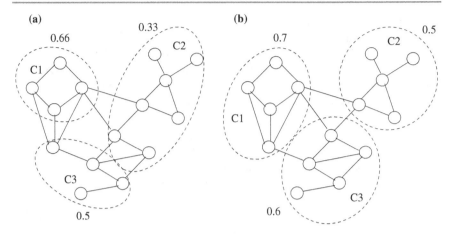

Fig. 11.1 Intra-cluster and inter-cluster densities example

The inter-cluster densities in Fig. 11.1a, b are 0.08 and 0.03 respectively, which again shows the clustering in (b) is better since we require this parameter to be as small as possible. The graph density in this example is $(2 \times 22)/(15 \times 14) = 0.21$ and based on the foregoing, we can conclude that a good clustering should provide a significantly higher intra-cluster density than the graph density, and the inter-cluster density should be significantly lower than the graph density.

When we are dealing with weighted graphs, we need to consider the total weights of edges in the cut set, as the internal and external edges, rather than the number of such edges. The density of an edge-weighted graph can be defined as the ratio of total edge weight to the maximum possible number of edges as follows:

$$\rho(G(V, E, w)) = \frac{2 \sum_{(u,v) \in E} w_{(u,v)}}{n(n-1)} \tag{11.5}$$

The intra-cluster density of a cluster C_i in such an edge-weighted graph can then be computed similarly to the unweighted graph but we sum the weights of edges inside the clusters and divide it by the maximum possible number of edges in C_i this time. The graph intra-cluster density is the average of intra-cluster densities of clusters as before and the general requirement is that this parameter should be significantly higher than the edge-weighted graph density. For inter-cluster density of an edge-weighted graph, we can compute the sum of weights of all edges between each pair of clusters and divide it by the maximum possible number of edges between clusters as in Eq. 11.4 by just replacing the number of edges with their total weight. We can then compare this value with the graph density as before and judge the quality of clustering.

11.2.1.2 Modularity

The modularity parameter proposed by Newman [35] is a more direct evaluation of the goodness of clustering than the above described procedures. Given an undirected and unweighted graph $G(V, E)$ which has a cluster set $C = \{C_1, .., C_k\}$, modularity Q is defined as follows [36]:

$$Q = \sum_{i=1}^{k} (e_{ii} - a_i^2) \qquad (11.6)$$

where e_{ii} is the percentages of edges in C_i, and a_i is the percentage of edges with at least one edge in C_i. We actually sum the differences of probabilities of an edge being in C_i and a random edge would exist in C_i. The maximum value of Q is 1 and a high value approaching 1 shows good clustering. For calculating Q conveniently, we can form a modularity matrix M which has an entry m_{ij} showing the percentage of edges between clusters i and j. The diagonal elements in this matrix represent the e_{ii} parameter in Eq. 11.6 and the sum of each row except the diagonal is equal to a_{ij} of the same equation. We will give a concrete example to clarify these concepts. Evaluating the modularity matrices M_1 and M_2 for Fig. 11.1a, b respectively yields:

$$M_1 = \begin{bmatrix} 0.18 & 0.09 & 0.14 \\ 0.09 & 0.32 & 0.14 \\ 0.14 & 0.14 & 0.14 \end{bmatrix} \quad M_2 = \begin{bmatrix} 0.32 & 0.05 & 0.09 \\ 0.05 & 0.23 & 0.05 \\ 0.09 & 0.05 & 0.27 \end{bmatrix}$$

For the first clustering, we can calculate the contributions to Q using M_1 from clusters C_1, C_2 and C_3 as 0.127, 0.267, and 0.060 giving a total Q value of 0.247. We can see straight away clustering structure in C_3 is worse than others as it has the lowest score. For M_2 matrix of clusters in (b), the contributions are 0.30, 0.22, and 0.25 providing a Q value of 0.77 which is significantly higher than the value obtained using M_1 and also closer to unity. Hence, we can conclude that the clustering in (b) is much more favorable than the clustering in (a). We will see in Sect. 11.4 that there is a clustering algorithm based on the modularity concept described.

11.2.2 Classification of Clustering Algorithms

There are many different ways to classify the clustering algorithms based on the method used. In our approach, we will focus on the methods used for clustering in biological networks and provide a taxonomy of clustering algorithms used for this purpose only as illustrated in Fig. 11.2. We have mostly included fundamental algorithms in each category that have distributed versions or can be distributed.

We classify the clustering algorithms in four basic categories as hierarchical, density-based, flow-based, and spectral algorithms. The hierarchical algorithms construct nested clusters at each step and they either start from each vertex being a single cluster and combine them into larger clusters at each step, or they may start from one

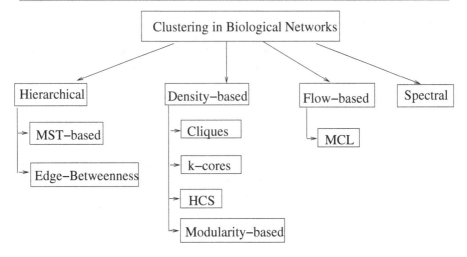

Fig. 11.2 A taxonomy of clustering algorithms in biological networks

cluster including all of the nodes and divide them into smaller clusters in each iteration [28]. The MST-based and edge-betweenness-based algorithms are examples of
the latter hierarchical methods. Density-based algorithms search for the dense parts
of the graph as possible clusters. Flow-based algorithms on the other hand are built
on the idea that the flow between nodes in a cluster should be higher than the rest of
the graph and the spectral clustering considers the spectral properties of the graph
while clustering.

We search for clusters in biological networks to understand their behavior, rather
than partitioning them. However, we will frequently need to partition a graph representing such a network for load balancing in a distributed memory computing system.
Our aim is to send a partition of a graph to a process in such a system so that parallel
processing can be achieved. The BFS-based partitioning algorithm of Sect. 7.5 can
be used for this purpose. In the next sections, we will investigate sample algorithms
of these methods in sequential and distributed versions in detail.

11.3 Hierarchical Clustering

We have described the basic hierarchical clustering methods in Sect. 7.3. We will now
investigate two graph-based hierarchical clustering approaches to discover dense
regions of biological networks.

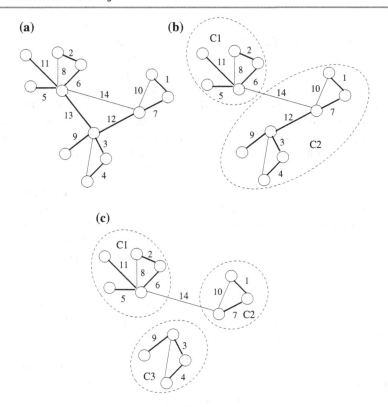

Fig. 11.3 MST-based clustering in a sample graph. MST is shown by *bold lines* and the edges are labeled with their weights. The highest weight edge in the MST has weight 13 and removed in the first step resulting in two clusters C_1 and C_2. The next iteration removes the edge with weight 12 and three clusters C_1, C_2, and C_3 are obtained

11.3.1   MST-Based Clustering

The general approach of MST-based clustering algorithms is to first construct an MST of the graph after which the heaviest weight edges from the MST are iteratively removed until the required number of clusters is obtained. The idea of this heuristic is that two nodes that are far apart should not be in the same cluster. Removing one edge in the first step will disconnect MST as MST is acyclic like any tree and will result in two clusters, hence we need to remove the heaviest $k - 1$ edges to get k clusters. Note that removing the heaviest edge may not result in a disconnected graph. Figure 11.3 displays the MST of a graph removing of two edges from which results in three clusters.

Instead of removing one edge at each iteration of the algorithm, we may start with a threshold edge weight value τ and remove all edges that have higher weights than τ in the first step which may result in a number of clusters. We can then check the quality Q of the clusters we obtain and continue if Q is lower than expected. This parameter can be the ratio of intra-cluster density to the inter-cluster density or it can simply be computed as the ratio of the total weight of intra-cluster edges in the current clusters to the total weight of inter-cluster edges. We may modify the value of τ as we proceed to refine the output clusters as a large τ value may result in many small clusters and a small value will generally give few large clusters. MST of a graph can be constructed using one of the greedy approaches as follows:

- *Prim's Algorithm*: This algorithm greedily includes an edge of minimum weight in MST among edges that are incident on the current MST vertices but not part of the current MST as we have seen in Sect. 3.6. Prim's algorithm requires $O(n^2)$ as it checks each vertex against all possible vertex connections but this time may be reduced to $O(m\log n)$ by using the binary heap data structure and to $O(m + n\log n)$ by Fibonacci heaps [13].
- *Kruskal's Algorithm*: Edges are sorted with respect to their weights and starting from the lightest weight edge, an edge is included in MST if it does not create a cycle with the existing MST edges. The time for this algorithms is dominated by the sorting of edges which is $O(m \log m)$ and if efficient algorithms such as union-find are used, it requires $O(m \log n)$ time.
- *Boruvka's Algorithm*: This algorithm is the first MST algorithm designed to construct an efficient electricity network for Moravia, dating back to 1926 [5]. It finds the lightest edges for each vertex and contracts these edges to obtain a simpler graph of components and then the process is repeated with the components of the new graph until an MST is obtained. It requires $O(m \log n)$ time to build the MST.

11.3.1.1 A Sequential Algorithm

We will now describe a simple sequential MST algorithm that does not require the number of clusters beforehand. It starts with an initial distance value τ and a cluster quality value Q and at each iteration of the algorithm, edges that have weights greater than the current value τ_i are deleted from the graph to obtain clusters. The new cluster quality Q_i is then computed and if this value is lower than the required quality of Q_{req}, another iteration is executed. Algorithm 11.1 shows the pseudocode for this algorithm [20]. *BFS_form* is a function that builds a BFS tree starting from the specified vertex and includes all vertices in this BFS tree in the specified cluster.

Algorithm 11.1 *Seq_MST_Clust*

1: **Input** : $G(V, E, w)$	▷ edge-weighted graph
2: $\quad \tau_i \leftarrow \tau_1, Q_{req} \leftarrow Q_1$	▷ initialize
3: **Output** : $C_1, ..., C_k$	▷ k clusters
4: **construct** MST of G	
5: $D[n, n] \leftarrow$ distances between points	
6: **int** $i \leftarrow 1$	
7: **while** $Q_i < Q_{req}$ **do**	
8: $\quad j \leftarrow 1$	
9: $\quad$ **for all** $(a, b) \in D$ such that $d(a, b) < \tau_i$ **do**	
10: $\quad\quad D[a, b] \leftarrow \infty$	
11: $\quad\quad BFS\_form(a, C_j)$	
12: $\quad\quad BFS\_form(b, C_{j+1})$	
13: $\quad\quad j \leftarrow j + 1$	
14: $\quad$ **end for**	
15: $\quad$ **compute** Q_i	
16: $\quad$ **adjust** τ_i if required	
17: $\quad i \leftarrow i + 1$	
18: **end while**	

11.3.1.2 Distributed Algorithms

Let us review the MST-based clustering problem; we need to first construct an MST of the graph, then we either remove the heaviest weight edge at each step or may remove a number of edges that have weights greater than a threshold value. The building of the MST dominates the time taken for the clustering algorithm and we have already reviewed ways of parallelizing this process in Sect. 7.5. We may partition the graph to processors of the distributed system and then implement Boruvka's algorithm in parallel as a first approach. The CLUMP algorithm takes a different approach by forming bipartite subgraphs and use Prim's algorithm in parallel in these graphs and combine the partial MSTs to get the final MST as described next.

CLUMP

Clustering through MST in parallel (CLUMP) is a clustering method designed to detect dense regions of biological data [38]. It is not particularly designed for biological networks, however, it uses representation of biological data as a weighted undirected graph $G(V, E)$ in which each data point is a node and an edge (u, v) connecting nodes u and v has a weight proportional to distance between these two points. This algorithm constructs an MST of the graph G and proceeds similarly to the MST-based clustering algorithms to find clusters. Since the most time-consuming part of an any MST-based clustering scheme is the construction of the MST, the following steps of CLUMP are proposed:

1. The original graph $G(V, E, w)$ is partitioned into $G_j(V_j, E_j), j = 1, \ldots, s$ where G_j is the subgraph induced by V_j, E_{ij} is the set of edges between V_i and V_j.
2. The bipartite graphs $B_{ij} = \{V_i \cup V_j, E_{ij}\}$ for all subgraphs formed in step 1 are constructed.
3. For each G_i; an MST T_{ii}, and for each B_{ij} an MST T_{ij} is constructed in parallel.
4. A new graph $G^0 = \bigcup T_{ij}, 1 \leq i \leq j \leq s$ is constructed by merging all MSTs from step 3.
5. The MST of the graph G^0 is constructed.

The authors showed that the MST of G^0 is the MST of the original graph G. The idea of this algorithm is to provide a speedup by parallel formation of MSTs in step 3 since the formation of G^0 in the last step is relatively less time consuming due to the sparse structure of this graph. Prim's algorithm was used to construct MSTs and the algorithm was evaluated using MPI and ANSI C. The authors also provided an online CLUMP server which uses an MySQL database for registered users. CLUMP is implemented for hierarchical classification of functionally equivalent genes for prokaryotes at multi-resolution levels and also for the analysis of the *Diverse Sitager Soil Metagenome*. The performance of this parallel clustering algorithm was found highly effective and practical during these experiments.

MST-based clustering is used in various applications including biological networks [45]. A review of parallel MST construction algorithms is provided in [34] and a distributed approach using MPI is reported in [17].

11.3.2 Edge-Betweenness-Based Clustering

As we have seen in Sect. 11.5, the vertex betweenness centrality $C_B(v)$ of a vertex v is the percentage of the shortest paths that pass through v. Similarly, the edge betweenness centrality $C_B(e)$ of an edge e is the ratio of the shortest paths that pass through e to total number of shortest paths. These two metrics are shown below:

$$C_B(v) = \sum_{s \neq t \neq v} \frac{\sigma_{st}(v)}{\sigma_{st}}, \qquad C_B(e) = \sum_{s \neq t \neq v} \frac{\sigma_{st}(e)}{\sigma_{st}} \qquad (11.7)$$

Girvan and Newman proposed an algorithm (GN algorithm) based on edge betweenness centrality to provide clusters of large networks which consists of the following steps [24].

1. Find edge betweenness values of all edges of the graph $G(V, E)$ representing the network.
2. Remove the edge with the highest edge betweenness value from the graph.
3. Recalculate edge betweennesses in the new graph.
4. Repeat steps 1 and 2 until a quality criteria is satisfied.

The general idea of this algorithm is that an edge e which has a higher edge betweenness value than other edges has a higher probability of joining two or more clusters as there are more shortest paths passing through it. In the extreme case, this edge could be a bridge of G in which case removing it will disconnect G. It is considered as a hierarchical divisive algorithm since it starts with a single cluster containing all vertices and iteratively divides clusters into smaller ones. The fundamental and most time consuming step in this algorithm is the computation of the edge betweenness values which can be performed using the algorithms described in Sect. 11.5.

11.3.2.1 A Distributed Edge-Betweenness Clustering Algorithm

GN algorithm has a high computational cost making it difficult to implement in large networks such as the PPI networks. Yang and Lonardi reported a parallel implementation of this algorithm (YL Algorithm) on a distributed cluster of computers and showed that a linear speedup is achieved up to 32 processors [46]. The quality of the clusters is validated using the modularity concept described in Sect. 11.2. All-pairs shortest paths for an unweighted graph G can be computed using the BFS algorithm for all nodes of G and the edge betweenness values can be obtained by summing all pair dependencies $\delta_{st}(v)$ over all traversals. The main idea of YL Algorithm is that the BFS algorithm can be executed independently in parallel on a distributed memory computer system. It first distributes a copy of the original graph G to k processors, however, each processor p_i executes BFS on its set of vertices V_i only. There is one supervisor processor p_s which controls the overall operation. Each processor p_i finds the pair dependencies for its vertices in V_i it is assigned and sends its results to p_s which in turn sums all of the partial dependencies and finds the edge e with the highest edge betweenness value. It then broadcasts the edge e to all processors which delete e from their local graph and continue with the next iteration until there are no edges left as shown in Algorithm 11.2 for the supervisor and each worker process. Modularity is used to find where to cut the output dendrogram. YL algorithm was implemented in C++ using MPI on five different PPI networks. The results showed it found clusters in these networks correctly with linear speedups up to 32 processors.

Algorithm 11.2 *YL_Alg*

1: **Input** : $G(V, E)$ ▷ undirected graph,
2: **Output** : edges $e_1, \ldots, e_m$ of G in reversal removal order
3: **if** I am root **then**
4:    **assign** vertex sets $V_1, \ldots, V_k$ to processors $p_1, \ldots, p_k$
5: **end if**
6: **while** there are edges left on processors **do**
7:    **if** I am root **then**
8:       **receive** $\delta_{st}(v)$
9:       **calculate** edge betweenness values for all edges
10:       **find** the edge e with the maximum value and **broadcast** e
11:    **else**
12:       **for all** $v \in V_i$ **parallel do**   ▷ worker process p_i does this part in parallel with others
13:          $BFS(G_i, v)$
14:          **send** all pair dependencies $\delta_{st}(v_i)$ to the root
15:       **end for**
16:       **receive** the edge e that has the highest betweenness
17:       **remove** e from the local graph partition G_i
18:       **synchronize**
19:    **end if**
20: **end while**

11.4 Density-Based Clustering

The dense parts of an unweighted graph have more edges than average and exhibit possible cluster structures in these regions. If we can find methods to discover these dense regions, we may detect clusters. Cliques are perfect clusters and detecting a clique does not require any comparison with the density in the rest of the graph. In many cases, however, we will be interested in finding denser regions of a graph with respect to other parts of it rather than absolute clique structures. We will first describe algorithms to find cliques in a graph and then review k-cores, HCS, and modularity-based algorithms with their distributed versions in this section.

11.4.1 Clique Algorithms

A clique of a graph G is a subset of its nodes which induce a complete graph as we saw in Sect. 3.6, and finding the maximum clique of a graph is NP-hard [23]. In the extreme case, detecting clusters in a graph G can be reduced to finding cliques of G. However, a graph representing a biological network may have only few cliques due to some links not being detected or deleted from it. For this reason, we would be interested in finding *clique-like* structures in a graph rather than full cliques to discover clusters. These structures can be classified as follows [20]:

Definition 11.1 (*k-clique:*) In a k-clique subgraph G' of G, the shortest path between any two vertices in G' is at most k. Paths may consist of vertices and edges external to G'.

Definition 11.2 (*quasi-clique:*) A quasi-clique is a subgraph G' of G where G' has at least $\gamma|G'||G'| - 1)/2$ edges. In other words, a quasi-clique of size m has γ fraction of the number of edges of the clique of the same size.

Definition 11.3 (*k-club:*) In a k-club subgraph G' of G, the shortest path between any two vertices which consists of vertices and edges in G', is at most k.

Definition 11.4 (*k-core:*) In a k-core subgraph G' of G, each vertex is connected to at least k other vertices in G'. A clique is a $(k - 1)$ core.

Definition 11.5 (*k-plex:*) A k-plex is a subgraph G' of G, each vertex has at most k connected to at least $n - k$ other vertices in G'. A clique is a 1-plex.

11.4.1.1 Bron and Kerbosch Algorithm

Bron and Kerbosch proposed a recursive backtracking algorithm (BK algorithm) to find all cliques of a graph [9]. This algorithm works using three disjoint sets of vertices; R which is the current expanding clique, P is the set of potential vertices connected to vertices in R, and X contains all of the vertices already processed. The algorithm recursively attempts to generate extensions to R from the vertices in P which do not contain any vertices in X. It consist of the following steps [9]:

1. Select a candidate.
2. Add the selected candidate to R.
3. Create new sets P and X from the old sets by removing all vertices not connected to R.
4. Call the extension operator on the newly formed sets.
5. Upon return, remove the selected candidate from R, add it to X.

In order to have a clique, set P should be empty, as otherwise R could be extended. Also, the set X should be empty to ensure R is maximal otherwise it may have been contained in another clique. Algorithm 11.3 shows the pseudocode of this algorithm based on the above steps. The time complexity of this algorithm was evaluated to be $O(4^n)$ according to the experimental observations of the authors. Bron and Kerbosch also provided a second version of this algorithm that uses pivots with an experimental time complexity of $O(3.14^n)$ where 3^n is the theoretical limit.

Blaar et al. implemented a parallel version of Bron and Kerbosch algorithm using thread pools in Java and provided test results using 8 processors [6]. Mohseni-Zadeh et al. provided a clustering method they called Cluster-C to cluster protein sequences based on the extraction of maximal cliques [32] and Jaber et al. implemented a parallel version of this algorithm using MPI [27]. Schmidt et al. provided a scalable parallel implementation of Bron and Kerbosch algorithm on a Cray XT supercomputer [41].

Algorithm 11.3 *Bron Kerbosch Algorithm*

1: **procedure** *BronKerbosch*(*R, P, X*)
2:   $P \leftarrow V$ includes all of the vertices and $R, X \leftarrow \emptyset$
3:   **if** $P = \emptyset \land X = \emptyset$ **then**
4:     **return** R as a maximal clique
5:   **else**
6:     **for all** $v \in P$ **do**
7:       *BronKerbosch*$(R \cup \{v\}, P \cap N(v), X \cap N(v))$
8:       $P \leftarrow P \setminus \{v\}$
9:       $X \leftarrow P \cup \{v\}$
10:     **end for**
11:   **end if**
12: **end procedure**

11.4.2 *k*-core Decomposition

Given a graph $G(V, E)$, a subgraph $G_k(V', E')$ of G induced by V' is a k-core of G if and only if $\forall v \in V' : \delta(v) \geq k$ and G_k is the maximum graph with this property. *Main core* of a graph G is the core of G with maximum order and the *coreness value* of a vertex v is the highest order of a core including v. The k-class of a vertex can be defined as the set of vertices which all have a degree of k [15]. Cores of a graph may not be connected and a smaller core is the subset of a larger core. Figure 11.4 displays 3 nested cores of a sample graph.

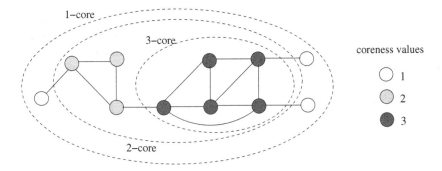

Fig. 11.4 Cores of a sample graph

The *k-core decomposition* of a graph G is to find the k-core subgraphs of G for all k which can therefore be reduced to finding coreness values of all vertices of G. Core decomposition has been used for complex network analysis [2] and to detect k-cores in PPI networks [1]. Detecting group structures such as *cliques, k-cliques, k-plexes,* and *k-clubs* are difficult and NP-hard in many cases, however, finding k-cores of a graph can be performed in polynomial time as we describe in the next section.

11.4.2.1 Batagelj and Zaversnik Algorithm

Batagelj and Zaversnik proposed a linear time algorithm (BZ algorithm) to find the core numbers of all vertices of a graph G [4] based on the property that removing all vertices of degree less than k from a graph with their incident edges recursively will result in a k-core. This algorithm first sorts the degrees of vertices in increasing order and inserts them in a queue Q. It then iteratively removes the first vertex v from the queue, and labels it with the core value which equals the current degree of v, and decrements the degree of each neighbor u of v, if u has a larger degree than v, to effectively delete the edge between them. The vertex u may not have a degree smaller than v as Q is sorted but u may have equal degree to v in which case we do not want to change its degree since u will be moved to a lower class. The pseudocode of this algorithm is shown in Algorithm 11.4 [4,20]. The algorithm ends when Q is empty and k-cores of G consist of vertices which have label values up to and including k. Batagelj and Zaversnik showed that the time complexity of this algorithm is $O(max(m, n))$, and time complexity is $O(m)$ in a connected network as $m \geq n - 1$ in such a case.

Algorithm 11.4 *BZ_Alg*

1: **Input** : $G(V, E)$
2: **Output** : core values of vertices
3: $Q \leftarrow$ sorted vertices of G in increasing weight
4: **while** $Q \neq \emptyset$ **do**
5:    $v \leftarrow$ front of Q
6:    $core(v) \leftarrow \delta(v)$
7:    **for all** $u \in N(v)$ **do**
8:       **if** $\delta(u) > \delta(v)$ **then**
9:          $\delta(u) \leftarrow \delta(u) - 1$
10:       **end if**
11:    **end for**
12:    **update** Q
13: **end while**

Execution steps of this algorithm in the sample graph of Fig. 11.5 is shown in Table 11.1.

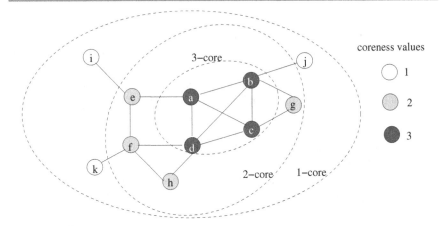

Fig. 11.5 Output of BZ algorithm on a sample graph

11.4.2.2   Molecular Complex Detection Algorithm

The Molecular Complex Detection (MCODE) Algorithm is used to discover protein complexes in large PPI networks [3]. It consists of vertex weighting, complex prediction, and optional post-processing to add or filter proteins in the output complexes. In the first step of this algorithm, the local vertex density of a vertex v is computed using the highest k-core of its neighborhood. We will repeat the definition of the clustering coefficient $cc(v)$ of a vertex v as follows:

$$cc(v) = \frac{2m_v}{n_v(n_v - 1)}, \tag{11.8}$$

where n_v is the number of neighbors of v and m_v is the existing number of edges between these neighbors. It basically shows how well connected the neighbors of v are. The MCODE algorithm defines the *core clustering coefficient* of a vertex v as the density of the highest k-core of the closed neighborhood of v. Using this parameter provides removal of many low-degree vertices seen in PPI networks due to the scale-free property, while emphasizing the high-degree nodes which we expect to see in the clusters. The weight of a vertex v, $w(v)$, is then assigned as the product of the core clustering coefficient $ccc(v)$ of vertex v and the highest k-core value k_{max} in the closed neighborhood of v as follows.

$$w(v) = ccc(v) \times k_{max} \tag{11.9}$$

The second step of the algorithm involves selecting the highest weighted vertex v as the seed vertex and recursively adding vertices in its neighborhood if their weights are above a threshold. The threshold named vertex weight percentage (VWP) is a predetermined percentage of the weight of the seed vertex v. When there are no more vertices that can be added to the complex, this sub-step is stopped and the process is repeated with the next highest and unvisited vertex. WWP parameter effectively specifies the density of the complex obtained, with a high threshold resulting in a

Table 11.1 BZ algorithm execution

Iterations	Queue sorted in ascending degree					Coreness values		
	1	2	3	4	5	$k = 1$	$k = 2$	$k = 3$
1	i, j, k	g, h	e	a, c, f	b, d	$\{i\}$	$\{\varnothing\}$	$\{\varnothing\}$
2	j, k	g, h, e	–	a, c, f	b, d	$\{i, j\}$	$\{\varnothing\}$	$\{\varnothing\}$
3	k	g, h, e	–	a, c, f, b	d	$\{i, j, k\}$	$\{\varnothing\}$	$\{\varnothing\}$
4	–	g, h, e	f	a, c, b	d	$\{i, j, k\}$	$\{g\}$	$\{\varnothing\}$
5	–	h, e	f, b, c	a	d	$\{i, j, k\}$	$\{g, h\}$	$\{\varnothing\}$
6	–	e, f	b, c	a, d	–	$\{i, j, k\}$	$\{g, h, e\}$	$\{\varnothing\}$
7	–	f	b, c, a	d	–	$\{i, j, k\}$	$\{g, h, e, f\}$	$\{\varnothing\}$
8	–	–	b, c, a, d	–	–	$\{i, j, k\}$	$\{g, h, e, f\}$	$\{b\}$
9	–	–	c, a, d	–	–	$\{i, j, k\}$	$\{g, h, e, f\}$	$\{b, c\}$
10	–	–	a, d	–	–	$\{i, j, k\}$	$\{g, h, e, f\}$	$\{b, c, a\}$
11	–	–	d	–	–	$\{i, j, k\}$	$\{g, h, e, f\}$	$\{b, c, a, d\}$

smaller and a denser complex and a low value results in the contrary. The last step is used to filter and modify the complexes. Complexes that do not have at least a 2-core are removed during filtering process and the optional *fluff* operation increases the size of the complexes according to the fluff parameter which is between 0.0 and 1.0. The time complexity of this algorithm is $O(nmh^3)$ where h is the vertex size of the average vertex neighborhood in G as shown in [3].

There is not a reported parallel or distributed version of this algorithm, however, the search of dense neighbors can be performed by the BFS method and this step can be employed in parallel using a suitable parallel BFS algorithm such as in [10].

11.4.2.3 A Distributed k-core Algorithm

Although the BZ algorithm is efficient for small graphs, distributed algorithms are needed to find k-cores in large biological networks such as PPI networks. The BZ algorithm has inherently serial processing as we need to find the vertex with the smallest degree globally and hence is difficult to parallelize. One very recent effort to provide a distributed k-core decomposition method was proposed by Montresor et al. [33]. This algorithm attempts to find coreness values of vertices in a graph G which provides k-cores of G indirectly. They considered two computational models; in *one-to-one* model, each computational node is responsible for one vertex and one node handles all processing for a number of vertices in *one-to-many* model. The latter is obtained by extending the first model. The main idea of this algorithm is based on the *locality* concept in which the coreness of a node u is the greatest k value where u has at least k neighbors, each belonging to k or larger cores. Based on this concept, a node can compute its coreness value using the coreness values of its neighbors. Each node u in this algorithm forms an estimate of its coreness value and sends this value to it neighbors and uses the value received from neighbors to recompute its estimate. After a number of periodic rounds, coreness values can be determined when no new estimates are generated. The time complexity was shown to be $O(n - s + 1)$ rounds where s is the number of nodes in the graph with minimal degree. The messages exchanged during the algorithm was shown to be $O(\Delta m)$ where Δ is the maximum degree of the graph. The authors have experimented one-to-one and one-to-many versions of this algorithm with both a simulator and real large data graphs and found it is efficient.

11.4.3 Highly Connected Subgraphs Algorithm

The highly connected subgraphs (HCS) algorithm proposed by Hartuv and Shamir [26] searches dense subgraphs with high connectivity rather than cliques in undirected unweighted graphs. The general idea of this algorithm is to consider a subgraph G' of n vertices of a graph G as highly connected if G' requires a minimum of $n/2$ edges to have it disconnected. In other words, the edge connectivity of G', $k_E(G')$ should be $n/2$ to accept it as a highly connected subgraph. The algorithm shown in

Algorithm 11.5 starts by first checking if G is highly connected, otherwise uses the minimum cut of G to partition G into H and H', and recursively runs HCS procedure on H and H' to discover highly connected subgraphs.

Algorithm 11.5 *HCS_Alg*

1: **procedure** *HCS(G)*
2:   **Input** : $G(V, E)$
3:   **Output** : highly connected clusters of G
4:   $(H, \bar{H}, C) \leftarrow MinCut(G)$
5:   **if** G is highly connected **then**
6:     **return**(G)
7:   **else**
8:     $HCS(H)$
9:     $HCS(\bar{H})$
10:   **end if**
11: **end procedure**

The execution of HCS algorithm is shown in a sample graph in Fig. 11.6 after which three clusters are discovered. HCS has a time complexity of $2N \times f(n, m)$ where N is the number of clusters discovered and $f(n, m)$ is the time complexity of finding a minimum cut in a graph that has n vertices and m edges. HCS has been successfully used to discover protein complexes, and cluster identification via connecting kernel (CLICK) algorithm is an adaptation of HCS algorithm for weighted graphs [43].

11.4.4 Modularity-Based Clustering

We have seen that the modularity parameter Q provides a good indication of the quality of the clustering in Sect. 12.2. The algorithm proposed by Girvan and Newman

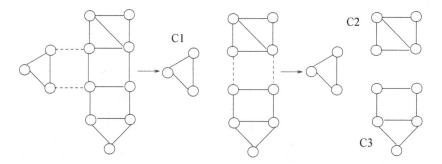

Fig. 11.6 HCS algorithm run on a sample graph. Three clusters C_1, C_2 and C_3 are discovered

(GNM algorithm) attempts to obtain clustering by increasing the value of Q as follows [35]:

1. Each node of the graph is a cluster initially.
2. Merge the two clusters that will increase the modularity Q by the largest amount.
3. If merges start reducing modularity, stop.

This algorithm can be classified as an agglomerative hierarchical clustering algorithm as it iteratively forms larger clusters and the output is a dendrogram as in such algorithms. Its time complexity is $O((m + n)n)$, or $O(n^2)$ on sparse graphs.

11.4.4.1 A Survey of Modularity-Based Algorithms

A review of modularity and methods for its maximization is presented by Fortuna [22]. Clauset et al. provided a method to find clusters using modularity in favorable time using the sparse structure of a graph [16]. They kept the clustering information using a binary tree in which every node is a cluster formed by combining its children. The time complexity of this algorithm is $O(mh \log n)$ where h is the height of the tree. The Louvain method proposed in [7] evaluates modularity by moving nodes between clusters. A new coarser graph is then formed where each node is a cluster. This greedy method optimizes modularity in two steps. The small communities are searched locally first, and these communities are then combined to form the new nodes of the network. These two steps are repeated until modularity is maximized.

Parallel and Distributed Algorithms

Parallel and distributed algorithms that find clusters using modularity are scarce. Gehweiler et al. proposed a distributed diffusive heuristic algorithm for clustering using modularity [25]. Riedy et al. proposed a massively parallel community detection algorithm for social networks based on Louvani method [39]. We will describe this algorithm in detail as it is one of the only parallel algorithms for this purpose. It consists of the following steps which are repeated until a termination condition is encountered:

1. Every edge of the graph is labeled with a score. If all edges have negative scores, exit.
2. Compute a weighted maximal matching using these scores.
3. Coarsen matched groups into a new group which are the nodes of the new graph.

In the first step, the change in optimization metric is evaluated if two adjacent clusters are merged and a score is associated with each edge. The second step involves selecting pairs of neighboring clusters merging of which will improve the quality of clustering using a greedy approximately maximum weight maximal matching and the selected clusters are contracted according to the matching in the final step. The time complexity of this algorithm is $O(mk)$ where k is the number of contraction steps. Each step of this algorithm is independent and can be performed in parallel. Reidy et al. implemented this algorithm in Cray XMT2 and Intel-based server platforms using OpenMP and obtained significant speedups with high performance and good data scalability.

LaSalle and Karypis recently provided a multithreaded modularity-based graph clustering algorithm using the multilevel paradigm [31]. Multilevel approaches for graph partitioning are popular due to the high quality partitions produced by them. These methods consist of *coarsening*, *initial clustering*, and *uncoarsening* phases. The initial graph G_0 is contracted into a series of smaller graphs $G_1, \ldots, G_s$ in the coarsening phase using some heuristics. The final graph G_s is then partitioned into a number of nonoverlapping partitions using a direct partitioning algorithm in the second phase. In the uncoarsening phase, a coarser graph is projected back to a finer graph followed by a cluster refinement procedure such as the Kernighan–Lin algorithm [29]. Maximal matching of edges is a frequently used heuristic during coarsening. The algorithm matches vertex u with its neighbor vertex v that provides maximum modularity in the coarsening phase of multilevel methods. In this study, the authors developed two modularity-based refinement methods called random boundary refinement and greedy boundary refinement which consider border vertices between clusters. This algorithm named *Nerstrand* has an overall time complexity of $O(m + n)$ which is the sum of complexities of three phases and the space complexity is shown to be also $O(m+n)$. The *Nerstrand* algorithm is implemented in shared memory multithreaded environment where the number of edges each thread works is balanced explicitly and the number of vertices for each thread is balanced implicitly. Each thread creates one or more initial clustering of its vertices and the best clustering is selected by reduction. Each thread then performs cluster projection over the vertices it is responsible. The authors implemented this algorithm in OpenMP environment and compared serial and multithreaded *Nerstrand* performances with other methods such as Louvain. They found serial *Nerstrand* performs similar or slightly better than Louvain method but parallel *Nerstand* finds clusters much faster than contemporary methods, 2.7–44.9 times faster.

11.4.4.2 A Distributed Algorithm Proposal

We now propose a distributed algorithm for modularity-based clustering. The general idea of our algorithm is to partition the modularity finding computation among k processes. Each process computes the merge that causes the maximum modularity in its partition and sends the pair that gives this value to the central process. Upon the gathering of local results in the central process, it finds the merge operation that results in maximum modularity change and broadcasts the cluster pair that has to be merged to all processes. This process continues until modularity does not change significantly anymore. We show below the main steps of this algorithm where we have k processes running on k processors and one of these processes, p_0, is designated as the root process that controls the overall execution of the algorithm as well as performing part of the algorithm. We will assume the graph $G(V, E)$ is already

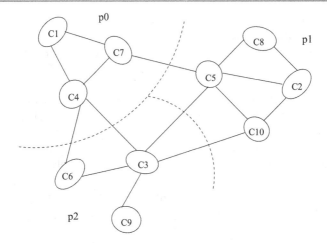

Fig. 11.7 Clusters for the distributed modularity-based clustering algorithm

partitioned into a number of clusters, however, we could have started by the original graph assuming each node is a cluster. The root performs the following steps:

1. Assume each cluster is a supernode and perform a BFS partition on the original graph to have k cluster partitions such that $C = \{C_1, \dots, C_k\}$.
2. **send** each partition to a process p_i.
3. **find** the best cluster pair in my partition.
4. **receive** best cluster pairs from each p_i.
5. **find** the pair C_x, C_y that gives the maximum modularity.
6. **broadcast** C_x, C_y to all processes.
7. **repeat** steps 3–5 until modularity starts reducing.

Figure 11.7 displays an example network that already has 10 clusters. The root process p_0 partitions this network by the BFS partitioning algorithm and sends the two cluster partitions to processes p_1 and p_2. In this example, p_1 computes the modularity values for combining operations $C_2 \cup C_8$, $C_2 \cup C_{10}$, $C_2 \cup C_5$, $C_5 \cup C_8$, $C_5 \cup C_{10}$, and $C_5 \cup C_7$, assuming for any cluster pair across the borders, the process that owns the lower identifier cluster is responsible to compute modularity. Further optimizations are possible such as in the case of local cluster operation in a process is decided to merge C_2 and C_{10} in p_1, the processes p_0 and p_2 do not need to compute their modularity values again as they are not affected.

11.5   Flow Simulation-Based Approaches

A different approach than the traditional graph clustering methods using density is considered in flow simulation-based methods. The goal in this case is to predict the regions in the graph where the flow will gather. The analogy of the graph is a water distribution network with nodes representing storages, and the edges as the pipes between them. If we pump water to such a network, flow will gather at nodes which have many pipes ending in them and hence in clusters. An effective way of simulating the flow in a network is by using random walks which is described in the next section.

11.5.1   Markov Clustering Algorithm

Markov Clustering Algorithm (MCL) is a fast clustering algorithm based on stochastic flow in graphs [18] and uses the following heuristics [20]:

1. Number of paths of length k between two nodes in a graph G is larger if they are in the same cluster and smaller if they are in different clusters.
2. A *random walk* starting from a vertex in a dense cluster of G will likely end in the same dense cluster of G.
3. Edges between clusters are likely to be incident on many shortest paths between all pairs of nodes.

The algorithm is based on random walks assuming by doing the random walks on the graph, we may be able to find where the flow gathers which shows where the clusters are. Given an undirected, unweighted graph $G(V, E)$ and its adjacency matrix A, this algorithm first forms a column-stochastic matrix M. The ith column of M shows the flows out of node v_i and the ith row contains the flows into v_i. The sum of column i of this matrix equals 1 as this is the sum of the probabilities of reaching any neighbor from vertex v_i, however, the sum of the rows may not add up to 1. The matrix M can be obtained by normalizing the columns of the adjacency matrix A as follows.

$$M(i, j) = \frac{A(i, j)}{\sum_{k=1}^{n} A(k, j)} \tag{11.10}$$

This operation is equal to $M = AD^{-1}$ where D is the diagonal matrix of the graph G. The MCL algorithm inputs the matrix M and performs two iterative operations on M called *expansion* and *inflation* and an additional *pruning* step as follows.

- **Expansion**: This operation simply involves taking the eth power of M as below:

$$M_{\exp} = M^e, \tag{11.11}$$

e being a small integer, usually 2. Based on the properties of M, $M_{\exp}$ shows the distribution of a random walk of length r from each vertex.

• **Inflation**: In this step, the rth power of each entry in M is computed and this value is normalized by dividing it to the sum of the rth power of column values as below.

$$M_{\inf}(i,j) = \frac{M(i,j)^r}{\sum_{k=1}^{n} M(k,j)^r} \tag{11.12}$$

The idea here is to emphasize the flow where it is large and to decrease it where it is small. This property makes this algorithm suitable for scale-free networks such as PPI networks, as these networks have few high-degree hubs and many low-degree nodes. As clusters are formed around these hubs, emphasizing them and deemphasizing the low-degree nodes removes extra processing around the sparse regions of the graph.

• **Pruning**: The entries which have significantly smaller values than the rest of the entries in that column are removed. This step reduces the number of nonzero column entries so that memory space requirements are decreased.

Algorithm 11.6 displays the pseudocode for MCL algorithm [42].

Algorithm 11.6 *MCL_Alg*

1: **Input** : $G(V, E)$ ▷ undirected unweighted graph
2:        expansion parameter e, inflation parameter r
3: **Output** : Clusters $C_1, \ldots, C_k$ of G
4: $A \leftarrow$ adjacency matrix of G
5: $M \leftarrow AD^{-1}$ ▷ initialize M
6: **repeat**
7:      $M_{exp} \leftarrow M^e$ ▷ expand
8:      **inflate** $M_{inf} \leftarrow M_{exp}$ using r ▷ inflate
9:      $M \leftarrow M_{exp}$
10: **until** M converges
11: **interpret** the resulting matrix M_{inf} to find clusters

---

After a number of iterations, there will be only one nonzero element at each column of M and the nodes that have flows to this node will be interpreted as a single cluster. The level of clustering can be modified by the parameter r, with the lower r resulting in a coarser clustering. The time complexity of this algorithm is $O(n^3)$ steps since multiplication of two $n \times n$ matrices takes n^3 time during expansion, and the inflation can be performed in $O(n^2)$ steps. The convergence of this algorithm has been shown experimentally only where the number of rounds required to converge was between 10–100 steps [3]. The MCL algorithm has been successfully implemented in biological networks in various studies [8,44], however, the scalability of MCL especially at the expansion step was questioned in [42]. Also, MCL was found to discover too many clusters in the same study and a modification to MCL by a multilevel algorithm was proposed.

11.5.2 Distributed Markov Clustering Algorithm Proposal

The MCL algorithm has two time consuming steps as expansion and inflation described above. We can see that these are matrix operations which can be performed independently on a distributed memory computing system whether a multiprocessor or totally autonomous nodes connected by a network. The expansion is basically a matrix multiplication operation in which many parallel algorithms exist. The inflation operation yields asynchronous operations and hence can be performed in a distributed system.

We will now sketch a distributed algorithm to perform MCL in parallel using m number of distributed memory processes $p_0, \ldots, p_{m-1}$. The supervisor process p_0 controls the overall flow of the algorithm and $m-1$ worker processes. The supervisor initializes the matrix M, broadcasts it to $m-1$ nodes which all perform multiplication of M by itself using row-wise 1-D partitioning and send back the partial results to the supervisor. This process now builds the M^2 matrix which can be partitioned again and sent to workers which will multiply part of it with their existing copy of M. This process is repeated for t times to conclude the expansion of the first iteration and for $t = 2$, finding M^2 will be sufficient. The supervisor can now send the expanded matrix M_{exp} by column partitioning it to $m-1$ processes each of which simply takes the eth power of each entry in columns, normalizes them and sends the resulting columns to the supervisor. The task of the supervisor now is to check whether M has converged and if this is not achieved a new iteration is started with the seed M_p. Algorithm 11.7 shows the pseudocode for the distributed MCL algorithm which can easily be implemented using MPI.

Algorithm 11.7 *DistMCL_Alg* Supervisor Process

1: **Input** : $G(V, E)$ ▷ undirected graph,
2:      expansion parameter e, inflation parameter r
3: **Output** : Clusters $C_1, \ldots, C_k$ of G
4: $A \leftarrow$ adjacency matrix of G
5: $M_p \leftarrow AD^{-1}$ ▷ initialize M
6: **repeat**
7:      **broadcast** M_p to $m - 1$ processes
8:      **compute** my partial product
9:      **gather** partial products from all workers   ▷ synchronize
10:     **build** M_{exp}
11:     **send** rows of M_{exp} to $m - 1$ processes
12:     **compute** my partial inflation
13:     **gather** partial products from all workers   ▷ synchronize
14:     **build** M_{exp}
15:     $M_p \leftarrow M_{exp}$
16: **until** M converges
17: **interpret** the resulting matrix M_{inf} to find clusters

---

Fig. 11.8 An example graph for distributed MCL algorithm

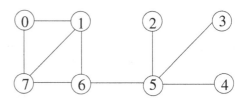

We will show the implementation of the distributed MCI algorithm using a simple example graph of Fig. 11.8 which will also show the detailed operation of the sequential algorithm.

The M matrix row partitioned by the supervisor for parallel processing using this graph will be:

$$
M =
\begin{bmatrix}
0 & 0.33 & 0 & 0 & 0 & 0 & 0 & 0.33 & p_0 \\
0.5 & 0 & 0 & 0 & 0 & 0 & 0.33 & 0.33 & \\
\hline
0 & 0 & 0 & 0 & 0 & 0.25 & 0 & 0 & p_1 \\
0 & 0 & 0 & 0 & 0 & 0.25 & 0 & 0 & \\
\hline
0 & 0 & 0 & 0 & 0 & 0.25 & 0 & 0 & p_2 \\
0 & 0 & 1 & 1 & 1 & 0 & 0.33 & 0 & \\
\hline
0 & 0.33 & 0 & 0 & 0 & 0.25 & 0 & 0.33 & p_3 \\
0.5 & 0.33 & 0 & 0 & 0 & 0 & 0.33 & 0 &
\end{bmatrix}
$$

Assuming we have 4 processors p_0, p_1, p_2 and p_3; and p_0 is the supervisor; the row partitioning of M will result in rows 2,3 to be sent to p_1; rows 4,5 to p_2 and 6,7 to p_3. When the partial products are returned to p_0, it will form M_{exp} shown below:

$$
M_{exp} =
\begin{bmatrix}
p_0 & & p_1 & & p_2 & & p_3 & \\
0.33 & 0.109 & 0 & 0 & 0 & 0.083 & 0.012 & 0.109 \\
0.165 & 0.383 & 0 & 0 & 0 & 0 & 0.083 & 0.274 \\
0 & 0 & 0.25 & 0.25 & 0.25 & 0 & 0.083 & 0 \\
0 & 0 & 0.25 & 0.25 & 0.25 & 0 & 0.083 & 0 \\
0 & 0 & 0.25 & 0.25 & 0.25 & 0 & 0.083 & 0 \\
0 & 0.109 & 0 & 0 & 0 & 0.159 & 0 & 0.109 \\
0.33 & 0.165 & 0.25 & 0.25 & 0.25 & 0 & 0.301 & 0.109 \\
0.165 & 0.274 & 0 & 0 & 0 & 0.083 & 0.109 & 0.383
\end{bmatrix}
$$

We then do a column partitioning of it and distribute it to processors p_1, p_2 and p_3 which will receive columns 2,3; 4,5; and 6,7 consecutively. After they perform inflation operation on their columns, they will return the inflated columns to p_0 which will form the M_{inf} matrix as below. Although some of the entries start diminishing as can be seen, convergence has not been detected yet, and p_0 will continue with the next iteration.

$$M_{inf} = \begin{bmatrix} 0.401 & 0.044 & 0 & 0 & 0 & 0.178 & 0.001 & 0.047 \\ 0.099 & 0.538 & 0 & 0 & 0 & 0 & 0.053 & 0.291 \\ 0 & 0 & 0.25 & 0.25 & 0.25 & 0 & 0.053 & \\ 0 & 0 & 0.25 & 0.25 & 0.25 & 0 & 0.053 & \\ 0 & 0 & 0.25 & 0.25 & 0.25 & 0 & 0.053 & \\ 0 & 0.044 & 0 & 0 & 0 & 0.643 & 0 & 0.047 \\ 0.401 & 0.099 & 0.25 & 0.25 & 0.25 & 0 & 0.695 & 0.047 \\ 0.099 & 0.275 & 0 & 0 & 0 & 0.178 & 0.092 & 0.570 \end{bmatrix}$$

As one of the few studies to provide parallel/distributed MCL algorithm, Busta-mam et al. implemented it using MPI with results showing improved performance [11]. They also provided another parallel version of MCL this time using graphic cards processors (GPUs) with many cores [12].

11.6 Spectral Clustering

Spectral clustering refers to a class of algorithms that use the algebraic properties of the graph representing a network. We have noted that the Laplacian matrix of a graph G is $L = D - A$ in unnormalized form, with D being the diagonal degree matrix which has d_i as the degree of the vertex i in its diagonal and A is the adjacency matrix. In normalized form, the Laplacian matrix $L = I - D^{-1/2}AD^{-1/2}$ and L has interesting properties that can be analyzed to find the connectivity information about a graph G. First of all, the eigenvalues of L are real as L is real and symmetric. The second eigenvalue is called the *Fiedler value* and the corresponding eigenvector for this eigenvalue, the *Fiedler vector* [21] provides connectivity information about the graph G. Using the Fiedler vector, we can partition a graph G into two balanced partitions in *spectral bisection* as follows [19]. We first construct the Fiedler vector and then compare each entry of this vector with a value s, if the entry $F[i] \leq s$ then the corresponding vertex of G, v_i, is put in partition 1 and otherwise it is placed in the second partition as shown in Algorithm 11.8. The variable s could simply be 0 or the median of the Fiedler vector. Figure 11.9 displays a simple graph that is partitioned using the value of 0.

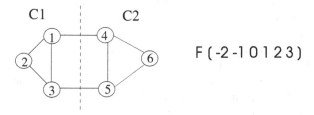

Fig. 11.9 Partitions formed using Fiedler vector. The first three elements Fiedler vector have values smaller or equal to zero and are put in the first partition and the rest are placed in the second

---

Algorithm 11.8 *Spect_Part*

1: **Input** : $A[n, n]$, $D[n, n]$ ▷ adjacency matrix and degree matrix of G
2: **Output** : V_1, V_2 ▷ two balanced partitions of G
3: $L \leftarrow D - A$
4: **calculate** Fiedler vector F of L
5: **for** $i = 1$ to n **do**
6:     **if** $F[i] \leq s$ **then**
7:         $V_1 \leftarrow V_1 \cup \{v_i\}$
8:     **else** $V_2 \leftarrow V_2 \cup \{v_i\}$
9:     **end if**
10: **end for**

Newman also proposed a method based on the spectral properties of the modularity matrix Q [37]. In this method, the eigenvector corresponding to the most positive eigenvalue of the modularity matrix is first found and the network is divided into two groups according to the signs of the elements of this vector. Spectral bisection provides two partitions and can be used to find a k-way partition of a graph when executed recursively. Spectral clustering however, is more general than spectral bisection and finds the clusters in a graph directly. This method is mainly designed to cluster n data points $x_1, \ldots, x_n$ but can be adapted for graphs as it builds a similarity matrix S which have entries s_{ij} showing how similar two data points x_i and x_j are. The spectral properties of this matrix are then investigated to find clusters of data points and the normalized Laplacian matrix, $L = I - D^{-1/2}SD^{-1/2}$ can then be constructed. A spectral clustering algorithm consist of the following steps [14]:

1. Construct similarity matrix S for n data points.
2. Compute the normalized Laplacian matrix of S.
3. Compute the first k eigenvectors of L and form the matrix V with columns as these eigenvectors.
4. Compute the normalized matrix M of V.
5. Use k-means algorithm to cluster n rows of M into k partitions.

The k-means algorithm is a widely used method to cluster data points as we reviewed in Sect. 7.3. The initial centers $c_1, \ldots, c_k$ can be chosen at random initially and the distance of data points to these centers are calculated and each data point is assigned to the cluster that it is closest. The spectral clustering algorithm described requires significant computation power and memory space for matrix operations and also to run the k-means algorithm due to the sizes of matrices involved. A simple approach would involve computation of the similarity matrix S using row partitioning. Finding eigenvectors can also be parallelized using parallel eigensolvers and the final step of using the k-means algorithm can also be parallelized [30]. A distributed algorithm based on the described parallel operations was proposed by Chen et al. and they experimented this algorithm with two large data sets using MPI and concluded it is scalable and provides significant speedups [14].

11.7 Chapter Notes

Graph clustering is a well-studied topic in computer science and there are numerous algorithms for this purpose. Our focus in this chapter was the classification and revision of fundamental sequential and distributed clustering algorithms in biological networks. We have also proposed two new distributed algorithms which can be implemented conveniently using a distributed programming environment such as MPI.

Two types of hierarchical clustering algorithms, MST-based and edge betweenness-based algorithms have found applications in clustering of biological networks more than other algorithms. Finding the MST of a graph using Boruvka's algorithm can be parallelized conveniently due to its nature. A different approach is taken in the CLUMP algorithm where the graph is partitioned into a number of subgraphs and bipartite graphs are formed. The MSTs for the partitions and the bipartite graphs are formed in parallel and then merged to find the MST of the whole graph. The edge-betweenness algorithm removes the edge with the highest betweenness value from the graph at each iteration to divide it into clusters. The algorithm proposed by Yang and Lonardi partitions the graph into processors which find pair dependencies by running BFS in parallel on their local partitions.

Density-based clustering algorithms aim to discover dense regions of a graph as these areas are potential clusters. Cliques are one example of such dense regions, however, clique-like structures such as k-cliques, k-cores, k-plexes, and k-clubs are more frequently found in biological networks than full cliques due to the erroneous measurements and dynamicity in such environments. Out of these structures, only k-cores of a graph can be found in linear time and therefore our focus was on sequential and distributed algorithms for k-core decomposition of graphs. The MCODE algorithm which uses a combination of clustering coefficient parameter and the k-core concept is successfully used in various PPI networks. There is not a reported parallel/distributed version of MCODE algorithm and this may be a potential research area as there are possibilities of independent operations such as finding weights of vertices. The k-core based algorithms are widely used to discover clusters in complex networks such as the biological networks and the Internet, and we reviewed a distributed k-core algorithm. The modularity concept provides a direct evaluation of the quality of clusters obtained and has formed the basis of various clustering algorithms used in social and biological networks. We proposed as simple distributed modularity-based algorithm that can be used for any complex network including biological networks.

Flow-based algorithms consider the flows in the networks and assume flows gather in clusters. An example algorithm with favorable performance that has been experimented in PPI networks is the MCL algorithm which we reviewed in sequential form and proposed its distributed version which can be implemented easily. Lastly, we described spectral bisection and clustering based on the Laplacian matrix of a graph and showed ways to implement distributed spectral clustering.

In summary, we can state most of these algorithms perform well in biological networks. However, each have merits and demerits and complexities of a number of

Fig. 11.10 Example graph
for Exercise 1

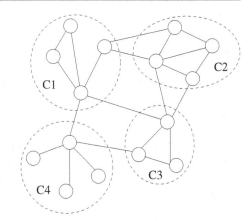

Fig. 11.11 Example graph
for Exercise 2

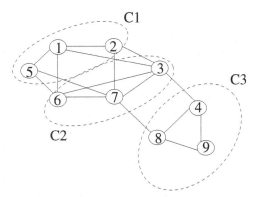

them have only been determined experimentally. Distributed algorithms for this task
are very rare and this may be a potential research area for researchers in this field.

Exercises

1. Find the intra-cluster and inter-cluster densities of the graph of Fig. 11.10. Do
   these values indicate good clustering? What can be done to improve this clustering?
2. Work out the modularity value in the example graph of Fig. 11.11 based on three
   clusters C_1, C_2, and C_3 and determine which merge operation is to be done to
   improve modularity. We need to check combining each cluster pair and decide
   on the pair that improves modularity by the largest amount.
3. Show the output of the BFS-based graph partitioning algorithm on the example
   graph of Fig. 11.12 in the first iteration, with the root vertex s. Then, partition the
   resulting graphs again to obtain four partitions and validate the partitions in terms
   of the number of vertices in each partition and the size of the minimum edge cut
   between them.

Fig. 11.12 Example graph
for Exercise 3

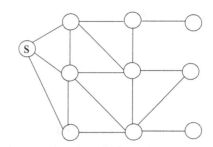

Fig. 11.13 Example graph
for Exercise 4

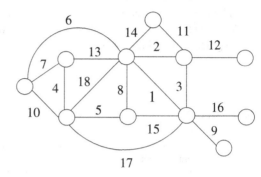

Fig. 11.14 Example graph
for Exercise 5

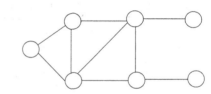

Fig. 11.15 Example graph
for Exercise 6

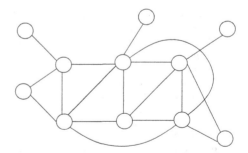

4. Work out the MST of the weighted graph of Fig. 11.13 using Boruvka's algorithm.
   In the second step, partition the graph into two processor p_1 and p_2 and provide a
   distributed algorithm in which p_1 and p_2 will form the MSTs in parallel. Show also
   the implementation of the distributed Boruvka algorithm in this sample graph.
5. Find the edge betweenness values for all edges in the graph of Fig. 11.14 and
   partition this graph into 3 clusters using Newman's edge betweenness algorithm.

Fig. 11.16 Example graph
for Exercise 7

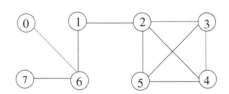

6. For the example graph of Fig. 11.15, implement Batagelj and Zaversnik algorithm
   to find the coreness values of all vertices. Show all iterations of this algorithm
   and compose the k-cores for this graph in the final step.
7. Work out the two iterations of Markov Clustering Algorithm (MCL) in the example graph of Fig. 11.16. Is there any display of the clustering structure?

References

1. Altaf-Ul-Amin Md et al (2003) Prediction of protein functions based on kcores of protein-
   protein interaction networks and amino acid sequences. Genome Inf 14:498–499
2. Alvarez-Hamelin JI, DallAsta L, Barrat A, Vespignani A (2006) How the k-core decomposition
   helps in understanding the internet topology. In: ISMA workshop on the internet topology
3. Bader GD, Hogue CWV (2003) An automated method for finding molecular complexes in
   large protein interaction networks. BMC Bioinform 4(2)
4. Batagelj V, Zaversnik M (2003) An O(m) algorithm for cores decomposition of networks.
   CoRR (Computing Research Repository), arXiv:0310049
5. Boruvka O (1926) About a certain minimal problem. Prce mor. prrodoved. spol. v Brne III (in
   Czech, German summary) 3:37–58
6. Blaar H, Karnstedt M, Lange T, Winter R (2005) Possibilities to solve the clique problem by
   thread parallelism using task pools. In: Proceedings of the 19th IEEE international parallel and
   distributed processing symposium (IPDPS05)—Workshop 5—Volume 06 in Germany
7. Blondel VD, Guillaume J-L, Lambiotte R, Lefebvre E (2008) Fast unfolding of communities
   in large networks. J Stat Mech: Theory Exp (10):P10008
8. Brohee S, van Helden J (2006) Evaluation of clustering algorithms for protein-protein interaction networks. BMC Bioinform 7:488. doi:10.1186/1471-2105-7-488
9. Bron C, Kerbosch J (1973) Algorithm 457: finding all cliques of an undirected graph. Commun
   ACM 16:575–577
10. Buluc A, Madduri K (2011) Parallel breadth-first search on distributed memory systems. CoRR
    arXiv:1104.4518
11. Bustamam A, Sehgal MS, Hamilton N, Wong S, Ragan MA, Burrage K (2009) An efficient
    parallel implementation of Markov clustering algorithm for large-scale protein-protein inter-
    action networks that uses MPI. In: Proceedings of the fifth IMT-GT international conference
    mathematics, statistics, and their applications (ICMSA), pp 94–101
12. Bustamam A, Burrage K, Hamilton NA (2012) Fast parallel Markov clustering in bioinformatics
    using massively parallel computing on GPU with CUDA and ELLPACK-R sparse format.
    IEEE/ACM Trans Comp Biol Bioinform 9(3):679–691
13. Cormen TH, Leiserson CE, Rivest RL, Stein C (2009) Introduction to algorithms, 3rd edn. The
    MIT Press
14. Chen W-Y, Song Y, Bai H, Lin C-J, Chang EY (2011) Parallel spectral clustering in distributed
    systems. IEEE Trans Pattern Anal Mach Intell 33(3):568–586

15. Cheng J, Ke Y, Chu S, Ozsu MT (2011) Efficient core decomposition in massive networks. In: ICDE'11 proceedings of the 2011 IEEE 27th international conference data engineering, pp 51–62

16. Clauset A, Newman ME, Moore C (2004) Finding community structure in very large networks. Phys Rev E 70(6):066111

17. Du Z, Lin F (2005) A novel approach for hierarchical clustering. Parallel Comput 31(5):523–527

18. Dongen SV (2000) Graph clustering by flow simulation. PhD Thesis, University of Utrecht, The Netherlands

19. Elsnern U (1997) Graph partitioning, a survey. Technical report, Technische Universitat Chemnitz

20. Erciyes K (2014) Complex networks: an algorithmic perspective. CRC Press, Taylor and Francis. SBN-13: 978-1466571662, ISBN-10: 1466571667, Chap. 8

21. Fiedler M (1989) Laplacian of graphs and algebraic connectivity. Comb Graph Theory 25:57–70

22. Fortunato S (2010) Community detection in graphs. Phys Rep 486(3):75–174

23. Garey MR, Johnson DS (1978) Computers and intractability: a guide to the theory ofNP-completeness. Freeman

24. Girvan M, Newman MEJ (2002) Community structure in social and biological networks. Proc Natl Acad Sci USA 99:7821–7826

25. Gehweiler J, Meyerhenke H (2010) A distributed diffusive heuristic for clustering a virtual P2P supercomputer. In: Proceedings of the 7th high-performance grid computing workshop (HGCW10) in conjunction with 24th international parallel and distributed processing symposium (IPDPS10). IEEE Computer Society

26. Hartuv E, Shamir R (2000) A clustering algorithm based on graph connectivity. Inf Process Lett 76(4):175–181

27. Jaber K, Rashid NA, Abdullah R (2009) The parallel maximal cliques algorithm for protein sequence clustering. Am J Appl Sci 6:1368–1372

28. Johnson SC (1967) Hierarchical clustering schemes. Psychometrika 2:241–254

29. Kernighan BW, Lin S (1970) An efficient heuristic procedure for partitioning graphs. Bell Syst Tech J 49(2):291–307

30. Kraj P, Sharma A, Garge N, Podolsky R, Richard A, Mcindoe RA (2008) ParaKMeans: implementation of a parallelized K-means algorithm suitable for general laboratory use. BMC Bioinform 9:200

31. LaSalle D, Karypis G (2015) Multi-threaded modularity based graph clustering using the multilevel paradigm. J Parallel Distrib Comput 76:66–80

32. Mohseni-Zadeh S, Brezelec P, Risler JL (2004) Cluster-C, an algorithm for the large-scale clustering of protein sequences based on the extraction of maximal cliques. Comput Biol Chem 28:211–218

33. Montresor A, Pellegrini FD, Miorandi D (2013) Distributed k-Core decomposition. IEEE Trans Parallel Distrib Syst 24(2):288–300

34. Murtagh F (2002) Clustering in massive data sets. Handbook of massive data sets, pp 501–543

35. Newman MEJ (2004) Fast algorithm for detecting community structure in networks. Phys Rev E 69:066133

36. Newman MEJ, Girvan M (2004) Finding and evaluating community structure in networks. Phys Rev E 69:026113

37. Newman MEJ (2006) Finding community structure in networks using the eigenvectors of matrices. Phys Rev E 74:036104

38. Olman V, Mao F, Wu H, Xu Y (2009) Parallel clustering algorithm for large data sets with applications in bioinformatics. IEEE/ACM Trans Comput Biol Bioinform 6:344–352

39. Riedy J, Bader DA, Meyerhenke H (2012) Scalable multi-threaded community detection in social networks. In: 2012 IEEE 26th international parallel and distributed processing symposium workshops and PhD forum (IPDPSW), IEEE, pp 1619–1628

40. Schaeffer SE (2007) Graph clustering. Comput Sci Rev 1:27–64
41. Schmidt MC, Samatova NF, Thomas K, Park B-H (2009) A scalable, parallel algorithm for maximal clique enumeration. J Parallel Distrib Comput 69:417–428
42. Satuluri V (2009) Scalable graph clustering using stochastic flows: applications to community discovery. In: Proceedings of the 15th ACM SIGKDD international conference on knowledge discovery and data mining, pp 737–746. Srinivasan Parthasarathy, Proceeding KDD'09
43. Sharan R, Shamir R (2000) CLICK: a clustering algorithm with applications to gene expression analysis. Proc Int Conf Intell Syst Mol Biol 8(307):16
44. Vlasblom J, Wodak SJ (200) Markov clustering versus affinity propagation for the partitioning of protein interaction graphs. BMC Bioinform 10:99
45. Xu X, Jager J, Kriegel H-P (1999) A fast parallel clustering algorithm for large spatial databases. Data Min Knowl Disc 3:263–290
46. Yang Q, Lonardi S (2007) A parallel edge-betweenness clustering tool for protein-protein interaction networks. Int J Data Min Bioinform (IJDMB) 1(3):241–247
47. Zadeh LA (1965) Fuzzy sets. Inf Control 8:338–353

Network Motif Search

12

12.1 Introduction

Network motifs are the building blocks of various biological networks such as the transcriptional regulation networks, protein–protein interaction networks and metabolic networks. Discovering network motifs gives insight to system level functions in such networks. Given a graph G that represents a biological network, a *network motif m* is a small recurrent and connected subgraph of G that is found in a greater frequency in G than expected in a random graph with a similar structure to G. A motif is assumed to have some biological significance and believed to perform a specific function in the network; it is widely accepted that there is a relation between its structure and its function in general.

Searching for network motifs provides analysis of the basic functions performed by a biological network and also helps to understand the evolutionary relationships between organisms. Conserved motifs in a PPI network allow protein–protein interaction prediction [1] and a conserved motif in two or more organisms may have similar functions in them. For instance, identical motifs found in transcriptional interaction network of *E. coli* and the yeast *S. cerevisiae*, and may that mean common functions are carried by these motifs [14]. Discovery of network motifs is a computationally difficult task as the number of possible motifs grows exponentially with the size of the motif, which is expressed by the number of vertices contained in it. The motif discovery process consists of three basic steps: discovery of frequently appearing subgraphs of the target graph representing the biological network; identification of topologically equivalent subgraphs of a given size in the network, which is the graph isomorphism problem in NP but not known to be NP-complete. The subgraph isomorphism problem, however, is NP-complete [7]. A set of random graphs with a similar topology needs to be constructed, and the motifs should be searched in these graphs to determine the statistical significance of the motifs found, as the last step. The statistical testing involves computing p-value and z-score of the motif. There are various algorithms for this purpose which can be broadly classified as exact counting in which every occurrence of a subgraph is discovered in the target

© Springer International Publishing Switzerland 2015
K. Erciyes, *Distributed and Sequential Algorithms for Bioinformatics*,
Computational Biology 23, DOI 10.1007/978-3-319-24966-7_12

graph, and sampling-based methods where a sample of the graph is searched. Exact methods are time-consuming and are limited in their motif size, whereas sampling-based methods provide approximate results with lower time complexities. Parallel and distributed algorithms for motif discovery are very scarce as we will see.

It would be right to note that the network motif concept has some criticism. It was argued in [2] that concepts such as the preferential attachment may lead to exhibition of motifs which may not have attributed functionality in them, and this was answered by Milo et al. [19] stating that the discovery of motifs should consider subgraph significance profiles as well as the frequency of the subgraph patterns as was also noted in [34].

In this chapter, we first define the motif discovery problem formally, classify algorithms for this purpose, and state parameters for measuring the significance of motif occurrence. We then briefly review and compare the main sequential motif discovery algorithms. Our emphasis is later on parallel and distributed algorithms for motif discovery where we review the few existing approaches and propose general guidelines for potential distributed algorithms.

12.2 Problem Statement

A network motif m in a graph G is a subgraph of G which appears significantly more than in a random graph of similar structure to G. Similarity in structure is typically considered as a graph G' that has the same degree sequence as G. We may be searching motifs in a directed graph or an undirected graph, depending on the application. Clearly, the number of possible motifs of a given size is much higher in directed motifs than undirected ones. Figure 12.1 shows all possible undirected motifs of sizes 3 and 4.

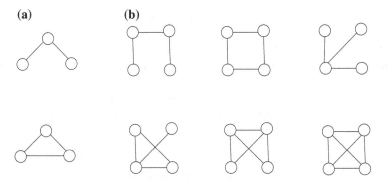

Fig. 12.1 All possible undirected network motifs of **a** size 3, **b** size 4

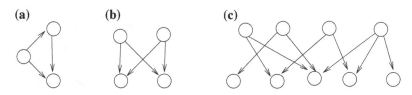

Fig. 12.2 Directed motifs found in biological networks, **a** Feed-forward-loop, **b** Bifan, and **c** Multi-input motifs

Certain motifs are found in abundance in some biological networks. For example, the feed-forward-loop (FFL) and bifan motifs are shown in Fig. 12.2 were discovered in transcriptional regulatory networks and neuronal connectivity networks. The FFL performs a basic function by controlling connections and signaling.

12.2.1   Methods of Motif Discovery

There are three subtasks involved in finding a motif of size k (m_k) in a target graph G as follows.

1. Finding all instances of m_k in G: This step can be done by exact counting methods or sampling-based methods. The former requires enumeration of all subgraphs which has high time complexity. Sampling-based methods implement the algorithm in a sampled subgraph of the target graph and find approximate solutions in a much less time than exact methods.
2. Determination of the isomorphic classes of the discovered subgraphs: Some of the found subgraphs may be equivalent which requires grouping of them into equivalent isomorphic classes.
3. Determination of the statistical significance of the discovered subgraphs: A set of random graphs that have similar structures to the target graph need to be generated and the steps above should be performed on this set. Statistical comparisons of the results in the target graph and in these graphs will then show whether the discovered subgraphs in the target graph are actual motifs.

There are two fundamental ways of investigating motifs in a network; first, we may be interested to search for all subgraphs of a given size k in a network and decide which of these subgraphs may be considered as motifs. This method is commonly called *network-centric* motif search. Second, we may need to know whether a certain subgraph m_k of size k can be considered as a motif in a graph G and search for m_k in G only. This approach is commonly referred to as *motif-centric* search, or sometimes *query-based* motif search or *single-subgraph method*. We will see there are algorithms for both cases. It would make sense first to search for all subgraphs of a given size k in a graph G_1 representing a biological network N_1 and conclude

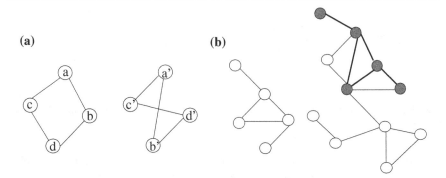

(a)   **(b)**

Fig. 12.3 a Two isomorphic graphs where vertex x is mapped to vertex x' for all vertices. **b** The small graph is isomorphic to the subgraph of the larger graph shown in bold. The subgraph may not be induced as shown in this example

that certain subgraphs of this size are motifs in G_1. We can then search for these motifs in another graph G_2 which represents a similar network N_2 to N_1, and may conclude that these networks have similar subgraphs, may have evolved from the same ancestors and also have similar functionality.

12.2.2 Relation to Graph Isomorphism

Informally, two isomorphic graphs are the same graph that looks different. Graph isomorphism can be defined as follows.

Definition 12.1 (*Graph Isomorphism*) Two graphs $G_1(V_1, E_1)$ and $G_2(V_2, E_2)$ are isomorphic if there is a one-to-one and onto function $f : V_1 \rightarrow V_2$ such that $(u, v) \in E_1 \Leftrightarrow (f(u), f(v) \in E_2)$.

If any two vertices u and v of the graph G_1 with an edge between them are mapped to vertices w and x vertices of G_2; there should be an edge between w and x. Given two graphs $G_1(V_1, E_1)$ and $G_2(V_2, E_2)$, with G_1 being a smaller graph then G_2, the *subgraph isomorphism problem* is to search for a subgraph G_3 of G_2 with maximal size that is isomorphic to G_1. Figure 12.3 displays these concepts. The decision version of this problem is to determine whether G_1 is isomorphic to any subgraph of G_2. It can be shown this problem is NP-complete by reduction from the clique problem, in which given an input graph G and an integer k, we check whether G contains a clique with k vertices.

Two graphs G_1 and G_2 with corresponding adjacency matrices A_1 and A_2 are isomorphic if there exists a permutation matrix P such that $A_2 = P A_1 P^T$. A permutation matrix is obtained by permutating rows and columns of the identity matrix

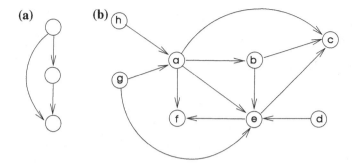

Fig. 12.4 Motif frequency concepts. **a** Feed-forward-loop and **b** A sample graph. The motifs found using F_1 are $\{(a, b, c), (a, e, f), (a, e, c), (b, e, c), (g, a, e)\}$; one of the possible sets using F_2 are $\{(a, b, c), (a, e, f)\}$; and using F_3 is any one of the motifs of F_1

I of the same size, and therefore has exactly a single 1 in each row and column. Ullmann's algorithm was an early effort to find subgraph or graph isomorphism between two graphs using the permutation matrix [31]. It progressively generates a permutation matrix to the size of the small graph, and the equation above is checked to find isomorphism. The *nauty* algorithm proposed by McKay [15,20] also finds subgraph isomorphism using vertex invariants such as k-cliques and the number of vertices at certain distances from a vertex and group theory and is used in network motif search algorithms such as FANMOD [33].

While searching for all subgraphs of a given size in the target network, we need to classify the found subgraphs into isomorphic classes which is the graph isomorphism problem in NP. If we know which motif to search in the target network, the problem is slightly different, we need to deal with the subgraph isomorphism problem in general which is NP-complete.

12.2.3   Frequency Concepts

The frequency of a motif m in a graph G is the number of times it matches to a subgraph in G. This frequency can be evaluated in three ways denoted by F_1, F_2, and F_3. F_1 shows all motifs of a given size allowing overlapping nodes and edges; F_2 shows edge disjoint motifs allowing node overlaps only. The F_3 frequency is the number of edge and vertex disjoint motifs. Figure 12.4 displays these frequency concepts. Various algorithms and tools for motif discovery may use different concepts, and in some cases overlapping nodes and edges may be required and in some other cases, disjoint motifs may be of interest. In general, F_1 is the most general and commonly used frequency parameter.

(a)   **(b)**

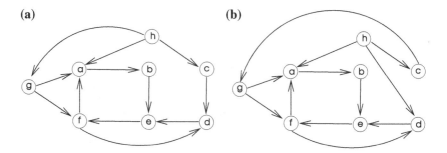

Fig. 12.5 Random graph generation using Monte-carlo method. **a** The original graph, **b** The random graph generated using edges (h, g) and (c, d). These edges are deleted and the newly formed edges are (h, d) and (c, g)

12.2.4   Random Graph Generation

The frequencies of a motif m should be computed in both the target graph G and a set of k random graphs $\mathcal{R} = R_1, \ldots, R_k$. In many motif searching methods, random graphs are typically generated by the Monte-Carlo algorithm, sometimes referred to as the *switching method*. At each iteration of this algorithm, two random edges (a, b) and (c, d) are selected and these edges are replaced by two news edges (a, d) and (b, c), preserving the in-degrees and out-degrees of the vertices. If there are multiple edges or loops are detected as a result of this modification, this step is discarded. This process is repeated sufficient enough to obtain a random graph G' which is quite different than G but has the same topological properties such as the number of vertices and edges, and degree sequence. Exchange of two edges using this method is shown in Fig. 12.5. Members of the set $\mathcal{R}$ is formed using this algorithm and the motif searching can then be performed on the graphs of this set to determine the statistical significance of the subgraphs found.

12.2.5   Statistical Significance

Once the set of k random graphs $\mathcal{R} = R_1, \ldots, R_k$ is formed, we need to apply the same steps of motif search procedure to all elements of $\mathcal{R}$ and analyze the results found in the target graph G and elements of $\mathcal{R}$ using statistical methods. The results will show the frequencies of the subgraphs in both cases. Commonly used statistical metrics for this purpose are the P-value, the Z-score and the *motif significance profile* which commonly use the frequency F_1, as described below [6]:

- *P*-**value**: The P-value of a motif m, $P(m)$ in a target graph G represents the probability of m to have an equal or greater number occurrences in a random network set $\mathcal{R} = R_1, \ldots, R_n$ than in G. A smaller value of $P(m)$ shows a more represented motif and a motif m is considered as statistically significant if $P(m) < 0.01$. This parameter is determined by finding the number of elements of $\mathcal{R}$ that have more frequency of m than in G. This number is divided by the size of the set $\mathcal{R}$ as follows.

$$P(m) = \frac{1}{n} \sum_{i=1}^{n} \sigma_{R_i}(m) \tag{12.1}$$

in which $\sigma_{R_i}(m)$ is set to 1 if the occurrence of motif m in the random network $R_i \in \mathcal{R}$ is higher and 0 if lower than in G.

- **Z-score**: If motif of m occurs F_m times in the network G, and $\overline{F}_r$ and σ_r^2 are the mean and variance frequencies of m in a set of random networks, and Z-score of a motif m_k, $Z(m)$, is the ratio of the difference between the frequency F_m of m in the target network and its average frequency $\overline{F}_r$ in a set of randomized networks, to the root of the standard deviation σ_r^2 of the $F_G(m)$ values in a set of randomized networks as shown below. A $Z(m)$ value that is greater than 2.0 means the motif is significant [11].

$$Z(m) = \frac{F_m - \overline{F}_r}{\sqrt{\sigma_r^2}} \tag{12.2}$$

- **Motif significance profile**: The motif significance profile vector (SP) is formed having elements as Z-scores of a set of motifs $m_1, m_2, \ldots, m_k$ and normalized to unity as shown in Eq. 12.3. It is used to compare various networks considering the motifs that exist in them.

$$SP(m_i) = \frac{Z(m_i)}{\sum_{i=1}^{n} Z(m_i)^2} \tag{12.3}$$

12.3   A Review of Sequential Motif Searching Algorithms

The type of algorithm depends on whether it is network centric or motif centric; and whether it is exact census or sampling based as shown in the classification of Fig. 12.6. Since the number of subgraphs grows exponentially both with the size of the network and the size of the subgraph investigated, the exact census algorithms demand high computational power and their performance degrade with the increased motif size. An efficient way to overcome this problem is to employ probabilistic algorithms where a number of subgraphs of required size are sampled in the target graph and in randomly generated graphs, and the algorithms are executed in these samples rather than the entire graphs. The accuracy of the method employed will increase with the number of samples used; however, we need to ensure that these subgraphs are qualitative representatives of the original graphs to provide the results with reasonable certainty. *Mfinder*, and ESU algorithms are exact motif centric algorithms with sampling-based versions, whereas *Kavosh* has only exact census version. Grochow–Kellis

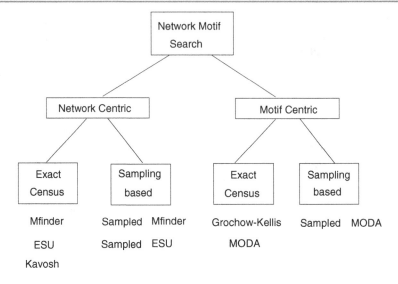

Fig. 12.6 Taxonomy of motif discovery algorithms

and MODA algorithms are motif centric with MODA also having a sampling-based version. We will briefly review all of these algorithms in the next sections.

12.3.1 Network Centric Algorithms

Network centric motif search evaluates all possible motifs in the target graph and groups them into isomorphic classes. The isomorphic testing can be done using different methods but the *nauty* algorithm [15,20] is commonly used in various applications. *Mfinder*, ESU and *Kavosh* algorithms all implement network centric methods as described next.

12.3.1.1 *Mfinder* Algorithm

Mfinder is an early motif search algorithm proposed by Milo et al. [11,17,18]. It uses frequency F_1 and can be used for directed and undirected networks. Given the target graph $G(V, E)$, the algorithm picks an edge $(u, v) \in E$ and enumerates all subgraphs of size k that contain (u, v). The vertices in the subgraph are then inserted into a hash-table to prevent the vertices in the pattern from being revisited. This procedure continues with other edges of G until all edges are visited. Visiting each edge of a subgraph results in detecting the subgraph as many times as the number of edges included in it. This redundancy results in usage of high memory space and high run times. For this reason, *mfinder* can be used efficiently only with motif sizes up to 5.

Mfinder **with Sampling**

Kashtan et al., provided a sampling-based version of *mfinder* called edge sampling algorithm (ESA) in which edges are selected randomly to grow subgraph patterns [12]. An edge (u, v) is randomly selected first and the edges around (u, v) are also randomly selected to form a motif of size k as shown in Algorithm 12.1. The subgraph returned has vertices contained in V_m and edges in E_m. While the size k of the subgraph is not reached, randomly selected edges that are neighbors to the selected edge but not included in the subgraph are included in E_m and the vertices in V_m. This algorithm will be called for each subgraph discovery.

Algorithm 12.1 *Mfinder_Sampling*

1: **procedure** ESA(G, k)
2:     **Input** : $G(V, E)$ ▷ directed or undirected graph
3:     **Output** : A subgraph of size k
4:     $V_m \leftarrow \emptyset$; $E_m \leftarrow \emptyset$
5:     **select** a random edge $(u, v) \in E$
6:     $V_m \leftarrow \{u, v\}$; $E_m \leftarrow (u, v)$
7:     **while** $|V_k| \neq k$ **do**
8:         $P \leftarrow$ neighbor edges of $\{u, v\}$
9:         $P \leftarrow P \setminus$ all edges between the vertices in V_m
10:         **if** $P = \emptyset$ **then exit**
11:         **end if**
12:         **select** a random edge $(x, y) \in P$
13:         $V_m \leftarrow V_m \cup \{x, y\}$
14:         $E_m \leftarrow E_m \cup (x, y)$
15:     **end while**
16:     **return** V_m, E_m
17: **end procedure**

12.3.1.2   Enumerate Subgraphs Algorithm

Enumerate Subgraph (ESU) algorithm is an efficient motif finding algorithm that enumerates all subgraphs by considering the nodes instead of the edges [34,35]. All nodes have identifiers which are used for symmetry breaking while forming the subgraphs. In this algorithm, the exclusive neighborhood of a node v with respect to $V' \subseteq V$ is defined as $N_{\text{excl}}(v, V') = N(\{v\} \setminus N(V'))$. Starting from a vertex v, a vertex u is added to the extended vertex set V_{ext} if its label is greater than that of v and u should not be a neighbor to any other vertex in V_{sub} other than the newly added vertex, V_{sub} holding the current included vertices in the subgraph under consideration. Algorithm 12.2 displays the code of this algorithm which calls *ExtSub* function for each vertex. This function performs recursive calls adding neighboring vertices to extend V_{sub} until the required motif size k is reached.

When a new node is selected for expansion, it is removed from the possible extensions, and its exclusive neighbors are added to the new possible extensions. This way, the algorithm ensures that each subgraph will be enumerated exactly once since the nonexclusive nodes will be considered in the execution of another recursion. Figure 12.7 displays the execution of this algorithm in a sample graph of six nodes labeled $1, \ldots, 5$, and the output consists of four triads and a triangle as shown. The FANMOD tool incorporates ESU algorithm uses the *nauty* algorithm to calculate isomorphisms of graphs [33].

Algorithm 12.2 *ESU*

1: **Input** : $G(V, E)$, **int** $1 \leq k \leq n$
2: **Output** : All k-size subgraphs of G
3: **for all** $v \in V$ **do**
4:      $V_{ext} \leftarrow \{u \in N(v) : label(u) > label(v)\}$
5:      $ExtSub(\{v\}, V_{ext}, v)$
6: **end for**
7: **return**
8: **procedure** EXTSUB(V_{sub}, V_{ext}, v)
9:      **if** $|V_{sub}| = k$ **then output** $G[V_{sub}]$
10:          **return**
11:      **end if**
12:      **while** $V_{ext} \neq \emptyset$ **do**
13:          $V_{ext} \leftarrow V_{ext} \setminus \{$a random vertex $w \in V_{ext}\}$
14:          $V'_{ext} \leftarrow V_{ext} \cup \{u \in N_{excl}(w, V_{sub}) : label(u) > label(v)\}$
15:          $ExtSub((V_{sub} \cup \{w\}, V'_{ext}, v))$
16:      **end while**
17:      **return**
18: **end procedure**

Randomized ESU Algorithm

As complete traversal of the ESU-tree is time-expensive, the ESU algorithm can be modified to traverse parts of the ESU-tree such that each leaf is reached with equal probability. In order to accomplish this, a probability $0 < p_d \leq 1$ is introduced for each depth $1 \leq d \leq 1$ in the ESU-tree. Based on this probability p_d, a decision is made for each child vertex v at depth d to traverse the subtree rooted at v. Implementation is provided by replacing line 5 with "With probability p_1 call $ExtSub(..)$" and line 15 with "With probability p_d, call $ExtSub(..)$" where $d = |V_{sub} + 1|$. This new algorithm is called RAND-ESU.

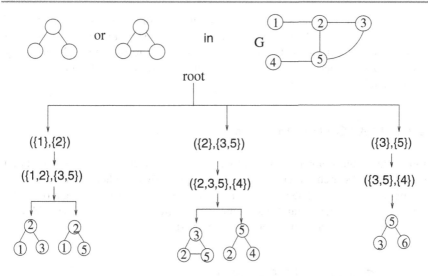

Fig. 12.7 Subgraphs found in a graph using the ESU algorithm

The author has implemented RAND-ESU and *mfinder* (ESA) using sampling in C++ with instances COLI (transcriptional network of *E. Coli* [30]), YEAST (transcriptional network of *S. Cereviciae* [18]), ELEGANS (neuronal network of *Caenorhabditis Elegans* [12]), and ythan (food web of the YTHAN estuary [36]). It was found that RAND-ESU is much faster than sampling-based *mfinder* reaching several orders of magnitude of better performance for graphs of size $k \geq 5$. The two versions of RAND-ESU called *fine-grain* and *coarse-grain* based on the value of the p_d were used in tests. The fine-grain RAND-ESU provided consistent sampling quality, whereas ESA had a very good sampling quality in some ranges. Another advantage of RAND-ESU is reported as being unbiased and ability to estimate the total number of subgraphs.

12.3.1.3   *Kavosh* **Algorithm**

Kavosh is another network motif search tool which can input directed and undirected graphs [10]. It has four main steps: enumeration, classification, random graph generation, and motif identification. In the first step, an unmarked node u is selected and a tree T_u with at most k depth is formed around this node to search for a motif of size k. The first level of T_u is u; the second level consists of neighbors $N(u)$ of u, and third level has nodes that are neighbors of nodes in $N(u)$. The higher level nodes in the tree are labeled similarly up to depth k. The revolving door algorithm [13] is used to discover all k-size motifs around the node u using T_u. Node u is then removed from the graph and the process is repeated for all other nodes. Classification is performed using the *nauty* algorithm. The random graphs are generated using the Monte-Carlo method, and Z-score and P-value parameters are used to determine the significance of motif frequencies. The authors tested the method in the metabolic pathway of the bacteria *E. coli* and the transcription network of *yeast S. cerevisiae*, a real social network and an electronic network, and compared their results to FANMOD. They concluded *Kavosh* has less memory usage and is faster than FANMOD for motif sizes 6, 7, and 8.

12.3.2 Motif Centric Algorithms

Motif centric algorithms input typically a single but sometimes a set of motifs of a given size k and search for these motifs in the input network rather than enumerating all subgraphs of size k. They are commonly used in situations where the motif to be searched is known beforehand. Two representative methods of this class are MODA and Grochow–Kellis algorithms as described next.

12.3.2.1 Grochow–Kellis Algorithm

The motif centric algorithm proposed by Grochow and Kellis (GK algorithm) proposed aims to overcome the limitations imposed by the network centric algorithms [8]. They propose the following features for improvement:

- Searching for a single subgraph is clearly less time-consuming than enumerating all subgraphs. In order to find all motifs of a given size, they additionally employ *geng* and *directd* tools [16]. In this sense, this algorithm is closely related to the subgraph isomorphism problem.
- Instead of finding all subgraphs of a given size and then checking for their isomorphism to the input query graph, the authors propose to use *mapping* to find matches of the query graph in the network.
- A method to find the subgraphs only once which results in writing the subgraph to disk as soon as it is discovered which provides efficient usage of memory.
- The isomorphism test is improved by considering the degrees of nodes and the degrees of each node's neighbors.
- The subgraphs are hashed based on their degree sequences which provides significant improvements in the required number of isomorphism tests.

The last two items are also applicable to network centric methods. The main idea of the algorithm is progressively mapping the desired query on the target graph. Grochow and Kellis proposed a novel symmetry-breaking technique to avoid repeated isomorphism tests. They also consider degrees of nodes and the degrees of their neighbors to improve isomorphism tests using subgraph hashing where subgraphs of a given size are hashed based on their degree sequences. Algorithm 12.3 shows the pseudocode for this algorithm as adapted from [8], where finding all of the instances of the query graph M in the target graph G is described.

The first step of the algorithm is calculating equivalence classes M_E of the query graph to map only one representative of each class. Then symmetry-breaking conditions C are calculated to avoid counting the same subgraph for multiple times. After this process, the algorithm tries to match every vertex g of the graph G into one of the vertices of M. If the algorithm finds a partial map f from M to G, the *IsoExt* function is called to find all isomorphic extensions of f satisfying the conditions C. This function recursively finds the most constrained neighbor of g and adds it into the mapping if it does not violate the symmetry conditions, and it has appropriate degree sequence until the whole query graph is mapped. GK algorithm can be used with *gtools* package [15] to generate all possible subgraphs of a given size and count the frequency of each subgraph.

Algorithm 12.3 *Grochow_Alg*

1: **procedure** FINDSUBINST($H(V_1, E_1), G(V_2, E_2)$)
2:    **Input**: M : query graph, G : target graph
3:    **Output**: all instances of M in G
4:    **find** Auth(M)   ▷ find automorphisms of H
5:    $M_E \leftarrow$ equivalence representatives of M
6:    $C \leftarrow$ symmetry-breaking conditions for M using M_E and Auth(M)
7:    **order** nodes of G by first by increasing degrees and then increasing neighbor degrees
8:    **for all** $u \in V_2$ **do**
9:       **for all** $v \in M[M_E]$ such that u can support v **do**
10:          $f \leftarrow$ partial map where $f(v) = u$
11:          $IsoExt(f, M, G[C(m)])$ ▷ find all isomorphic extensions of f until symmetry
12:          **add** the images of these maps to instances
13:       **end for**
14:       $G \leftarrow G - \{u\}$
15:    **end for**
16:    **return** the set of all instances
17: **end procedure**

The authors implemented this algorithm in a PPI network [9] and a transcription network of *S. cerevisiae* [5], and compared its performance to other methods of motif discovery. They showed that GK algorithm provides an exponential time improvement when compared with *mfinder* when subgraphs up to size 7 are counted. Similarly, the memory space was used exponentially less than *mfinder* as the memory requirement of the algorithm is proportional to the size of the query graph, whereas the compared algorithm requires space proportional to the number of subgraphs of a given size. The main advantage of GK algorithm is reported as having a symmetry-breaking method which ensures each subgraph instance is discovered exactly once. Using this method, the authors report of finding significant structures of 15 and 20 nodes in the PPI network and the regulatory network of *S. cerevisiae* and also the re-discovery of the cellular transcription machinery, as a 29-node cluster of 15-node motifs, based solely on the structure of the protein interaction network.

12.3.2.2 MODA

MODA is another motif centric algorithm proposed by Omidi et al. [21]. It uses *pattern growth* method which is the general term referring to extending subgraphs until the desired size is reached. At each step of the algorithm, the frequency of the query graph is searched in the target network. This algorithm makes use of information about previous subgraph query searches. If the current query is a super-graph of a previous query, the previous mapping can be used. A hierarchical structure is needed to implement the subgraph and super-graph relationships between the queries. The expansion tree T_k structure introduced for this purpose has query graphs as nodes and as the depth of T_k increases, the query graphs become more complete. The expansion tree T_k for size k has the following properties [21].

1. Level of the root is assumed to be zero.
2. Each node of the tree except the root contains a graph of size k.
3. Each node at ith level includes a graph of size k with $k - 2 + i$ edges.
4. Number of nodes at the first level is equal to the number of nonisomorphic trees of size k.
5. Graphs at nodes are nonisomorphic to all other graphs in T_k.
6. Each node except the root is a subgraph of its child.
7. There is only one node at level $(k^2 - 3k + 4)/2$ which contains the complete graph K_k.
8. Longest path from the root node to a leaf contains $(k^2 - 3k + 4)/2$ edges.

Figure 12.8 displays the expansion tree T_4 for 4-node graphs. An edge is added at each level to form a child graph. All of the graphs at each level are nonisomorphic to prevent redundancy. The depth of the tree T_k is the depth of the node with the complete graph of k nodes (K_k).

The subgraph frequency calculation uses the expansion tree T_k in a bottom-up direction and employs two methods called *mapping* and *enumerating*. A BFS traversal is performed on T_k to fetch the graphs stored at is nodes. First, the trees at first level of T_k are fetched, and all mappings from these trees to the network are formed by the mapping module which are saved for future reference. In the second step, the query graphs in the second level are fetched and processed by an enumerating module to compute the appearance numbers of these graphs. This process continues until the complete graph K_k is reached. The subgraph frequencies computed using *mapping* and *enumeration* modules are checked against a threshold Δ and any exceeding values are added to the *frequent subgraph list*. The pseudocode of the MODA subgraph frequency finding algorithm is shown in Algorithm 12.4 as adapted from [21].

Algorithm 12.4 *MODA*

1: **Input** : $G(V, E)$, **int** $1 \leq k \leq n$, Δ : threshold value
2: **Output** : L : List of frequent k-size subgraphs of G
3: **repeat**
4:     $G'(V', E') \leftarrow Get\_Next\_BFS(T_k)$ ▷
5:     **if** $|E'| = k - 1$ **then**
6:         $Mapping Module(G', G)$
7:     **else**
8:         $Enumerating Module(G', G, T_k)$
9:     **end if**
10:    save F_2
11:    **if** $|F_G| > \Delta$ **then**
12:        $L \leftarrow L \cup G'$ ▷ add subgraph to the list
13:    **end if**
14: **until** $|E'| = (k - 1)/2$
15: **procedure** MAPPINGMODULE(G, G')
16:    **for** $i=1$ to m **do**
17:        **select** $v \in G$ with a probability proportional to $deg(v)$
18:        **for all** $u \in G'$ **do**
19:           **if** $k_v \geq k_u$ **then**
20:             $f(u) \leftarrow v$
21:             **find** all isomorphic mappings
22:             **add** mappings to F_G
23:           **end if**
24:        **end for**
25:        $G \leftarrow G - \{v\}$
26:    **end for**
27:    **return** and **save** F_G
28: **end procedure**
29: **procedure** ENUMERATINGMODULE(G', G, T_k)
30:    $F_G \leftarrow \emptyset$
31:    $H \leftarrow Get\_Parent(G', T_k)$
32:    **get** F_H from memory
33:    $(u, v) \leftarrow E(G') - E(H)$
34:    **for all** $f \in F_H$ **do**
35:        **if** $f(u), f(v) \in G$ **and** $< f(u), f(v) >$ violates the corresponding conditions **then**
36:           **add** f to $F_{G'}$
37:        **end if**
38:    **end for**
39:    **return** F_G
40: **end procedure**

Get_Next_BFS routine is used to implement a BFS traversal of the expansion tree T_k. If the node $v \in T_k$ under consideration by this procedure is a tree, a call to *Mapping Module* is made. Otherwise, *Enumerating Module* procedure is activated which in turn fetches the graphs of the parent nodes of v from memory and checks to find if the new edge of T_k exists in the target graph

Fig. 12.8 The expansion
tree T_4 for 4-node graphs
(adapted from [21])

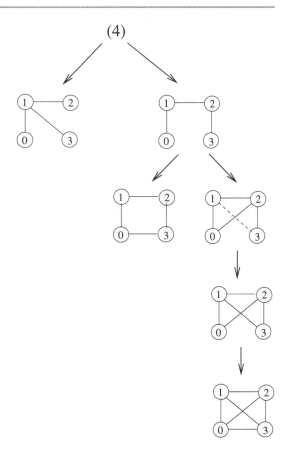

The value of m in the mapping module is a predefined value to show the number of samples of the network. All possible mappings are computed between lines 18 and 24. All isomorphic mappings are found in line 21 by the branch and bound method as in GK algorithm. The enumeration module starts by finding the parent node of G and loading its parent set F_H found in the previous step. The new edge added to G by its parent is determined in line 33 and the mapping to support the selected criteria is decided in lines 34–38.

MODA with Sampling

Since the mapping module of MODA uses a lot of time, the sampling version of MODA is proposed for better efficiency [21]. In this version, the time-consuming mapping module is modified so as to select only a sample of nodes from the network. Sampling should provide reliable results, that is, the number of mappings for the input query must not vary significantly by the different runs of the algorithm. The number of subgraphs that a node is included is proportional to its degree as shown by experiments. Therefore, selecting a node with a probability related to its degree for sampling will provide the required reliability. The sampling MODA based on this concept is shown in Algorithm 12.5 [23].

Algorithm 12.5 $MODA\_Sample$

1: **Input** : network graph $G_1(V_1, E_1)$, query graph $G_2(V_2, E_2)$
2: **Output** : approximate frequency and mapping of G'
3: **procedure** MAPSAMPLE(G_2)
4:     **for** $i = 1$ to *number of samples* **do**
5:         **choose** $v \in V$ with probability $deg(u)$
6:         **for all** $u \in V_2$ **do**
7:             **if** $deg(v) \geq deg(u)$ **then**
8:                 $Grochow(G_2)$ with $f(u) = v$
9:             **end if**
10:            $V \leftarrow V \setminus \{v\}$
11:        **end for**
12:    **end for**
13: **end procedure**

MODA requires substantial memory space as all k-subgraphs are stored in memory. Sampling is more efficient at the cost of decreased accuracy. In [21], both sampling and nonsampling versions of the algorithm were experimented and compared with GK algorithm, *mfinder*, FANMOD and FPF [28]. The tests were carried with *E. coli* transcription network data. Comparing MODA with sampling-based MODA shows sampling MODA is faster and can detect motifs up to size of 9. When all algorithms were tested in terms of time taken for subgraph sizes, MODA turned out to be second fastest after FANMOD, and the size of the subgraph reached was 9 with MODA whereas all others reached the subgraph size 8 as maximum.

12.4   Distributed Motif Discovery

Parallel and distributed algorithms for motif discovery are scarce. Before we review few reported studies of distributed motif search, we will describe general ways of parallelizing the three steps of motif search.

12.4.1   A General Framework

The main steps of motif search and possible parallel processing in these steps as follows.

1. *Subgraph enumeration*: For network centric motif search, we can partition the
   network graph and distribute it evenly to k processes $p_o, \ldots, p_{k-1}$. The root
   processes p_0 can perform the partition and for edges incident across the partitions,
   ghost vertices concept in which border vertices are duplicated in each partition
   can be used. Each p_i has information about these vertices and using a simple
   labeling scheme, the processing of the boundary edges can be done, for example,
   by the lowest identifier processes that has the ghost vertex.

   Alternatively, we can distribute all of the network graph to all processes and the
   root process sends a set of specific motifs of a given size to each process p_i which
   performs a motif centric search for the specific motifs it receives. It then reports
   the results to the root which can combine all of the outputs from all processes.

   For motif centric approach, we can simply partition and distribute the graph
   and have each process p_i search for the specific motif in its partition. Similar
   strategies can be employed for sampling-based motif searches. For example, the
   whole graph can be distributed to all processes, and each process can perform its
   own sampling and searching motifs.
2. *Finding isomorphic graphs*: This step can be performed in parallel as a contin-
   uation of the first step. For the case of the partitioned graph, each process can
   also find the isomorphic classes of the motifs it has discovered using an algorithm
   such as *nauty*. If the graph is sent as a whole to each process, the same isomorphic
   classification can be done on the motifs found.
3. *Evaluating statistical significance*: In this case, we need first to generate m number
   of randomly graphs $R_1, \ldots, R_m$ using an algorithm such as Markov chain. We
   need then to find subgraphs as in step 1 and find the frequencies in the original
   graph and in these randomly graphs. In the simplest case, we can have each of
   the k processes generate m/k random graphs and perform the rest of processing
   sequentially in the random graphs it owns. The results can then be sent to the root
   process which can then combine all of them to produce the final output received
   and report the final statistical significance.

12.4.2 Review of Distributed Motif Searching Algorithms

The reported studies for distributed motif discovery are due to Wang et al., Schatz
et al., and Riberio et al., as described next .

12.4.3 Wang et al.'s Algorithm

Wang et al., presented one of the first studies for exact parallel motif discovery [32].
The main idea of this method is to define neighborhoods of nodes and limit the search
space by searching for motifs in these neighborhoods and perform these in parallel.
The neighborhood of a node v, $Nbr(v)$, consists of nodes v' such that $d(v', v) \leq k-1$
for a given integer k. The first algorithm proposed constructs a BFS tree $T(V', E')$
from a node v at depth $k-1$. The role of each node in $Nbr(v)$ is also defined as

uncle, nephew, or *hidden-uncle* node. The level of a node u in BFS tree is $level(u)$, and $uncle(u)$ is a node v that has the same level as the parent of u with an edge between u and v, that is $(u, v) \in E$, but it is not the parent of u. The nephew of u is the node w which is the child of the uncle node of u, as in the family relationships. The uncles, nephews, and hidden-uncles are identified using the first algorithm.

The second algorithm adds more edges called the *virtual edges* to the existing edges in $Nbr(v)$. If two nodes u and v both have an edge to another node w where $level(u) = level(v) = level(w) + 1$ with $(u, v) \notin E$, a virtual edge is added between u and v. Virtual edges are also added between nodes and their hidden-uncles. The auxiliary neighborhood $Aux\_Nbr(v)$ of a node v is obtained this way. The third algorithm searches all k-connected subgraphs in $Aux\_Nbr(v)$. These three algorithms are used to find motifs in parallel in a number of processes $p_0, \dots, p_{k-1}$ with p_0 as the supervisor and the rest of the processes as the workers. The p_0 process sorts the nodes according to their degrees and broadcasts this list to all workers in the initial step as shown below. It then partitions the graph and sends the partitions to each process each of which runs the three algorithms described for the nodes in their partitions.

1. **Input**: Target graph $G(V, E)$
2.   Process set $P = p_0, \dots, p_{k-1}$
3. **Output**: Discovered motifs set $M = m_1, \dots, m_t$
4. **If** $p_i = p_0$ **then** (I am the supervisor)
5.   **initialize** all $v \in V$ as unmarked and unvisited.
6. $V' \leftarrow$ sorted nodes of G w.r.t. their degrees
7.   **broadcast** V' to all processes $p_1, \dots, p_{k-1}$
8.   **partition** nodes of G and send each partition to a process p_i
9.   **collect** all the results from the worker processes to M'
10.   **check** subgraph isomorphism in M' to get M
11. **Else** (I am a worker)
12.   **For** each node u in my partition
13.   Implement Alg. 1 to get $Nbr(u)$
14.   Implement Alg. 2 to $Nbr(u)$ to get $Aux\_Nbr(u)$
15.   Implement Alg. 3 to $Aux\_Nbr(u)$ to get M_i
16.   **EndFor**
17.   send M_i to p_0 (send the found motifs to supervisor)
18. **EndIf**

The results are gathered at the supervisor at the end of processing which then checks isomorphism and decides on the motifs. This method has to be implemented for a number of random graphs to determine the significance of the discovered motifs. The authors applied this procedure for data obtained for interactions between transcription factors and operans in *E. coli* network. The MPI library with C programming interface was used in a 32-node cluster with two processors in each node. The exhaustive search, random sampling, and the parallel algorithm were tested in the *E. coli* transcriptional network, and the running times and precision of all three methods were recorded. The parallel method proposed outperforms the other two in time for motifs up to size 4 but is slower than random sampling method for larger sizes. Precision of the parallel method is shown to be the best among all methods. When both parameters are considered, parallel algorithm is reported to have the best performance.

12.4.4 Schatz et al.'s Algorithm

Schatz et al., provided a distributed version of Grochow–Kellis algorithm by query parallelization and network partitioning [27]. The former refers to distribution of query set element to a set of processes. For example, there are 13 possible motifs for the size 4, and these can be distributed to four processes along with a copy of the target graph G. Each process p_i can then search for the motifs it is assigned in G. They implemented two scheduling policies as *naive* and *first-fit*. All possible subgraphs are computed and then distributed evenly to processes in the naive policy which may result in unbalanced nodes. The first-fit method is dynamic where the supervisor distributes the queries to current idle processes.

The network is partitioned into overlapping graphs in network parallelization and each process p_i is responsible for discovering motifs in its partition. They implemented a variation of hierarchical clustering where a partition centered at a node v includes all nodes that are at most r hops from v. Therefore, each partition contains every query of radius r. The partitions that are subsets of other partitions are removed resulting in a set of nonredundant and overlapping partitions. These partitions can then be distributed to processes which search for the queries in their partitions. This method was implemented in a 64 node cluster each with a dual core processor for detecting motifs in the yeast PPI network. In query parallelization, it was observed that the first-fit performed better than the naive approach and network partitioning resulted in linear speedups in many cases.

12.4.5 Riberio et al.'s Algorithms

Riberio et al., provided three methods for distributed motif discovery. The first algorithm makes use of a special data structure called *g-trie*, the second one is an attempt to parallelize the ESU algorithm we have reviewed, and the third one is a method to parallelize all three steps of the motif discovery process using the ESU algorithm as an example. We will first describe the *g-trie* structure with a sequential algorithm based on this structure and then review the distributed algorithms.

12.4.5.1 G-Tries Algorithm
The argument by Riberio is that the network centric exact subgraph enumeration methods spend a lot of time to find all subgraphs in the target network, and the random networks most of which may not be useful. Also, the sampling-based algorithms

Fig. 12.9 A prefix trie for
four words

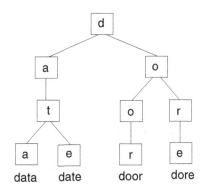

search the same graph for different queries which may be avoided. The idea is to
design an algorithm that is motif centric but does not use a single query motif but is
specialized in finding a set of input motifs. Similar to the discovery of DNA/RNA
and protein sequence motifs we have reviewed in sequences in Chap. 8, the notion of
searching repeated words in a sequence can be used here by searching for repeated
subgraph patterns. In a prefix trie, a path from the root to a leaf corresponds to a
prefix of the input word as shown in Fig. 12.9.

A prefix trie can be searched in linear time for the dictionary operation of finding
a word. It is also efficient in storing data as common prefixes are stored only once
rather than once for each word. The authors have applied these concepts for graphs
by introducing the Graph reTRIEval (*G-trie*) data structure for motif search which
is basically a prefix trie which has graphs as nodes instead of letters. Two types of
data are stored at each node of a *g-trie*: information about a single-graph vertex and
its incident edges to ancestor nodes, and a boolean variable to show whether it is
the last vertex of a graph. A *g-trie* is formed by iteratively inserting subgraphs to
it. Figure 12.10 shows the *g-trie* for subgraphs of up to size 4. Each path in a *g-trie*
corresponds to a unique subgraph.

G-Tries for Motif Discovery

In the first step of motif discovery, the *g-trie* data structure is built to search the set of
subgraphs in the target network G. The next step involves finding the *g-tries* graphs in
G and the random networks, and finally, statistical significance of the results obtained
should be determined. While searching for subgraphs in G, the isomorphism tests
are carried concurrently so that additional tests are not needed. Two approaches are
proposed in [23] to use *g-tries* for motif discovery:

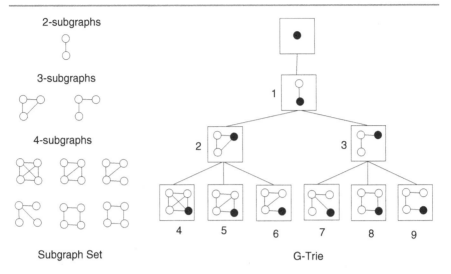

Fig. 12.10 A G-Trie for subgraphs of up to size 4 (adapted from [23])

- *G-trie use only*: All possible subgraphs of a given size are generated, for example by the *gtools* package [16],and then these are inserted in a *g-trie* structure. Then, *g-trie* matching algorithm is applied to match nodes of the *g-trie* structure to the target network *G*. For the discovered subgraphs, a new *g-trie* is formed and *g-trie* matching is implemented in the random networks with this new g-trie.
- *Hybrid approach*: The *g-trie* structure is formed only for the discovered subgraphs in *G* and *g-trie* matching is applied to random networks with this structure. Enumerating subgraphs in *G* can be performed by another network-centric algorithm such as ESU.

An important issue about subgraph matching is breaking subgraph symmetries. Ribeiro et al., proposed symmetry-breaking rules similar to ones proposed by Grochow–Kellis. They also provided their matching algorithm with symmetry-breaking version.

Parallel G-tries

Riberio et al., provided a parallel implementation of the g-tries-based motif search [23,26]. In both approaches of using g-tries for motif discovery, a subgraph census is achieved. The whole process of motif discovery using *g-tries* can be considered as a large space tree. Two synchronizations in this process are when the census on the target network is computed and when census is done for all networks [23]. The first synchronization is needed as only *g-tries* for the discovered subgraphs will be used. Statistical computations require completion of all frequencies which necessitates the second synchronization step. The *g-tries* discovery applies a recursive matching algorithm and different calls to this procedure can be performed independently, therefore parallelism can be achieved at this level. Each call is termed a *work unit* to be distributed to processes. A receiver-initiated dynamic load balancing strategy is employed to distribute the work units evenly to processes.

Parallel ESU

Riberio et al., selected ESU for parallelization as it is an efficient algorithm for sub-graph enumeration [26]. The procedure $ExtSub$ in this algorithm (Algorithm 12.2) recursively extends the subgraph and the main idea of parallelization is to distribute this procedure call to a set of processors. Observing each call to $ExtSub$ is independent, it is called a *primary* work unit to be distributed to processors. A load balancing strategy is needed as the amount of work by these work units vary greatly. After experimenting with several methods, the authors chose the supervisor-worker model where the workers ask for work units from the supervisor which sends them these until all work units are finished. The supervisor merges all of the results and computes the frequencies of the subgraphs at the completion of work units. The ordering of the work units on the supervisor's list of works plays an important role on the overall performance of the system. The selected strategy is called *largest processing time first* (LPTF) which was tested to be effective. There is still the problem due to the power-law distribution in biological networks which is displayed by few very high-degree nodes and many low-degree ones. Executing the recursive calls to find subgraphs around very large degree nodes requires high computation times, and this source of imbalance is to be addressed. The authors devised a strategy to prevent this situation by giving higher labels to these hubs than ordinary nodes with lower degrees. This results in fewer neighbors with higher degree than the originating node. The load balancing scheme incorporating the described features along with a hierarchical data aggregation procedure as an important component of the method proposed; it is called *adaptive parallel enumeration* (APE).

Ribeiro et al., implemented the methods presented in a 192-core cluster distributed memory system with message passing using OpenMPI library. Various biological networks such as neural, gene, metabolic, and protein interaction networks were tested. Comparison of sequential *g-tries* with its parallel version showed the parallel version is scalable up to 128 processors. Tests for parallel ESU showed linear speedups were obtained for gene and metabolic networks but neural and PPI network tests did not result in linear performance. The authors conclude APE consumes a lot of time during the final aggregation of results.

12.4.5.2 Parallel Network Motif Discovery

Wang's algorithm uses graph partitioning [32], Schatz's algorithm basically performs query distribution [27] and previous works of Riberio employ parallelization of recursive calls in the tree structure [25,26]. In a more recent study, Riberio et al., provided an extended and more generalized form of parallel motif discovery, again using the ESU algorithm [22] as in their previous work. Their earlier work involved parallelizing a single subgraph census only, this time, they parallelized all of the steps of the motif discovery process and introduced a distributed and dynamic receiver-initiated load balancing policy. They argue the existence of the following parallelization opportunities, similar to what we have discussed as general guidelines for parallelism.

- *Census parallelization*: The network is evenly partitioned to overlapping subgraphs, and each partition is distributed to a process which searches for motifs in its partition. The recursive tree search procedures can be distributed to processes and the queries can be divided among the processes.
- *Parallelization of random network processing*: Generation of random networks and search for motifs in these networks can be distributed to processes.
- *Statistical Computations*: The determination of the statistical significance of the found motifs can be done in parallel by the processes.

A single call to $ExtSub$ procedure is called an ESU work unit (EWU) which is the call to $ExtSub$ with added network identifier. The main idea of the parallelization is dividing EWUs among a number of processors. The parallel algorithm consists of three main steps:

1. *Pre-processing phase*: All initialization including the initial work queue for each process is performed.
2. *Work phase*: Two algorithms are used to analyze the network and find frequencies of subgraphs.
3. *Aggregation phase*: The subgraph frequencies found by processes are gathered and their significance is computed.

In the work phase, two policies called *master-worker strategy* and *distributed strategy* are designed. In the former, the master performs load balancing and other processes are involved in the work phase. The master waits for requests from workers and sends them any available work. In the distributed scheme, each worker has a work queue and if work units in this queue are finished, it asks for work from other workers.

Tests were carried out in the 192 core system using the same software environment as in [26]. The foodweb, social, neural, metabolic, protein, and power networks were used to experiment the method designed. The authors reported speedups of their parallel ESU method using up to 128 cores. First, they showed hierarchical method performs better than naive method in aggregation phase. Second, they compared master-worker and distributed control strategies for parallelization of whole motif discovery process. In both methods, they achieve nearly linear speedups up to 128 cores but distributed control performs better because all cores are involved in computation without wasting additional power. Finally, they compared the speedups of distributed control using 128 cores when number of random networks varies from 0 to 1000. The algorithm scales well, but speedup decreases as the number of random networks increases, as expected.

12.5 Chapter Notes

We started this chapter by first describing the network motifs which are the building blocks of biological networks. Motifs are assumed to have biological significance by carrying out special functions. Discovery of network motifs is a fundamental research area in bioinformatics as it provides insight into the functioning of the network. Network motifs may also be used to compare biological networks as similar motifs found in two or more networks may mean similar functions are performed in them by the same motifs. The network motif search problem consists of subgraph enumeration, graph isomorphism, and determining statistical significance steps. We need to enumerate all of the subgraphs in exact enumeration, or a part of the network in the sampling-based enumeration. The discovered subgraphs need to be classified into isomorphic classes in the second step which is a problem in NP. Finally, we need to asses the statistical significance of the found subgraphs by generating similar random graph to the original graph and applying the same procedures to these random graphs.

We reviewed some of the sequential algorithms for motif discovery. An early study to discover motifs is *mfinder* algorithm which enumerates all subgraphs. It has limited usage as it is biased, requires large memory space and may find the same motif more than once. The sampling version of *mfinder* selects edges with some probability to enlarge subgraphs [11]. The ESU algorithm was presented by Wernicke which handles some of the problems of *mfinder*; it is unbiased and discovers motifs only once [34]. Kavosh uses levels of a tree to form subgraphs and achieves discovery of motifs of sizes 8 and 9 [10]. NeMoFINDER uses trees to find subgraphs and is designed for PPI networks [4]. Motif Analysis and Visualization Tool (MAVisto) is a network centric algorithm that uses a pattern growth method [29]. Grochow–Kellis [8] and MODA [21] algorithms are motif centric, MODA also has a sampling version. Grochow–Kellis algorithm uses mapping with symmetry breaking to discover motifs and MODA also uses mapping with a pattern growth tree. Exact enumeration algorithms get slower as the size of the motif grows. FANMOD that uses RAND-ESU and *Kavosh* are more efficient than others and *Kavosh* can find motifs of larger sizes than FANMOD. FANMOD is faster than MODA; however, MODA can discover larger motifs [37].

We then described general approaches for parallel motif discovery in distributed memory architectures and then reviewed few sample algorithms reported in literature. Wang's algorithm [32] partitions the target graph for parallel operation and Schatz provided a distributed version of Grochow–Kellis algorithm by mapping the queries to a set of processors [27]. Riberio et al., provided three methods for parallel motif discovery. The first algorithm is based on a novel data structure called *g-trie* [24,25],

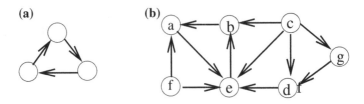

Fig. 12.11 Example graph for Exercise 1

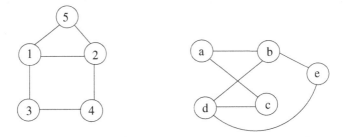

Fig. 12.12 Example graph for Exercise 2

the second algorithm is the parallel implementation of the RAND-ESU algorithm with a load balancing strategy [26], and the third method is a general framework [22]. Although the three steps of motif discovery are promising for parallel/distributed processing, parallel and distributed algorithms are scarce in this area necessitating more studies.

Exercises

1. Find the motif shown in Fig. 12.11a in the graph shown in (b) of the same figure using F_1, F_2 and F_3 frequency concepts.
2. Show that the two graphs shown in Fig. 12.12 are isomorphic by finding a permutation matrix P that transforms one to the other.
3. Find the 3-node network motifs in the graph of Fig. 12.13 using the *mfinder* algorithm.

Fig. 12.13 Example graph
for Exercise 3

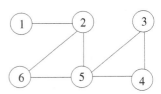

Fig. 12.14 Example graph
for Exercise 4

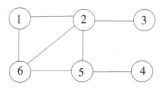

4. Find the 3-node network motifs in the graph of Fig. 12.14 using the *ESU* algorithm.
5. Compare exact census algorithms with approximate census algorithms in terms of their performances and the precision of the results obtained in both cases.

References

1. Albert I, Albert R (2004) Conserved network motifs allow protein-protein interaction prediction. Bioinformatics 20(18):3346–3352
2. Artzy-Randrup Y, Fleishman SJ, Ben-Tal N, Stone L (2004) Comment on network motifs: simple building blocks of complex networks and superfamilies of designed and evolved networks. Science 305(5687):1007
3. Battiti R, Mascia F (2007) An algorithm portfolio for the subgraph isomorphism problem. In: Engineering stochastic local search algorithms. Designing, Implementing and analyzing effective heuristics, Springer, Berlin Heidelberg, pp 106–120
4. Chen J, Hsu M, Lee L, Ng SK (2006) NeMofinder: genome-wide protein-protein interactions with meso-scale network motifs. In: Proceedings of 12th ACM SIGKDD international conference knowledge discovery and data mining (KDD'06), pp 106–115
5. Costanzo MC, Crawford ME, Hirschman JE, Kranz JE, Olsen P, Robertson LS, Skrzypek MS, Braun BR, Hopkins KL, Kondu P, Lengieza C, Lew-Smith JE, Tillberg M, Garrels JI (2001) Ypd(tm), pombepd(tm), and wormpd(tm): model organism volumes of the bioknowledge(tm) library, an integrated resource for protein information. Nucleic Acids Res 29:75–79
6. Erciyes K (2014) Complex networks: an algorithmic perspective. CRC Press, Taylor and Francis, pp 172. ISBN 978-1-4471-5172-2
7. Garey MR, Johnson DS (1979) Computers and intractability: a guide to the theory of NP-completeness. W. H. Freeman
8. Grochow J, Kellis M (2007) Network motif discovery using subgraph enumeration and symmetry-breaking. In: Proceedings of 11th annual international conference research in computational molecular biology (RECOMB'07), pp 92–106
9. Han J-DJ, N. Bertin N, Hao T, Goldberg DS, Berriz GF, Zhang LV, Dupuy D, Walhout AJM, Cusick ME, Roth FP, Vidal M, (2004) Evidence for dynamically organized modularity in the yeast protein-protein interaction network. Nature 430(6995):88–93
10. Kashani ZR, Ahrabian H, Elahi E, Nowzari-Dalini A, Ansari ES, Asadi S, Mohammadi S, Schreiber F, Masoudi-Nejad A (2009) Kavosh: a new algorithm for finding network motifs. BMC Bioinform 10(318)
11. Kashtan N, Itzkovitz S, Milo R, Alon U (2002) Mfinder tool guide. Technical report, Department of Molecular Cell Biology and Computer Science and Applied Mathematics, Weizman Institute of Science

12. Kashtan N, Itzkovitz S, Milo R, Alon U (2004) Efficient sampling algorithm for estimating sub-graph concentrations and detecting network motifs. Bioinformatics 20:1746–1758
13. Kreher D, Stinson D (1998) Combinatorial algorithms: generation, enumeration and search. CRC Press
14. Lee TI et al (2002) Transcriptional regulatory networks in Saccharomyces cerevisiae. Science 298(5594):799–804
15. McKay BD (1981) Practical graph isomorphism. In: 10th Manitoba conference on numerical mathematics and computing, Congressus Numerantium, vol 30, pp 45–87
16. McKay BD (1998) Isomorph-free exhaustive generation. J Algorithms 26:306–324
17. Mfinder. http://www.weizmann.ac.il/mcb/UriAlon/index.html
18. Milo R, Shen-Orr S, Itzkovitz S, Kashtan N, Chklovskii D, Alon U (2002) Network motifs: simple building blocks of complex networks. Science 298(5594):824–827
19. Milo R, Kashtan N, Levitt R, Alon U (2004) Response to comment on network motifs: simple building blocks of complex networks and superfamilies of designed and evolved networks. Science 305(5687):1007
20. Nauty User's Guide, Computer Science Dept. Australian National University
21. Omidi S, Schreiber F, Masoudi-Nejad A (2009) MODA: an efficient algorithm for network motif discovery in biological networks. Genes Genet Syst 84:385–395
22. Ribeiro P, Silva F, Lopes L (2012) Parallel discovery of network motifs. J Parallel Distrib Comput 72(2):144–154
23. Ribeiro P, Efficient and scalable algorithms for network motifs discovery. Ph.D. Thesis Doctoral Programme in Computer Science. Faculty of Science of the University of Porto
24. Ribeiro P, Silva F (2010) Efficient subgraph frequency estimation with g-tries. Algorithms Bioinform 238–249
25. Ribeiro P, Silva F (2010) G-tries: an efficient data structure for discovering network motifs. In: Proceedings of 2010 ACM symposium on applied computing, pp 1559–1566
26. Ribeiro P, Silva F, Lopes L (2010) A parallel algorithm for counting subgraphs in complex networks. In: 3rd international conference on biomedical engineering systems and technologies, Springer, pp 380–393
27. Schatz M, Cooper-Balis E, Bazinet A (2008) Parallel network motif finding. Techinical report, University of Maryland Insitute for Advanced Computer Studies,
28. Schreiber F, Schwbbermeyer H (2005) Frequency concepts and pattern detection for the analysis of motifs in networks. In: Transactions on Computational Systems Biology LNBI 3737, Springer, Berlin Heidelberg, pp 89–104
29. Schreiber F, Schwbbermeyer H (2005) MAVisto: a tool for the exploration of network motifs. Bioinformatics 21:3572–3574
30. Shen-Orr SS, Milo R, Mangan S, Alon U (2020) Network motifs in the transcriptional regulation network of Escherichia Coli. Nat Gen 31(1):64–68
31. Ullmann JR (1976) An algorithm for subgraph isomorphism. J ACM 23(1):31–42
32. Wang T, Touchman JW, Zhang W, Suh EB, Xue G (2005) A parallel algorithm for extracting transcription regulatory network motifs. In Proceedings of the IEEE international symposium on bioinformatics and bioengineering, IEEE Computer Society Press, Los Alamitos, CA, USA, pp 193–200
33. Wernicke S (2005) A faster algorithm for detecting network motifs. In: Proceedings of 5th WABI-05, vol 3692, Springer, pp 165–177
34. Wernicke S (2006) Efficient detection of network motifs. IEEE/ACM Trans Comput Biol Bioinform 3(4):347–359
35. Wernicke S, Rasche F (2006) FANMOD: a tool for fast network motif detection. Bioinformatics 22(9):1152–1153
36. Williams RJ, Martinez ND (2000) Simple rules yield complex food webs. Nature 404:180–183
37. Wong E, Baur B, Quader S, Huang C-H (2011) Biological network motif detection: principles and practice. Brief Bioinform. doi:10.1093/bib/bbr033

Network Alignment

13

13.1 Introduction

Network alignment is the process of finding structural similarities between two or more networks by comparing them. Informally, we attempt to fit one network into another as much as possible during an alignment. Alignment of networks between different species and developing models [37] provides us with the similarity information that can be used to deduce phylogenetic relations among them which will help to understand the evolution process. Additionally, we can detect functionally preserved modules (subnetworks) in different species and can elaborate on their functions.

Analogous to sequence alignment which searches similar sequences between two or more DNA/RNA or protein sequences, network alignment aims to detect subgraphs conserved in two or more biological networks. Biological networks such as PPI networks are noisy, making it difficult to take exact measurements, moreover, the edges in a PPI network are added and deleted due to evolutionary process making it more difficult to find exactly conserved modules. The graph alignment problem is also closely related to the graph [12] and subgraph isomorphism problem as the network motif search and is NP-hard [13] which necessitates the use of heuristics. Network alignment can be performed pairwise or multiple, and local or global. In pairwise alignment, two networks are aligned whereas multiple alignment considers more than two networks. Multiple alignment in general is more difficult than pairwise alignment due to increased computation. Local alignment algorithms compare subnetworks of the given networks whereas global alignment is used to compare two or more networks as a whole, resulting in one-to-one node similarity scores. Global alignment is commonly used for comparison of similar species, however, comparing the organisms of different species usually involves local alignment methods. *Node similarity* refers to comparing nodes of the network based on their internal structure such as the amino acid sequences of the protein nodes in PPI networks. *Topological similarity* is evaluated by using network structure only. It is a common practice to use both of these metrics with different weighting parameters to align networks as we will see.

© Springer International Publishing Switzerland 2015
K. Erciyes, *Distributed and Sequential Algorithms for Bioinformatics*,
Computational Biology 23, DOI 10.1007/978-3-319-24966-7_13

Local network alignment is closely related to the subgraph isomorphism problem where a small graph is searched within a larger graph. However, it is more general than subgraph isomorphism since we search for similar subgraphs rather than exact ones as in the former. Subgraph isomorphism is NP-hard, and hence, the related alignment problem is also intractable necessitating the use of approximation algorithms and more frequently heuristics to solve this computationally difficult problem. Once we have a heuristic algorithm that provides results in favorable time, we need to also evaluate the goodness of the alignment achieved.

In this chapter, we will first define the alignment problem formally and then show its relation to other problems such as subgraph isomorphism and bipartite matching in graphs.We will then take a closer look at aligning PPI networks as these are the target biological networks of most of the studies in practice. We will then review sequential alignment algorithms by describing the fundamental ones briefly and show sample distributed network alignment algorithms.

13.2 Problem Statement

Given a set of graphs $\mathcal{G} = G_1, \ldots, G_k$ representing k biological networks, our aim in network alignment is to compare these networks as a whole and find conserved subnetworks between these networks. If these networks represent different species, there is a correlation between the conservation of evolutionary structures and their functional significance. In other words, if we can discover that two or more species have similar subnetworks, we can assume that these subnetworks have some important functionality. We can further infer phylogenetic relationships between these species. The data obtained from the fundamental application of PPI networks have high noise and the investigation of similar networks rather than exact networks is more meaningful. From another angle, we may be interested to know how similar two or more networks in total are as in global alignment.

An *alignment graph* first introduced by Ogata et al. [29] and then Kelley et al. [16], has nodes representing k similar proteins for k species as a single node and edges of this graph show the conserved interactions between the PPIs of the species. Construction of such a graph between species shows us which modules or subnetworks are conserved across the species and is basically a local alignment method. It has been implemented in various studies to find conserved pathways [16] and complexes [19]. Figure 13.1 displays a sample alignment graph of the PPI networks of three species.

13.2.1 Relation to Graph Isomorphism

Given two graphs $G_1(V_1, E_1)$ and $G_2(V_2, E_2)$, they are isomorphic if there is a one-to-one and onto function $f : V_1 \rightarrow V_2$ such that $(u, v) \in E_1 \Leftrightarrow (f(u), f(v) \in E_2)$. Given two graphs $G_1(V_1, E_1)$ and $G_2(V_2, E_2)$, with G_1 being a smaller graph then

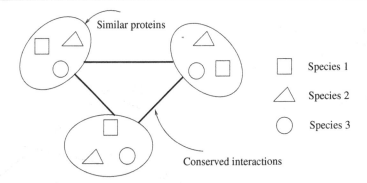

Fig. 13.1 An alignment graph of the PPI networks of 3 species. The proteins inside the clusters have similar structures. Conserved interactions or pathways exist in all of the species

G_2, the *subgraph isomorphism problem* is to search for a graph $G_3 \subset G_2$ with maximal size that is isomorphic to G_1. We have seen in Chap. 13 that graph isomorphism problem is closely related to network motif search. In network alignment, we basically search for similar subgraphs in two or more biological networks and this process is as hard as subgraph isomorphism. Various algorithms for subgraph isomorphism exist [4,23,28,40] and finding the largest common subgraph of two networks for comparison is basically an extension of the subgraph isomorphism method. The exponential time complexity of this problem have resulted in many approximate and heuristic approaches to evaluate the alignment of biological networks.

13.2.2 Relation to Bipartite Graph Matching

A matching in a graph $G(V, E)$ is a subset of its edges that do not share any endpoints and this concept can be used conveniently for network alignment of two graphs. Global network alignment can be performed by mapping vertices of two graphs $G_1(V_1, E_1)$ and $G_2(V_2, E_2)$, or more graphs, and is usually performed in two steps. In the first step, a similarity matrix R is constructed which has an entry r_{ij} showing the similarity score between the vertices i and j, one from each graph. A complete weighted bipartite graph $G_B(V_B, E_B, w)$ where $V_B = V_1 \cup V_2$ and $E_B = \{(v_i, v_j) : \forall v_i \in V_1, v_j \in V_2\}$ is then constructed. We then search for a maximal weighted matching in G_B as our aim is to find a mapping between V_1 and V_2 with the maximum score. A weighted matching in a weighted bipartite graph is shown in Fig. 13.2. We can therefore use any maximal weighted bipartite matching algorithm to discover node pairs with high similarities as we will see in Sect. 13.5.

13.2.3 Evaluation of Alignment Quality

The *topological similarity* between the graphs under consideration is based on the structure of the graphs. The *node similarity* on the other hand, searches the similarity

Fig. 13.2 An weighted
matching in a bipartite graph
formed between the vertices
of graphs G_1 and G_2. Using
this matching, we can align
the vertex pairs
$(a, s), (b, q)(c, p)(d, r)$

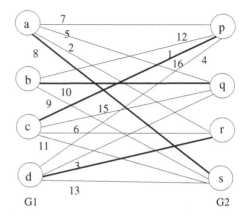

between the structures of the nodes of the graph. In a PPI network for example, a
node structure is identified by the amino acids and their sequence that constitute
that node, however, node similarity may also take topological properties of nodes
into consideration such as their degrees or the similarity of their neighbors. *Edge
correctness* is a parameter to evaluate topological similarity defined as follows [38].

Definition 13.1 (*Edge Correctness*) Given two simple graphs G_1 and G_2, Edge
Correctness (EC) is:

$$EC(G_1, G_2, f) = \frac{|f(E_1) \cap E_2|}{|E_1|}, \tag{13.1}$$

where f is the function that maps edges of the graph G_1 to G_2. Edge correctness
basically shows the percentage of correctly aligned edges and helps to evaluate
the quality of the alignment achieved. Patro and Kingford proposed the induced
conserved structure (ICS) score metric which extends the EC concept [30]. This
parameter is specified as follows:

$$ICS(G_1, G_2, f) = \frac{|f(E_1) \cap E_2|}{|E_{G_2[f(V_1)]}|}, \tag{13.2}$$

where the denominator is the size of the edges induced in the second graph G_2 by
the mapped vertices. The goal is to align a sparse region of G_1 to a sparse region of
G_2, or align their dense regions using ICS score to improve accuracy [9]. As another
measure of the quality of the alignment, genetic ontology (GO) consistency of the
aligned proteins is defined as the sum of the following over all aligned pairs [1].

$$GOC(u, v) = \frac{|GO(u) \cap GO(v)|}{|GO(u) \cup GO(v)|}, \tag{13.3}$$

for an aligned pair of nodes $u \in V_1$ and $v \in V_2$, where GO(u) shows the GO terms
associated with the protein u that are at a distance 5 from the root of the GO hierarchy.

A different alignment metric is the assessment of the size of the largest connected component (LCC) shared by the input graphs. A larger LCC shows greater similarity among the input graphs. The quality of the alignment achieved can be assessed by statistical methods and by comparing the alignment with the alignment of random networks of the same sizes [32].

The sequence alignment methods we have seen in Chap. 7 can be conveniently used to find node similarities. The BLAST algorithm [2] is frequently used for this purpose and Kuchaiev et al. proposed a scoring function between the nodes of the graphs which evaluates the number of small connected graphs called *graphlets* each node is included and therefore uses topological properties of nodes [20,21]. Various algorithms consider both topological and node similarities while assessing similarities between networks and this is especially meaningful in PPI networks where the function of a protein is dependent on both its position in the network and its amino acid sequence.

13.2.4 Network Alignment Methods

The methods for network alignment vary in their choice of the type of alignment as well as in their approach to solve this problem. In general, they can have one of the following attributes:

1. Pairwise or Multiple Alignment
2. Node or Topological similarity information
3. Local or Global Alignment

Two networks are aligned in pairwise alignment and more than two networks are aligned in multiple alignment which is computationally harder than the former. Path-BLAST [16], MaWISh [19], GRAAL [20], H-GRAAL [26] and IsoRank [38] are examples of pairwise alignment algorithms. Extended PathBLAST [36], Extended IsoRank [39], and Graemlin [11] algorithms all provide multiple network alignment. Functional or node similarity information looks at the information other than network topology to evaluate similarity between the nodes whereas only topological information is used to determine topological similarity between the nodes. In practice, a combination of both by different weighting schemes is used to decide on the similarity scores between node pairs.

Local alignment methods attempt to map small subnetworks between two or more species. For example, PPI networks of two species may have similar subnetworks but they may not be evaluated as similar by a global alignment method. The logic here is very similar to what we have discussed while comparing local and global sequence alignment in Chap. 6. Example local alignment algorithms are NetAlign [24] MaWIsh [19], PathBLAST [16], and NetworkBLAST [35]. Global network alignment (GNA) aims to find the best overall alignment from a given set of input graphs by detecting the maximum common subgraph between these networks. In this method, each node of an input graph representing a network is either matched

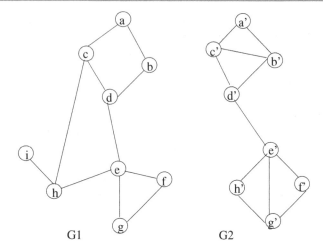

Fig. 13.3 An example GNA alignment between graphs G_1 and G_2 where a node x in G_1 is mapped to a node x' in G_2. Edge (c, h) is deleted and edges (c', b') and (g', h') are added in G_2; gap node i is not mapped

to some other node of another network or not which is usually shown by a gap. The global sequence alignment aims to compare genomic sequences to discover variations between species, and GNA similarly is used to understand the differences and similarities across species [38]. GRAAL [20], H-GRAAl [26], MI-GRAAl [21], C-GRAAL [27] IsoRAnk [38], and Graemlin [11] all provide global alignment. Figure 13.3 displays a GNA between two graphs G_1 and G_2. Aligning two graphs only on node similarity can be performed by the Hungarian algorithm [22] in $O(n^3)$ time.

13.3   Review of Sequential Network Alignment Algorithms

In this section, we provide a brief review of existing algorithms and tools for network alignment including recent ones.

13.3.1   PathBlast

The PathBLAST is a local network alignment and search tool that compares PPI networks across species to identify protein pathways and complexes that have been conserved by evolution [16]. It uses a heuristic algorithm with a scoring function related to the probability of the existence of a path. It searches for high-scoring alignments between pairs of PPI paths by combining the protein sequence similarity information with the network topology to detect paths of high scores. The query path input to PathBLAST consists of a sequence of several proteins and similar pathways in the target network are searched by pairing these proteins with putative orthologs occurring in the same order in the second path. The pathways are combined to form a global alignment graph with either direct links, gaps or mismatches.

The goal of PathBLAST is to identify conserved pathways which can be performed efficiently by dynamic programming for directed acyclic graphs. However, the PPI networks are not directed or acyclic and for this reason, the algorithm employed eliminates cycles by imposing random ordering of the vertices and then performing dynamic programming which is repeated for a number of times. The time complexity of the algorithm is $O(L!n)$ to find conserved paths of length L in a network of order n. Its main disadvantage is being computationally expensive and the search is restricted to specific topology. PathBLAST has a Web page that can be accessed [15] where the user can specify a protein interaction path as a query and select a target PPI network from the database. The output from the PathBLAST is a ranked list of paths that match the query in the target PPI network. It also provides a graphical view of the paths.

13.3.2   IsoRank

IsoRank algorithm is proposed by Singh et al. to find GNA between two PPI networks [38]. It uses both sequence similarity and local connectivity information for alignment, and is based on the PageRank algorithm. It consists of two stages. It first assigns a score between each node pairs, one from each network using eigenvalues, network, and sequence data; and builds the score matrix R. In the second step, highly matching nodes from R are extracted to find the GNA. The general idea in the construction of matrix R is that two nodes i and j are a good match if their neighbors also match well. After computing R, the node matchings which provide a maximum sum of scores is determined by using maximum bipartite graph matching. The bipartite graph G_B represents a network at each side and the weights of edges are the entries of R. The maximum-weight matching of G_B provides the matched nodes and the remaining nodes are the gap nodes. The authors have tested IsoRank for GNA of $S.$ $cerevisiae$ and $D.$ $melanogaster$ PPI networks and found a common subgraph of 1420 edges.

13.3.3   MaWIsh

Maximum-weight-induced subgraph (MaWIsh) proposed by Koyuturk et al. is a pairwise local alignment algorithm for PPI networks [19]. It uses a mathematical model to extend the concepts of match, mismatch, and gap in sequence alignment to that of match, mismatch, and duplication in network alignment. The scoring function to rank the similarities between the graphs accounts for evolutionary events. The similarity score is based on protein sequence similarity and is calculated by BLAST.

MaWIsh attempts to identify conserved multi-protein complexes by searching for clique-like structures. These structures are expected to contain at least one hub node with a high degree. Having found a hub in a clique-like structure, it greedily extends these subgraphs. The time complexity of MaWIsh is $O(n_1 n_2)$, where n_1 and n_2 are the sizes of the two graphs being compared. The main disadvantage of MaWIsh is that it looks for a specific topological property (cliques). MaWIsh was implemented to find network alignments of the PPI networks of yeast, fly and the worm successfully.

13.3.4 GRAAL

Graph Aligner (GRAAL) is a global alignment algorithm that uses topological similarity information only [20]. It uses graphlets which are small, connected, non-isomorphic subgraphs of a given size. Having two graphs $G_1(V_1, E_1)$ and $G_2(V_2, E_2)$ representing two PPI networks, it produces a set of ordered pairs (u, v) with $u \in V_1$ and $v \in V_2$, by matching them using the *graphlet degree signature similarity*. Graphlet degree signatures are computed for each node in each graph by finding the number of graphlets up to size 4 they are included, and then assigning a score that reflects this number. The scores of the nodes in the graphs are then compared to find similarities. GRAAL algorithm first finds a single seed pair of nodes with high graphlet degree signature similarity and afterwards enlarges the alignment radially around the seed using a greedy algorithm.

H-GRAAL [26] is a version of GRAAL that uses the Hungarian algorithm. The MI-GRAAL (Matching Integrative *GRAAL*) algorithm makes use of graphlet degree signature similarity, local clustering coefficient differences, degree differences, eccentricity similarity, and node similarity based on BLAST [21], and C-GRAAL (Common neighbors-based GRAAL) algorithm [27] is based on the idea that the neighbors of the mapped nodes in two graphs to be aligned should have mapped neighbors.

13.3.5 Recent Algorithms

Natalie [10] is a tool for pairwise global network alignment and uses the Lagrangian relaxation method proposed by Klau [17]. The GHOST is a spectral pairwise global network alignment method proposed by Patro and Kingsford [30]. It uses the spectral signature of a node which is related to the normalized Laplacian for subgraphs of various radii centered around that node. SPINAL (scalable protein interaction network alignment) proposed by Aladag and Erten [1] first performs a coarse-grained alignment and obtains similarity scores. The second phase of this algorithm involves a fine-grained alignment using the scores obtained in the first phase.

13.4 Distributed Network Alignment

Let us now review the basic steps typically followed in GNA:

1. A similarity matrix R is constructed with each entry r_{ij} showing the similarity of node i in the first network to node j in the second network.
2. A matching algorithm matches node pairs $\{i, j\}$, one from each network, by maximizing the similarity scores obtained from R.

The similarity matrix R is formed by computing the similarities between node pairs using topological and node similarities. There are various methods to evaluate the similarities between the nodes of the input graphs. A general approach followed in some studies is to consider the similarities of the neighbors when determining similarity where two nodes are considered similar if their neighbors are also similar using some metric. Kollias et al. provided an algorithm called *network similarity decomposition* (NSD) to compute similarities of nodes [18]. The computation of the similarity matrix R differs in local and global alignment methods. Local alignment methods attempt to find similarities between subgraphs of the input networks, therefore, dissimilarity between other subnetworks of the networks can be allowed [16, 19]. Subgraph isomorphism is frequently employed in local alignment methods such as in [41].

In PPI networks, the sequence similarities computed by the BLAST algorithm can be used as one parameter showing similarity. Also topological information about the degrees of nodes, and the degrees of their neighbors can also be combined to compute R. In the second step, we need to find the best matching nodes that give the total highest score. Given two graphs $G_1(V_1, E_1)$ and $G_2(V_2, E_2)$, we can form a complete weighted bipartite graph $G_B(V_1 \cup V_2, E_B, w)$ and the weights on edges $(i, j) \in E_B$ represent the entries r_{ij} of the score matrix R. Our aim is to find the maximal weighted matching in this graph.

Although the computation of matrix R is reported as the dominant cost in studies such as IsoRank, selecting a small set of similarity metrics have resulted in less complexity of this first step [18]. The attention recently is more diverted to the second step of GNA which can then be viewed as the maximal weighted bipartite graph matching (MWBGM) problem also referred to as the *assignment problem*. We will now review the fundamental MWBGM algorithms first sequentially and will look at ways of providing their distributed versions.

13.4.1 A Distributed Greedy Approximation Algorithm Proposal

A sequential greedy algorithm for MWBM problem can be designed which selects the heaviest-weight edge (u, v) from the active edge set and includes it in the matching. It then iteratively removes all edges incident to u and v from the active edge set and continues until there are no edges left as shown in Algorithm 13.1.

Algorithm 13.1 *Seq_MWBM*

1: **Input** : $G(V_1 \cup V_2, E)$ ▷ undirected bipartite graph
2: **Output** : Matching M ▷ matched node pairs set
3: $E_W \leftarrow E$ ▷ initialize the working edge set
4: $M \leftarrow \varnothing$
5: **while** $E_W \neq \varnothing$ **do**
6:    **select** the heaviest edge $(u, v) \in E_B$
7:       $M \leftarrow M \cup \{(u, v)\}$ ▷ include edge (u, v) in matching
8:       $E_W \leftarrow E_W \setminus \{ \text{all } (u, x) \in E_W \text{ and } (v, y) \in E_W \}$ ▷ remove all incident edges to u and v
9: **end while**

This algorithm as shown has $O(m \log m)$ complexity as sorting m elements of a list and then matching. The execution of this algorithm is shown in Fig. 13.4a for a simple weighted bipartite graph with 3×3 nodes. The size of the matching found this way is 18 which is less than the size of the maximal weighted matching of size 21 in (b). It can be shown this algorithm has an approximation ratio of 1/2 [3].

There are various methods to parallelize this greedy algorithm [8,25]. We will now describe a simple parallel version of this algorithm to be executed on the nodes of a distributed memory system, based on parallel sorting. The general idea is to partition the rows of the similarity matrix $R[n, n]$ among k processes and a designated process called the supervisor sends the assigned rows of R to processes such that each process p_i gets n/k rows. Each p_i then sorts the weighted edges in its partition that are incident to nodes in its partition and sends the sorted list to the root process p_0. This process merges the sorted lists to get the globally sorted list and includes edges in matching starting from the heaviest edge by obeying the matching principle, that is, cover an edge (u, v) as matched if neither u nor v are matched. Algorithm 13.2 shows the pseudocode for this algorithm proposed.

Fig. 13.4 a Output of the sequential greedy weighted matching algorithm. **b** The maximal weighted matching of the same graph

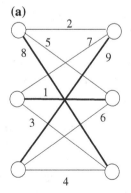
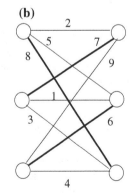

Algorithm 13.2 $GreedyMatch\_Alg$ Supervisor Process

1: **Input** : Similarity matrix R of $G(V_1 \cup V_2, E)$
2: $\qquad P = \{p_0, ..., p_{k-1}\}$
3: **Output** : Matching M
4: **for** $i=2$ to $k-1$ **do**
5: $\qquad$ **send** rows $((i-1)k/n)) + 1$ to ik/n of R to p_i
6: **end for**
7: **sort** my edges into L_0
8: **for** $i=1$ to $k-1$ **do**
9: $\qquad$ **receive** sorted list L_i from processor p_i
10: **end for**
11: **merge** sorted lists $L_i, i = 0, \ldots, k-1$ to one sorted list L_N
12: **while** $L_N \neq \emptyset$ **do**
13: $\qquad$ **remove** edge (u, v) from the front of L_N
14: $\qquad$ **if** u and v obey matching rule **then**
15: $\qquad\qquad M \leftarrow M \cup (u, v)$
16: $\qquad$ **end if**
17: **end while**

We will illustrate this concept by a simple example where our aim is to find GNA between two graphs G_1 with nodes a, b, c, d and G_2 which has p, q, r, s and there are two available processes p_0 and p_1. Let us assume the similarity matrix R, with rows as the G_1 partition and columns as G_2, is formed initially after which we partition the rows equally among the two processes as below:

$$R = \begin{bmatrix} 1 & 8 & 13 & 2 & | \; p_0 \\ 15 & 14 & 6 & 7 & | \\ \hline 3 & 12 & 4 & 16 & | \; p_1 \\ 11 & 10 & 5 & 9 & | \end{bmatrix}$$

Let us further assume p_0 is the supervisor and starts the matching process by sending the third and fourth rows of matrix R to p_1. The two processes both sort the edges that are incident to their vertices in decreasing weights. In this case, p_0 builds the list L_0 with elements $\{(b, p), 15\}, \{(b, q), 14\}, \{(a, r), 13\}, \{(a, q), 8\}, \{(b, s), 7\}, \{(b, r), 6\}, \{(a, s), 2\}, \{(a, p), 1\}$. Similarly, p_1 forms the list L_1 which contains $\{(c, s), 16\}, \{(c, q), 12\}, \{(d, p), 11\}, \{(d, q), 10\}, \{(d, s), 9\}, \{(d, r), 5\}, \{(c, r), 4\}, \{(c, p), 3\}$ and sends this list to p_0. The root p_0 then merges these two sorted lists and forms the new sorted list L_N. In the final step of this algorithm, p_0 starts to include the edges in the matching from the front of L_N as long as the edges obey the matching rule, that is, the endpoints of an edge are not included in a previous matching. In this example, the edges included in the matching in decreasing weight are $\{(c, s), 16\}, \{(b, p), 15\}, \{(a, r), 13\}$ and $\{(d, q), 10\}$ as shown in Fig. 13.5 giving a total weight of 54.

Fig. 13.5 Implementation of
the distributed greedy
bipartite matching algorithm
in a sample graph

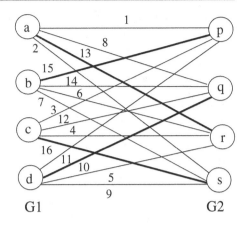

Analysis

Each process p_i receives $n \times n/k$ data items and sorting this data requires $(n^2/k \log n^2/k)$ time. The supervisor p_0 receives the sorted lists each of which has a size of n^2/k and it needs to perform the merging of k lists to find the globally sorted list. Merging two sorted lists of sizes l_1 and l_2 takes $O(l_1 + l_2)$ time. Thus, merging of k lists with n^2/k size each, takes $O(n^2)$ time and $O(n)$ time for the maximal matching in the last step. Therefore, total time for parallel execution has $(n^2/k \log n^2/k + n^2)$ time. The sequential case requires $O(m \log m)$ time to sort edges which is $n^2 \log n^2$ since a full bipartite graph with n nodes on each partition has n^2 edges. The speedup obtained is the ratio of the sequential algorithm to the distributed one which is $(k \log n)/((k/2) + \log n)$ ignoring the $\log k$ term. This indicates a slow increase of time with increased k. For example, for a large full bipartite graph with $2^{16} \times 2^{16}$ nodes and running the distributed algorithm with 4, 8, and 16 processors provides speedups of 3.6, 6.4, and 10.7, respectively, all without considering the interprocess communication costs.

13.4.2 Distributed Hoepman's Algorithm

An improvement to the greedy algorithm for maximal weighted matching which has $O(m \log m)$ time complexity was proposed by Preis to reduce the complexity to $O(m)$ with the same approximation ratio as the greedy algorithm [31]. This algorithm searches locally highest-weighted edges instead of globally ordering and selecting them. Given a graph $G(V, E, w)$ with weighted edges, Preis algorithm finds dominating edges and adds them to the matching. Hoepman provided a distributed version of Preis algorithm where each vertex of G is a computational node and participates in finding the approximated maximal weighted matching of G [14]. Later on, Manne et al. provided a sequential version of the distributed version of Hoepman's algorithm and showed how to parallelize this sequential algorithm [25].

We will first describe the sequential version of Hoepman's algorithm which is shown in Algorithm 13.3 as adapted from [25]. The general idea is to look for the heaviest-weight incident edges locally in the first phase. If an edge (u, v) on a vertex u is the heaviest-weight edge connected to v and, if the vertex v at the other end of this edge has (u, v) as its heaviest-weight incident edge; (u, v) is included in the matching. The matched vertices are added to the set D for processing in the second phase.

Algorithm 13.3 $Seq\_Hoepman$

1: **Input** : $G(V, E, w)$ ▷ undirected weighted graph
2: **Output** : Matching M ▷ matched node pairs set
3: $M \leftarrow \emptyset$ ▷ initialize the matching set and neighbors
4: **for all** $u \in V$ **do**
5:      $C(v) \leftarrow \emptyset$
6:      $neighs_v \leftarrow \Gamma(v)$
7: **end for**
8: **for all** $u \in V$ **do**   ▷ first phase
9:      $v \leftarrow H(u)$ ▷ find the heaviest edge
10:     $C(u) \leftarrow v$
11:     **if** $C(v) = u$ **then**   ▷ if it is heaviest for the opposite node, include in matching
12:         $D \leftarrow D \cup \{u\} \cup \{v\}$
13:         $M \leftarrow M \cup \{(u, v)\}$
14:     **end if**
15: **end for**
16: **while** $D \neq \emptyset$ **do**   ▷ second phase
17:     **select** $u \in D$
18:     $D \leftarrow D \setminus \{u\}$
19:     $v \leftarrow H(u)$
20:     **for all** $(w \in neighs_u \setminus v)$ that is not matched **do**
21:         $neighs_v \leftarrow neighs_v \setminus \{u\}$
22:         $C(w) \leftarrow H(w)$
23:         **if** $C(C(w)) = w$ **then**
24:             $D \leftarrow D \cup \{w\} \cup \{C(w)\}$
25:             $M \leftarrow M \cup \{(w, C(w))\}$
26:         **end if**
27:     **end for**
28: **end while**

We now search for other remaining edges to complete maximal matching by checking vertices in the dominated vertex set D. For each vertex in this set, its unmatched neighbors are found; and for each such neighbor, the basic rule in phase 1 is implemented. This processing continues for all vertices in D. Figure 13.6 shows an implementation of this algorithm in a simple example bipartite graph. Although the algorithm is for a general graph, we can use it for a bipartite graph. As we need to consider each edge once, the complexity of this algorithm is $O(m)$ as in Hoepman's algorithm.

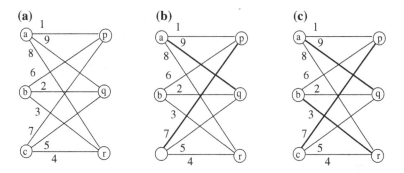

Fig. 13.6 Running of Hoepman's algorithm. **a** The original graph, **b** First phase of the algorithm processing nodes a, b and c in sequence, **c** Output from the second phase of the algorithm

A parallel formation of this algorithm on a distributed memory computer system can be achieved as shown in [25]. Each process has n/k vertices for a total of k processes, using block partitioning of the distance matrix. In the implementation, ghost vertices are used to handle the inter-partition edges. Each process p_i runs the first phase of the algorithm for the vertices it has and then the second phase is executed. The interleaved operation of local matching and communication is continued until the set D on each process is empty. The time complexity and the approximation ratio are the same as in Hoepman's algorithm.

13.4.3   Distributed Auction Algorithms

The strategy employed in a real-world auction can be used as the basis of MWBM algorithms. The idea in these *auction-based* algorithms is to assign buyers to the most valuable objects. In this model, given a weighted bipartite graph $G(V_1 \cup V_2, E, w)$, $i \in V_1$ is considered as a buyer, $j \in V_2$ is an object and the edge (i, j) between them is the cots of buying object j. Each buyer i bids for only one object.

The basic auction algorithm was proposed by Bertsekas to solve the assignment problem [6]. It has a polynomial time complexity and is also suitable for distributed processing. We will first describe the sequential auction algorithm and then provide its distributed version for maximal weighted bipartite graph matching. The bipartite graph $G(V_1 \cup V_2, E)$ we will consider has n nodes in each partition for a total of $2n$ nodes. The benefit of matching $u \in V_1$ to $v \in V_2$ is labeled as w_{uv} which is the weight of edge (u, v). Our aim is to find a matching between the nodes of V_1 and V_2 such that total weight $\sum_{u \in V_1, v \in V_2} w_{uv}$ is maximized. This algorithm considers nodes of V_1 as economic agents acting for their interest and V_2 as objects that can be bought. Each object j has a price p_j and the person receiving this object pays this price. Algorithm 13.4 shows the operation of the sequential auction algorithm as adapted from [34] which consists of three phases as initialization, bidding and assignment.

Algorithm 13.4 *Auction_Alg* Supervisor Process

1: **Input** : $G(V_1 \cup V_2, E, w)$ ▷ undirected weighted bipartite graph
2: **Output** : Matching M
3: I : set of buyers
4: $M \leftarrow \emptyset$
5: **initialize** ϵ
6: **for all** $j=1$ to n_2 **do**   ▷ initialize prices for objects
7:     $p_j \leftarrow 0$
8: **end for**
9: **while** $I \neq \emptyset$ **do**   ▷ start auction
10:     **select** $i \in I$ ▷ pick a free buyer
11:     $j_i \leftarrow max_j(w_{ij} - p_j)$ ▷ find the best object for this buyer
12:     $u_i \leftarrow w - ij(i) - p_j$ ▷ save its profit
13:     **if** $u_i > 0$ **then**
14:         $v_i \leftarrow max(w_{ij} - P_j)$ ▷ save the second best profit
15:         $p_j \leftarrow p_j + u_i - v_i + \epsilon$
16:         $M \leftarrow M \cup (i, j); I \leftarrow I \setminus \{i\}$ ▷ assign buyer
17:         $M \leftarrow M \setminus (k, j); I \leftarrow I \cup \{k\}$ ▷ free previous owner if exists
18:         **update** ϵ
19:     **else**
20:         $I \leftarrow I \setminus \{i\}$
21:     **end if**
22: **end while**

There are several studies aimed at finding parallel versions of the auction algorithm. Synchronous and asynchronous forms of the parallel algorithm was implemented in shared memory computer architectures in [5]. The algorithms presented in [7,33,34] all provide parallel auction algorithms that run on distributed memory architectures. We will take a closer look at the algorithm of [34] in which the three stages of the auction algorithm are parallelized. The bipartite graph is first distributed among the k processes $p_1, \ldots, p_k$ at the start of the algorithm. Each process p_i is responsible for a number of vertices and initializes its data structures and computes the bids for free buyers in its vertex set. All free buyers compute a bid in a single iteration and then the local prices are exchanged. A buyer becomes the owner of the object if it has the highest global bid after the message exchanges. Sathe et al. implemented this parallel algorithm using MPI in a distributed memory architecture and showed speedups obtained [34].

13.5 Chapter Notes

We have reviewed the network alignment problem in biological networks, described distributed maximal weighted bipartite graph matching algorithms that can be used for network alignment and proposed a new distributed greedy bipartite matching algorithm in this chapter.

Alignment of biological networks has few implications. We can deduce phylogenetic relationships between two or more species by learning how similar their PPI networks are, for example. Also, functionally similar subnetworks of organisms can be identified. The network alignment can be viewed from two perspectives in general; we try to find similarities between different subnetworks of two or more networks in local alignment to inspect conserved modules in them. These modules presumably are conserved as they have some important functionality. In the global alignment of networks, our aim is to map one network graph to another such that each vertex of one graph is mapped to another vertex in the other graph or not mapped at all. Global alignment provides us information about the similarity between the networks as a whole. There are various algorithms for both and we described a commonly used method for global alignment which consists of two steps. The similarity matrix which has an element r_{ij} showing the similarity between nodes i and j from each of the input graphs is first computed. There are various methods to compute this matrix, using node similarity, topological similarity or both as more commonly implemented. In the second step, a full bipartite graph having nodes of each input graph as one of its partitions is constructed with edges having weights equal to the entries in the similarity matrix. In this case, the global network alignment is reduced to finding maximal weighted matching of this graph. This problem has been studied extensively for general graphs and our focus in this chapter was on algorithms that can be distributed conveniently on a number of processing elements. One such algorithm is the simple greedy algorithm which has an approximation ratio of 1/2 and we described a distributed version of this algorithm that can be implemented in a distributed system using MPI. The auction algorithm considers the problem as an auction with buyers and objects, and has favorable complexity. We described this problem and reviewed a distributed version of it designed with network alignment goal.

Network alignment continues to be one of the fundamental problems in biological networks. Other than the investigation of relationships between the species, it can also aid to discover disease pathways by finding conserved subnetwork modules. There are only few distributed algorithms for network alignment and hence it is a potential topic for further research.

Exercises

1. Find the maximum common subgraph between the two graphs shown in Fig. 13.7 by visual inspection.
2. The networks in Fig. 13.8 are to be aligned. Provide a possible alignment between these two networks and work out the edge correction value.

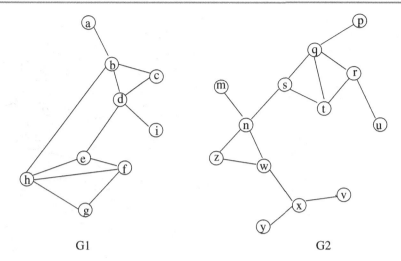

Fig. 13.7 Example graph for Exercise 1

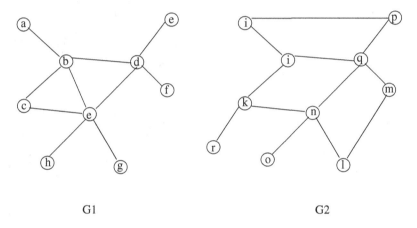

Fig. 13.8 Example graph for Exercise 2

3. Show the iteration steps of the greedy weighted bipartite matching algorithm on the graph of Fig. 13.9. Find also the maximum weighted matching in this graph and work out the approximation ratio of the greedy algorithm.
4. Show a possible execution steps of the distributed version of the greedy algorithm on the graph of Fig. 13.10.
5. Hoepman's algorithm in sequential form is to be executed to find MWBM in the graph of Fig. 13.10. Show the running of this algorithm and work out the approximation ratio with respect to the maximal weighted matching for this graph.
6. Compare the sequential network alignment tools IsoRank, PathFinder, GRAAL and MaWIsh in terms of their application domains and performances.

Fig. 13.9 Example graph for
Exercise 3

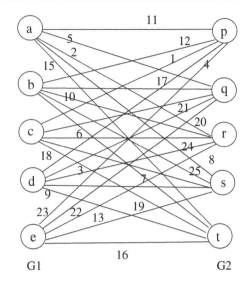

Fig. 13.10 Example graph
for Exercise 3

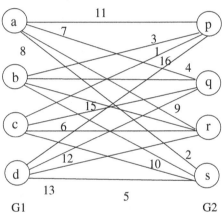

7. Compare the three distributed weighted bipartite graph matching algorithms,
   namely, the greedy approach, parallel Hoepman's algorithm on distributed memory computers and the distributed auction algorithm in terms of their complexities
   and the quality of the alignment achieved.

References

1. Aladag AE, Erten C (2013) SPINAL: scalable protein interaction network alignment. Bioinformatics 29(7):917–924
2. Altschul S, Gish W, Miller W, Myers E, Lipman D (1990) Basic local alignment search tool. J
   Mol Biol 215(3):403–410

3. Avis D (1983) A survey of heuristics for the weighted matching problem. Networks 13(4):475–493
4. Battiti R, Mascia F (2007) Engineering stochastic local search algorithms. designing, implementing and analyzing effective heuristics, An algorithm portfolio for the subgraph isomorphism problem. Springer, Berlin, pp 106–120
5. Bertsekas DP, Castannon DA (1991) Parallel synchronous and asynchronous implementations of the auction algorithm. Parallel Comput 17:707–732
6. Bertsekas DP (1992) Auction algorithms for network flow problems: a tutorial introduction. Comput Optim Appl 1:7–66
7. Bus L, Tvrdik P (2009) Towards auction algorithms for large dense assignment problems. Comput Optim Appl 43(3):411–436
8. Catalyurek UV, Dobrian F, Gebremedhin AH, Halappanavar M, Pothen A (2011) Distributed-memory parallel algorithms for matching and coloring. In: 2011 international symposium on parallel and distributed processing, workshops and Ph.D. forum (IPDPSW), workshop on parallel computing and optimization (PCO11), IEEE Press, pp 1966–1975
9. Clark C, Kalita J (2014) A comparison of algorithms for the pairwise alignment of biological networks. Bioinformatics 30(16):2351–2359
10. El-Kebir M, Heringa J, Klau GW (2011) Lagrangian relaxation applied to sparse global network alignment. In: Proceedings of 6th IAPR international conference on pattern recognition in bioinformatics (PRIB'11), Springer, pp 225-236
11. Flannick J, Novak A, Srinivasan BS, McAdams HH, Batzoglou S (2006) Graemlin: general and robust alignment of multiple large interaction networks. Genome Res 16:1169–1181
12. Fortin S (1996) The graph isomorphism problem. Technical Report TR 96-20, Department of Computer Science, The University of Alberta
13. Garey MR, Johnson DS (1979) Computers and intractability: a guide to the theory of NP-completeness. W.H. Freeman, New York
14. Hoepman JH (2004) Simple distributed weighted matchings. arXiv:cs/0410047v1
15. http://www.pathblast.org
16. Kelley BP, Sharan R, Karp RM, Sittler T, Root DE, Stockwell BR, Ideker T (2003) Conserved pathways within bacteria and yeast as revealed by global protein network alignment. Proc PNAS 100(20):11394–11399
17. Klau GW (2009) A new graph-based method for pairwise global network alignment. BMC Bioinform 10(Suppl 1):S59
18. Kollias G, Mohammadi S, Grama A (2012) Network Similarity Decomposition (NSD): a fast and scalable approach to network alignment. IEEE Trans Knowl Data Eng 24(12):2232–2243
19. Koyuturk M, Kim Y, Topkara U, Subramaniam S, Szpankowski W, Grama A (2006) Pairwise alignment of protein interaction networks. J Comput Biol 13(2):182–199
20. Kuchaiev O, Milenkovic T, Memisevic V, Hayes W, Przulj N (2010) Topological network alignment uncovers biological function and phylogeny. J Royal Soc Interface 7(50):1341–1354
21. Kuchaiev O, Przulj N (2011) Integrative network alignment reveals large regions of global network similarity in yeast and human. Bioinformatics 27(10):1390–1396
22. Kuhn HW (1955) The Hungarian method for the assignment problem. Naval Res Logistic Q 2:83–97
23. Kuramochi M, G. Karypis G (2001) Frequent subgraph discovery. In: Proceedings of 2001 IEEE international conference on data mining, IEEE Computer Society, pp 313–320
24. Liang Z, Xu M, Teng M, Niu L (2006) Comparison of protein interaction networks reveals species conservation and divergence. BMC Bioinf. 7(1):457
25. Manne F, Bisseling RH, A parallel approximation algorithm for the weighted maximum matching problem. In: Wyrzykowski R, Karczewski K, Dongarra J, Wasniewski J (eds) Proceedings of seventh international conference on parallel processing and applied mathematics (PPAM 2007), Lecture notes in computer science, vol 4967. Springer, pp 708–717

26. Milenkovic T et al (2010) Optimal network alignment with graphlet degree vectors. Cancer Inf. 9:121–137

27. Memievic V, Pruzlj N (2012) C-GRAAL: common-neighbors-based global GRaph ALignment of biological networks. Integr. Biol. 4(7):734–743

28. Messmer BT, Bunke H (1996) Subgraph isomorphism detection in polynomial time on pre-processed model graphs. Recent developments in computer vision. Springer, Berlin, pp 373–382

29. Ogata H, Fujibuchi W, Goto S, Kanehisa M (2000) A heuristic graph comparison algorithm and its application to detect functionally related enzyme clusters. Nucleic Acids Res 28:4021–4028

30. Patro R, Kingsford C (2012) Global network alignment using multiscale spectral signatures. Bioinformatics 28(23):3105–3114

31. Preis R (1999) Linear time 2-approximation algorithm for maximum weighted matching in general graphs. In: C. Meinel, S. Tison (eds) STACS99 Proceeedings 16th annual conference theoretical aspects of computer science, Lecture notes in computer science, vol 1563. Springer, New York, pp 259–269

32. Przulj N (2005) Graph theory analysis of protein-protein interactions. In: Igor J, Dennis W (eds) A chapter in knowledge discovery in proteomics. CRC Press

33. Riedyn J (2010) Making static pivoting scalable and dependable. Ph.D. Thesis, EECS Department, University of California, Berkeley

34. Sathe M, Schenk O, Burkhart H (2012) An auction-based weighted matching implementation on massively parallel architectures. Parallel Comput 38(12):595–614

35. Sharan R, Suthram S, Kelley RM, Kuhn T, McCuine S, Uetz P, Sittler T, Karp RM, Ideker T (2005) Conserved patterns of protein interaction in multiple species. Proc Natl Acad Sci USA 102:1974–1979

36. Sharan R et al (2005) Identification of protein complexes by comparative analysis of yeast and bacterial protein interaction data. J Comput Biol 12:835–846

37. Sharan R, Ideker T (2006) Modeling cellular machinery through biological network comparison. Nat Biotechnol 24(4):427–433

38. Singh R, Xu J, Berger B (2007) Pairwise global alignment of protein interaction networks by matching neighborhood topology. In: Research in computational molecular biology, Springer, pp 16-31

39. Singh R, Xu J, Berger B (2008) Global alignment of multiple protein interaction networks with application to functional orthology detection. PNAS 105(35):12763–12768

40. Ullmann JR (1976) An algorithm for subgraph isomorphism. J ACM 23(1):31–42

41. Yan X, Han J (2002) Gspan: graph-based substructure pattern mining. In: Proceedings of IEEE international conference on data mining, pp 721–724

Phylogenetics

14

14.1 Introduction

One of the fundamental assumptions in biology is that all living organisms share common ancestors. *Phylogeny* is the study of this evolutionary relationships among organisms. *Phylogenetics* is the study of these associations through molecular sequence and morphological data and aims to reconstruct evolutionary dependencies among the organisms. The earlier attempts to discover phylogenetic relationships between organisms by biologists involved using only morphological features. Data of living organisms under consideration in current research activities is a combination of DNA and protein amino acid sequences, and sometimes morphological characters. Finding these relations between organisms has many implications, for example, the prediction of disease transmission patterns using phylogenetics can be achieved to understand the spread of contagious diseases [6]. Phylogenetics can also be implemented in medicine to learn the origins of diseases and analyze the disease resistance mechanisms in other organisms to design therapy and cure in humans. For example, the evolution of viral pathogens such as flu viruses can be analyzed using phylogeny to design new vaccines.

Mutations are the driving forces behind evolution. A mutation event in an organism will be carried to offsprings. Many mutations are harmless and some result in improved traits, yet some are harmful and sometimes lethal to the organism. Mutations accumulated over time may result in the generation of new species. The evolutionary relatedness among the organisms may be depicted in a *phylogenetic tree* structure which shows the ancestor/descendant relationships between the living organisms under consideration in a tree topology. The leaves of this tree represent the living organisms and the interior nodes are the hypothetical ancestors.

However, a phylogenetic tree may not be adequate to represent the evolutionary relatedness between all organisms. A horizontal gene transfer is the transfer of genetic material between different and sometimes distantly related organisms and a phylogenetic tree does not represent such an event. A vertical gene transfer on the other hand involves the transfer of genetic material to descendants and is the basis of

© Springer International Publishing Switzerland 2015

K. Erciyes, *Distributed and Sequential Algorithms for Bioinformatics*,
Computational Biology 23, DOI 10.1007/978-3-319-24966-7_14

tree construction. *Phylogenetic networks* model reticulate evolutionary event such as horizontal gene transfer and are represented by a directed acyclic graph.

In this chapter, we will first describe methods and algorithms to construct phylogenetic trees with emphasis on distributed algorithms and then briefly review phylogenetic networks which have been the focus of recent studies.

14.2 Terminology

We will describe briefly the fundamental terms used in phylogeny as follows:

- *Taxonomy*: The classification and naming of organisms.
- *Species*: A class of organisms which have the ability to interbreed. A *subspecies* is a subgroup of a species.
- *operational taxonomic units* (OTUs): These are the leaves of the phylogenetic tree and consist of living organisms. They are also called *taxon* (taxa as plural).
- *Hypothetical taxonomic units* (HTUs): The internal nodes of the phylogeny. These are hypothetical as the name suggests and cannot be verified. Since there are differences between OTUs, it is not possible to distinguish an HTU from its child if it has only one child. It is therefore assumed that each OTU has at least two children.
- *Homology*: A feature that appears similar in two or more taxa with a common ancestor.
- *Reticulation*: Joining of separate branches of a phylogenetic tree, usually through hybridization or through horizontal gene transfer.
- *Clade*: A group of organisms that have descended from a particular ancestor.
- *Cladogram*: A tree that shows branching relationships between objects. Branches may not be proportional to the time passed.
- *Monophyly*: A clade where all the OTUs are derived including the descendants of a single common ancestor. It may include all or part of the descendants.
- *Phylogram*: A tree that shows inferred relationships among entities. It is different than a cladogram as the branches are drawn proportional to the sizes of mutations.
- *Speciation*: The evolutionary process causing new biological species.
- *Outgroup*: A group of OTUs assumed to be outside the clade of the OTU under consideration.
- *Molecular clock*: When present, it is assumed that the rate of mutational changes is constant by time.
- *Phenetics*: Taxonomy method based on the morphological or genetical similarity. It mainly considers observable similarities and differences.
- *Phenotype*: The set of measurable or detectable physical or behavioral features of an individual.

- *Orthologous*: Genes of common ancestors, passed from ancestors to descendants.
- *Paralogous*: Similar genes formed as a result of duplication during evolution.
- Xenologs: Genes formed as a result of horizontal transfers between organisms, for example, injection by a virus.

14.3 Phylogenetic Trees

A phylogenetic tree shows the ancestor–descendant relationships of organisms and although it does not show the reticulate events, it is still widely used to infer the similarities and differences of species and hence we will investigate its properties, and sequential and distributed algorithms to form these trees

A phylogenetic tree can be *rooted* or *unrooted*. In rooted trees, a designated node is the root of the tree and it serves as the common ancestors of all nodes. Hierarchy does not exist in unrooted trees but the distances and the relations between nodes can be determined. There are $1 \times 3 \times 5...(2n - 3) = (2n - 3)!!$ rooted trees and $(2n - 5)!!$ unrooted trees which have n taxa as their leaves. An unrooted tree can be transformed into a rooted tree by inserting a new node which will serve as the root. An unrooted tree can be generated from a rooted tree by discarding the labeling of the root. An internal node of a rooted tree represents the hypothetical most recent common ancestor (MRCA) of all nodes in its subtree. Leaves of this tree called $taxa$ represent the organisms where data is derived from. The length of edges represents the evolutionary time between the species. The time increases from the root, if this is available, to the leaves. The intermediate nodes represent the ancestors in the past. Figure 14.1 shows an unrooted and a rooted phlylogenetic tree.

The amino acid or DNA/RNA sequence similarities of the organisms to be analyzed can be found using multiple sequence alignment methods. A method/algorithm can then be selected to construct a phylogenetic tree and the structure obtained can be evaluated as the final step. The three fundamental methods to construct phylogenetic trees are *distance methods*, *maximum parsimony* and *maximum likelihood*. In distance methods, evolutionary distances are computed for all taxa and these distances are used to build trees. The maximum parsimony method on the other hand searches

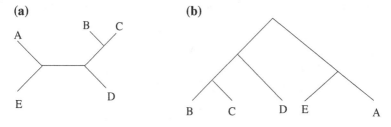

Fig. 14.1 a An unrooted phylogenetic tree. **b** The rooted version of the same tree

a tree that minimizes the number of changes to data, and the maximum likelihood method searches all trees selecting the most probable tree that explains the input data.

14.3.1 Distance-Based Algorithms

The input to a distance-based method is the distance matrix $D[n, n]$ for n taxa where d_{ij} is the distance between taxa i and j. Our aim is to construct a weighted-edge tree T where each leaf represents a single taxon and the distance between leaves i and j is approximately equal to the entry d_{ij} of D. We need to define the important properties of the phylogenetic trees as below:

Definition 14.1 (*metric tree*) A metric tree $T(V, E, w)$ is a rooted or unrooted tree where $w : E \rightarrow \mathbb{R}^+$

The metric trees have three main properties. All edges of a metric tree have nonnegative weights, and $d_{ij} = d_{ji}$ for any two nodes $i, j \in V$, implying symmetry. Also, for any three nodes i, j, k in a metric tree, $d_{ik} \leq d_{ij} + d_{jk}$ showing triangle inequality.

Definition 14.2 (*additive tree*) A tree is called additive if for all taxa pairs, the distance between them is the sum of the edge weights of the path between them.

Figure 14.2 shows such an additive tree and its associated distance matrix.

Definition 14.3 (*ultrametric tree*) A rooted additive tree is called ultrametric if the distance between any of its two leaves i and j and any of their common ancestors k is equal ($d_{ik} = d_{jk}$). Therefore, for any two leaves i and j and the root r of an ultrametric tree, $d(i, r) = d(j, r)$.

The distances of such a tree are said to be *ultrametric* if, for any triplet of sequences the distances are either all equal, or two are equal and the remaining one is smaller. This condition holds for distances derived from a tree with molecular clock. In other words, the sum of times down a path to the leaves from any node is the same, whatever

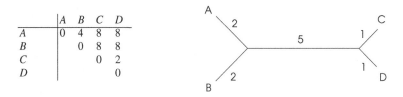

	A	B	C	D
A	0	4	8	8
B		0	8	8
C			0	2
D				0

Fig. 14.2 An additive tree and its associated distance matrix for taxa A, B, C, and D

the choice of path. This means that the divergence of sequences can be assumed to occur at the same constant rate at all points in the tree. The edge lengths in the resulting tree can therefore be viewed as times measured by a molecular clock with a constant rate. The ultrametric tree property implies additivity but not vice versa. If the weights of the edges of a rooted metric tree represent the times, all of the leaves should be equidistant from the root of an ultrametric tree as this distance represents the time elapsed since the MRCA which should be the same for all taxa. In short, ultrametric tree property implies the existence of a *molecular clock* which means a constant rate of evolution.

14.3.1.1 Sequential UPGMA

Unweighted pair group method using arithmetic averages (UPGMA) is a simple method to construct phylogenetic trees. It is an agglomerative hierarchical clustering algorithm which assumes that the tree is ultrametric and starts considering each taxon as a cluster initially. At each iteration of the algorithm, the closest clusters are found and merged into a new cluster and are placed in the tree. Then the distances of each element of the matrix D are calculated and D is updated. This process continues until there are two clusters left. The general idea of this algorithm is that the two taxa that have a low distance between them should be placed close to each other in the tree. It uses the distance metrics between clusters described below.

Definition 14.4 (*distance between two clusters*) Given two disjoint clusters C_i and C_j, the distance d_{ij} between them is the sum of the distances between the sequences in clusters normalized by the size of the clusters as follows:

$$d_{ij} = \frac{\sum_{u \in C_i} \sum_{v \in C_j} d_{uv}}{|C_i||C_j|} \tag{14.1}$$

We are computing the average distance between the two clusters in this case. We may, however, use the single-link distance for simplicity, which is the shortest distance between the clusters. The following lemma provides an efficient way to calculate distances between two clusters one of which already consists of two clusters.

Lemma 14.1 (efficiency) *Let us assume a cluster C_k consists of two clusters C_i and C_j such that $C_k = C_i \cup C_j$. The distance of C_k to another cluster C_l can be defined as follows:*

$$d_{kl} = \frac{d_{il}|C_i| + d_{jl}|C_j|}{|C_i| + |C_j|} \tag{14.2}$$

The proof is trivial and this lemma helps us to find the distance between two clusters in linear time. The UPGMA algorithm uses the above-defined two distance metrics to find the distances between the newly formed cluster to all other clusters. The steps of UPGMA are as follows:

	A	B	C	D
A	0	6	6	6
B		0	3	3
C			0	1
D				0

Fig. 14.3 UPGMA first iteration

1. Each node i is assigned to cluster C_i initially.
2. **Repeat**
    a. Find the two clusters C_i, C_j where d_{ij} is minimal. If there is more than one pair of clusters with this minimal value, pick one randomly.
    b. Define a new cluster $C_k = C_i \cup C_j$ and update distances of all clusters to C_k using Eq. 14.1 or Eq. 14.2.
    c. Form the new node k with two children C_i and C_j and place it at height $d_{ij}/2$.
3. **Until** two clusters i and j remain
4. Place the root at height $d_{ij}/2$

We will show a simple example where the distance matrix for four taxa A, B, C, and D is as shown in Fig. 14.3. The minimum value in this matrix is between taxa C and D. We therefore form a HTU labeled X which is equidistant to these nodes as depicted in the partial phylogenetic tree. These two nodes are then included in the cluster (C, D) formed in Fig. 14.3.

The distance of the remaining nodes to this cluster is then calculated and the distance matrix is updated as shown in Fig. 14.4. The distance from nodes A and B to the cluster is their average distance to the nodes in the cluster as follows:

$$d_{A,(CD)} = \frac{d_{AC} + d_{AD}}{2} = 6$$

$$d_{B,(CD)} = \frac{d_{BC} + d_{BD}}{2} = 3$$

We now find that distance between B and (CD) is the smallest and merge these two clusters and provide a new HTU labeled Y which is at half distance between them as depicted in Fig. 14.4.

We are now left with two clusters A and (B, C, D) and calculate the distance between them using Eq. 14.2 as below and the loop terminates.

$$d_{A,(B,C,D)} = \frac{d_{A,B}|B| + d_{A,(C,D)}|(C, D)|}{|B| + |(C, D)|} = \frac{(6 \times 1) + (6 \times 2)}{(1 + 2)} = 6$$

The root Z is placed to give a height of the half of the distance between these two clusters and the algorithm finishes giving the final tree shown in Fig. 14.5.

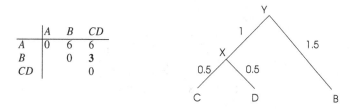

	A	B	CD
A	0	6	6
B		0	3
CD			0

Fig. 14.4 UPGMA second iteration

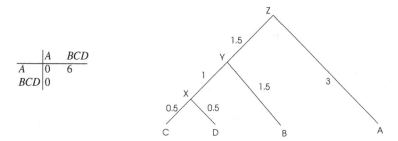

	A	BCD
A	0	6
BCD	0	

Fig. 14.5 The final phylogenetic tree

The time taken to find the smallest value element is $O(n^2)$ since the distance matrix has n^2 elements and there is a total of $n-1$ iterations of the algorithm for n taxa. The total time therefore is $O(n^3)$. The UPGMA has the disadvantage of assuming an ultrametric and also additive tree.

14.3.1.2 Distributed UPGMA Proposal

We can sketch a simple distributed algorithm to construct a phylogenetic tree using the UPGMA method. As can be seen, the UPGMA method is very similar to the single-link agglomerative hierarchical clustering algorithm. We can therefore use the same algorithmic procedure of Algorithm 7.6 of Sect. 7.5. We have k processes and the supervisor process p_0 partitions the rows of the distance matrix D and send them to the processes. Each process p_i then finds the minimum distance between the clusters it owns and sends this to p_0 which finds the global minimum distance. It consecutively broadcasts the minimum distance to the processes. The process(es) which have the marked minimum distance clusters merge them and all related processes update their distances to the new cluster formed. The speedup obtained in this algorithm is approximately equal to the number of processors, discarding the communication overheads.

We will describe the operation of the distributed algorithm for a small set of six taxa $t_1, \ldots, t_6$ to be processed by three processes p_0, p_1, and p_2. The distance matrix D row partitioned between these processes is shown in Fig. 14.6. In the first step, the minimum distance for p_1 is between t_3 and t_5 (arbitrarily); and for p_0 and p_2

	t_1	t_2	t_3	t_4	t_5	t_6	
t_1	0	18	2	18	1	18	p_0
t_2	18	0	18	4	18	4	
t_3	2	18	0	18	2	18	p_1
t_4	2	18	0	18	2	18	
t_5	1	18	2	18	0	18	p_2
t_6	18	4	18	3	18	0	

Fig. 14.6 Distributed UPGMA first round. The minimum distance is between t_1 and t_5 which are clustered to form the two leaves of the phylogenetic tree

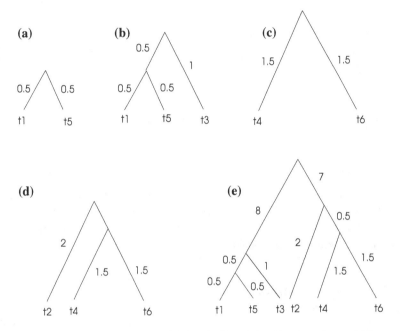

Fig. 14.7 The partial trees constructed with the distributed algorithm in each round. The trees in (**a**),...,(**e**) correspond to the rounds 1,...,5. The tree in (**e**) is final

is between t_1 and t_5. Therefore, the global minimum distance is between the taxa t_1 and t_5 and these are grouped to form the leaves of the tree in the first step.

This procedure continues for five rounds at the end of which, the final tree shown in Fig. 14.7 is constructed.

14.3.1.3 Sequential Neighbor Joining Algorithm

The neighbor joining (NJ) algorithm was first proposed by Saitou and Nei [34], and later modified by Studier and Kepler [38]. It is a hierarchical clustering algorithm like UPGMA, however, does not assume the ultrametric tree property and produces an unrooted phylogenetic tree.

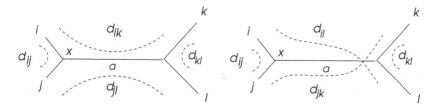

Fig. 14.8 The distances in an additive tree

The additive tree condition meant that for any two leaves, the distance between them is the sum of edge weights of the path between them. We need a method to check if a tree is additive or not by inspecting the distance matrix. We can now state the *four-point condition* between four taxa.

Definition 14.5 (*four-point condition*) Given four taxa i, j, k, and l, the four-point condition holds if two of the possible sums $d_{il} + d_{jk}$, $d_{ik} + d_{jl}$ and $d_{ij} + d_{kl}$ are equal and the third one is smaller than this sum.

As can be seen in Fig. 14.8, the possible distances between four taxa can be specified as follows:

$$d_{il} + d_{jk} = T + 2a$$
$$d_{ik} + d_{jl} = T + 2a$$
$$d_{ij} + d_{kl} = T$$

where T is the sum of the distances of the leaves to their ancestors. This would mean that the larger sum should appear twice in these three sums. A distance matrix $D[n, n]$ is additive if and only if the four-point condition holds for all of its four elements. We can check this condition for the distance matrix of Fig. 14.2. The three sums in this case are as follows:

$$d_{AB} + d_{CD} = 4 + 2 = 6$$
$$d_{AC} + d_{BD} = 8 + 8 = 16$$
$$d_{AD} + d_{BC} = 8 + 8 = 16$$

The two sums are equal and the third sum is smaller than this sum, so this matrix is additive. The NJ algorithm assumes additive tree property but this tree does not have to be ultrametric. It has a similar structure to the UPGMA and takes the distance matrix D which has an entry d_{ij} showing the distance between clusters i and j as input, and merges the closest clusters at each iteration to form a new cluster. However, the distance computations are different than UPGMA where the general idea of this algorithm is to merge the two clusters that are close (minimize the distance between the two clusters) but also farthest from the rest of the clusters (maximize their distance to all other clusters). UPGMA did not aim for the latter. This method starts with a starlike tree as shown in Fig. 14.9 and iteratively forms clusters as shown.

(a)   (b)   (c)

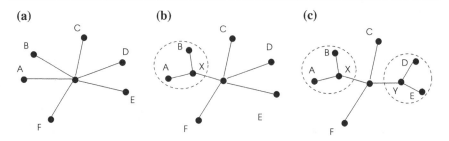

Fig. 14.9 Constructing a tree using NJ algorithm. **a** A star with six taxa (A, B, C, D, E, and F) as leaves is formed. **b** The two leaves A and B that have the minimum of the distance between them but are farthest to all other clusters are joined to form a new cluster and the distances to this new cluster from all clusters are calculated. **c** The two closest to each other and farthest to all other clusters are D and E and they are joined and the process is repeated

Let us now define the *separation* u_i of a cluster i from the rest of clusters as follows:

$$u_i = \frac{1}{n-2} \sum_j d_{ij} \tag{14.3}$$

where n is the number of clusters. The aim in NJ algorithm is then to find the two clusters that have the minimum distance between them and also have the highest distance to all other clusters. In other words, it searches for the cluster pair i and j with the minimum value of $d_{ij} - u_i - u_j$. These two clusters are replaced by a single node x in the tree and the distance of all other clusters to x are calculated. The distance between the node x representing the merged clusters i and j and the other clusters k and l in this case is as follows.

$$d_{xk} = \frac{1}{2}(d_{ik} + d_{jl} - d_{ij}) \tag{14.4}$$

This is exactly how the distance of the newly formed cluster (ij) to all other clusters is computed in the algorithm. We are now ready to review the NJ algorithm which consists of the following steps:

1. Each node i is assigned as a cluster initially.
2. **Repeat**
   a. For each cluster i, compute the separation to each cluster j as $u_i = \frac{1}{n-2} \sum d_{ij}$.
   b. Find the two clusters i, j where $d_{ij} - u_i - u_j$ is minimal.
   c. Define a new cluster (ij) with a node x representing this cluster. Calculate branch lengths from i and j to x as follows:

$$d_{ix} = \frac{1}{2}(d_{ij} + u_i - u_j), \qquad d_{jx} = \frac{1}{2}(d_{ij} + u_j - u_i) \tag{14.5}$$

Fig. 14.10 NJ algorithm first iteration. The separation values are determined and a subtree for taxa A and B is found

	A	B	C	D
A	0	9	15	16
B		0	10	19
C			0	25
D				0

	u_i
A	(9+15+16)/2=20
B	(9+10+19)/2=19
C	(15+10+25)/2=25
D	(16+19+25)/2=30

d. Compute the distance between the new cluster (i, j) and each cluster k using additive property of Eq. 14.4 as

$$d_{(ij)k} = \frac{1}{2}(d_{ik} + d_{jl} - d_{ij}) \qquad (14.6)$$

e. Remove clusters i and j from the distance matrix and replace them by a single cluster

3. **Until** two clusters i and j remain
4. Connect the two clusters i and j by a branch length d_{ij}

Finding the smallest distance takes n^2 time in the step 2b of the algorithm and the main loop is executed n times resulting in $O(n^3)$ time complexity. Let us illustrate this algorithm using an example; we are given a distance matrix D between four taxa $A, B, C,$ and D and then the initial separation values u_i for each taxon are calculated using Eq. 14.5 as shown in Fig. 14.10.

We can now find the distance values from each taxon to the others using the step 2b of the algorithm. We have two minimum values and we randomly select leaves B and C to form the first cluster as shown. The distance between the common ancestor X of B and C is calculated according to Eq. 14.5 and we find $d_{BX} = \frac{1}{2}(10+19-25) = 2$ and $d_{AX} = \frac{1}{2}(10+25-19) = 8$. This is represented as asymmetric tree branches as shown in Fig. 14.11.

We can now calculate the distances of all other taxa (A and D) to this new node X using Eq. 14.6. In this case, $d_{AX} = \frac{1}{2}(9 + 15 - 10) = 7$ and $d_{DX} = \frac{1}{2}(19 + 25 - 10) = 17$. We can now form the new distance matrix between taxa A and D and intermediate node X and find the new separation values in the second iteration of the algorithm as shown in Fig. 14.12.

The minimum values are all equal and we arbitrarily select A and D to cluster and these are merged into a new cluster which actually terminates the iteration as we have two clusters (A, D and B, C) now. We continue with the evaluation of the distances of the two children nodes A and D to the new ancestral node Y using Eq. 14.5. The distances are $d_{AY} = \frac{1}{2}(16 + 23 - 33) = 3$ and $d_{DY} = \frac{1}{2}(16 + 33 - 23) = 13$.

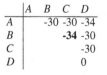

	A	B	C	D
A		-30	-30	-34
B			-34	-30
C				-30
D				0

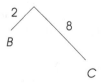

Fig. 14.11 NJ algorithm first iteration. The distance values are determined

Fig. 14.12 NJ algorithm
second iteration

	A	D	X
A	0	16	7
D		0	17
X			0

	u_i
A	16+7=23
D	16+17=33
X	7+17=24

Fig. 14.13 NJ algorithm
third iteration

	A	D	X
A		-40	-40
D			-40
X			

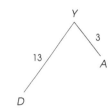

Fig. 14.14 The final
phylogenetic tree constructed
by the NJ algorithm

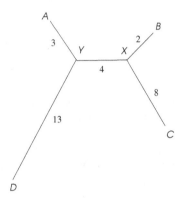

Selecting any other two clusters would result in the same phylogenetic tree in Fig. 14.13.

The distance between the two intermediate nodes X and Y can be calculated using Eq. 14.6. giving $d_{XY} = \frac{1}{2}(9 + 15 - 10) = 4$ and all of the subtrees we have built can be combined to give the final tree of Fig. 14.14 which represents the initial distance matrix precisely.

Distributed Neighbor Joining Algorithms

The operation of UPGMA and NJ algorithms is similar as they are both agglomerative hierarchical clustering algorithms, however, the calculation of distances are different. Therefore, we can proceed similar to the parallelization of the UPGMA in order to provide a parallel NJ algorithm to be executed on a distributed memory computer system. Our idea in this algorithm is again to partition n rows of the distance matrix to p processors and the algorithm is similar in structure to distributed UPGMA algorithm. The speedup obtained will be the same as the number of processors in this case.

Ruzgar and Erciyes combined NJ algorithm with fuzzy clustering to construct phylogenetic trees of Y-DNA haplogroup G of individuals [33]. Their algorithm consists of the following steps:

1. Calculate the distances between n data points in the set $P = \{p_1, \ldots, p_n\}$
2. Cluster the data points using three sample algorithms into m clusters $C_1, \ldots, C_m$. The algorithms used are a reference clustering algorithm, fuzzy C-means (FCM) [9], and fuzzy neighborhood density-based spatial clustering of applications with noise (FN-DBSCAN) [30] which provided the best grouping of data.
3. For each cluster C_i, implement NJ algorithm to construct phylogenetic trees and designate roots to trees formed.
4. Implement NJ algorithm to construct a phylogenetic tree consisting of cluster roots only.
5. Merge the phylogenetic trees obtained in steps 3 and 4 to obtain the whole tree.

Although they have not implemented the distributed version of this algorithm, description of the approach was provided. The clustering can be performed by a root process p_0 after which it distributes the clusters to individual processes in a distributed system. Each process p_i then constructs the NJ tree for data in its cluster (Step 3) and the tree topology formed is sent to p_0. The root then constructs the NJ tree for the root nodes and merges the whole tree structure into a single tree. They showed that the results obtained using this method were highly similar to the results of running of the original nonclustered NJ algorithm.

14.3.2 Maximum Parsimony

The basic idea of the *maximum parsimony* method is based on the philosophical idea of Ockham called *Ockham's razor*, that the best hypothesis to explain a complex process is the one that requires fewest assumptions. Adapting this idea to constructing phylogenetic trees, we need to find the tree that has the lowest number of mutations to explain the input taxa. We need to test a number of trees that provide the taxa at leaves and count the number of mutations in each of them. The one with the lowest number should be best to represent the real evolutionary process according to this method. There may be more than one solution giving the same number of mutations. The parsimony problem can be inspected at two levels as *small parsimony* problem and *large parsimony* problem which are described next.

14.3.2.1 The Small Parsimony Problem

The small parsimony problem is the process of finding a labeling of the ancestral nodes of a given tree which has the minimum number of changes along its edges. A tree topology is determined beforehand and our aim is to find the labeling of the vertices of this tree to yield the lowest score. This problem can be solved in linear time by a simple algorithm due to Fitch.

Fitch's Algorithm

Fitch's algorithm is a dynamic algorithm to compute the minimum number of mutations in a tree to explain the given taxa [13]. The state of an intermediate node is determined by the states of its children. The algorithm first starts from the leaves of the tree and for each parent of a leaf, it assigns a label which is the intersection of the labels of its children if this intersection is not empty. Otherwise, the labeling of the parent is the union of its children. The idea is to label a parent with the common elements of the children only if the children have some common elements. All of the uncommon elements must then have occurred for each child. For totally unrelated children, the parent must have had all of the uncommon elements to pass them to the children. It relabels the intermediate nodes starting from the root in the second phase. Formally, the algorithm consists of the following steps:

1. For each leaf v, $S_v \leftarrow X_v$ where X_v is the character label of v
2. $C \leftarrow 0$, the parsimony score is initialized
3. Starting from the leaves, traverse the tree upwards to the root. For any internal node u with children v and w,

$$S_u = \begin{cases} S_v \cap S_w & \text{if } S_v \cap S_w \neq \emptyset \\ S_v \cup S_w; \quad C \leftarrow C + 1 & \text{otherwise} \end{cases}$$

4. We need to finalize the X_v for each node v now. We start from the root and traverse the tree downward to the leaves. For each node u with a parent v

$$S_u = \begin{cases} S_u \leftarrow S_v & \text{if } S_u \in S_v \\ \text{arbitrarily assign any } y \in S_v \text{ to } S_u & \text{otherwise} \end{cases}$$

Figure 14.15 displays implementation of this algorithm in an example taxa set G, G, A, G, and C. The trees at (a) and (b) have the same score of 2 but the tree of (c) has a score of three which means we should choose either the tree of (a) or (b). The tree at (d) is obtained by running the algorithm from the root to the leaves of (a) in the second step. For n taxa using an alphabet of size m, time taken for this algorithm is $O(nm)$.

Weighted Parsimony and Sankoff's Algorithm

Fitch's algorithm assumed that all mutations have equal weights of unity which is not realistic. As was stated, the nucleotide A pairs with T and G with C in the DNA. A *transition* is defined as a point mutation that changes a purine nucleotide to another purine (A $\leftrightarrow$ G) or a pyrimidine nucleotide to another pyrimidine (C $\leftrightarrow$ T). A *transversion* on the other hand, changes a purine to pyrimidine base such as (A $\leftrightarrow$ C) or (G $\leftrightarrow$ T). Approximately, two out of three single nucleotide polymorphisms (SNPs) are transitions. It is therefore reasonable to assign different weights to different mutations, a higher weight for transversion than a transition as commonly done.

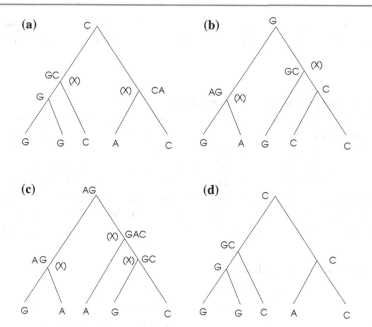

Fig. 14.15 Fitch's algorithm implementation for five taxa: G, G, A, A, and C. The mutations are marked with a $\times$. **a** A possible tree with a Finch score of 2, **b** Another tree with the same score, **c** A tree with a score of 3, **d** The final labeling of the first tree

A weight is assigned for each mutation in weighted parsimony. Sankoff's algorithm is a generalization of Fitch's algorithm for weighted parsimony [35]. A cost c_{ij} is assigned as the cost of going from state i to state j of a character. Let $S_j(i)$ be the smallest weighted number of steps needed to evolve the subtree at or above node j, with the node j in state i. This algorithm traverses the tree and consists of the bottom-up phase of labeling nodes from leaves to the root and top-down refinement phase of determining the final labels as described in [12]:

1. *Bottom-up Phase*
   a. $S_j(i) \leftarrow 0$ if node j has state i. Otherwise, $S_j(i) \leftarrow \infty$.
   b. Then doing a postorder traversal of tree, for a node a with children l and r

$$S_a(i) = min_j[c_{ij} + S_l(j)] + min_k[c_{ik} + S_r(k)]$$

2. *Top-down Phase*: After the optimal score of the root vertex is computed, we move down the tree and select the state that produced the minimal cost for a vertex in the bottom-up phase.

Processing of one node in the first phase takes $O(l^2)$ time where l is the number of different states of a character. For n input taxa with length m, time required is $O(nml^2)$.

14.3.2.2   The Large Parsimony Problem

Small parsimony problem involved finding labels for the internal nodes of a *given* tree topology which minimizes the parsimony score. In the large parsimony problem, our input is n taxa sequences and we are asked to find the topology and labeling of a tree with the minimum parsimony score. This task involves searching all $(2n - 3)!!$ rooted or $(2n - 5)!!$ unrooted possible trees which is intractable. This problem is NP-hard and various methods which incorporate heuristics are used to find approximate solutions. We will describe a 2-approximation algorithm and two ways to search for a solution to this task as *branch and bound*, and *nearest-neighbor interchange* methods.

A 2-Approximation Algorithm

A simple approximation algorithm which makes use of the MST property can be implemented to find an approximate parsimonious tree of input taxa. This algorithm first finds the Hamming distances among all input taxa and then builds a complete graph K_n which has the n taxa as its vertices and the distance between taxa i and j is labeled as the weight of the edge (i, j) between them. The minimum number of mutations that covers all of the nodes in this graph is in fact the MST of the graph. The problem of finding the most parsimonious tree is therefore reduced to finding the MST of K_n which can be performed in linear time using one of the Prim's, Kruskal's, or Boruvka's algorithms. We will illustrate this algorithm by an example with five input taxa P, Q, R, S, and T as shown in Table 14.1 with their Hamming distances.

We can now form the complete graph K_5 and then find the MST of this graph. We proceed by building the tree starting from a node of the MST as the root and then joining the nodes of the MST as displayed in Fig. 14.16. The approximation ratio of this algorithm is 2 as shown in [39]. Finding the Hamming distance between two taxa of length k requires k comparisons and there will be $O(n^2)$ such comparisons for n input data. The runtime for this step therefore is $O(n^2 k)$. Constructing the MST tree requires $O(n^2 \log n)$ for the complete graph with $O(n^2)$ edges. Hence, total time taken is $O(n^2(k + \log n))$.

Table 14.1   The input taxa and their Hamming distances

Taxa	DNA Sequence					P	Q	R	S	T
P	A	G	A	T		P	0	3	3	3	1
Q	C	T	A	G		Q		0	3	3	3
R	G	C	A	T		R			0	1	2
S	G	G	A	T		S				0	2
T	A	T	A	T		T					0

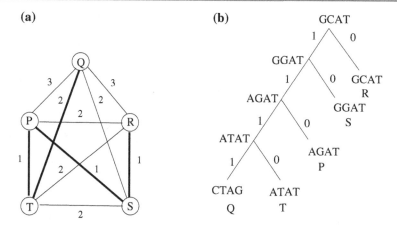

Fig. 14.16 **a** The MST of K_5 shown by bold lines, using the Hamming distances between taxa, **b** phylogenetic tree formed using the MST. The parsimony score is 5

Branch and Bound Method

The branch and bound (B&B) method in general restricts the search space of an algorithm by carefully pruning the search tree. If a partial solution cannot lead to a solution better than the already existing ones, it is abandoned. The large parsimony problem is a good premise for this method and it was first used for this purpose by Hendy and Penny [18]. It evaluates all possible trees and discards paths of the search tree if these will not lead to optimal trees.

We will now briefly describe how to grow a tree. In order to obtain a tree T_{i+1} representing taxa $t_1, \ldots, t_{i+1}$, we add an extra edge end point of which it is labeled with the taxon t_{i+1} in all possible ways. Figure 14.17 shows how to enlarge a tree with three leaves first to four leaves, and then to five leaves.

We can detect, for example, adding edge with taxon E in the middle 4-leaf tree results in a worse parsimanous tree than the best-known score and abandon all of the subtrees of that branch. The steps of a general B&B method for large parsimony can be stated as follows:

1. Construct an initial tree T_0 using a suitable algorithm such as NJ.
2. The upperbound $B \leftarrow$ parsimony score of T_0.
3. Start building new trees by adding edges. If adding an edge e to an existing partial tree T' results in a parsimony score that is equal or larger than B, all of the subtrees below e are abandoned.
4. If the parsimony score B' of the new tree is smaller than B, $B \leftarrow B'$
5. Report the best tree (one with the lowest B) obtained so far.

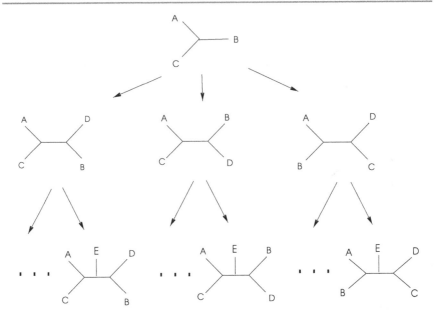

Fig. 14.17 Enlarging an initial tree with three leaves A, B, and C. The new edge with taxon D is placed between all possible taxa; A and B, A, and C, and B and C resulting in three more trees. Adding the new edge for taxon E results in 15 more trees some of which are shown

Nearest-Neighbor Interchange Method

Nearest neighbor interchange (NNI) is a heuristic method proposed by Robinson [31] to search for a maximum parsimonious tree. We start with an initial tree $T(V, E)$ which may be constructed using a suitable algorithm such as NJ algorithm and investigate subtrees of T that can be interchanged. An internal edge $(u, v) \in V$ in a tree is selected and subtrees connected to this edge are swapped to obtain different tree topologies. The *nni distance* between two trees is the minimum number of swap operations needed to transform one tree to another [7, 14]. Each tree formed is evaluated using Finch's or Sankoff's algorithm and the tree with the lowest score is selected. The neighborhood relationship in a tree can be defined in various ways. Figure 14.18 displays 4 subtrees A, B, C, and D positioned around an internal edge (u, v). The two possible tree configurations formed by swapping subtrees of the first one are shown.

14.3.2.3 Parallel Maximum Parsimony

Since the algorithms of Fitch and Sankoff already provide linear time solutions to the small parsimony problem, our focus on parallelization will be on the large parsimony which is also more general as there is no input tree. For an exhaustive search of all possible trees, we can simply generate all of these trees and assign

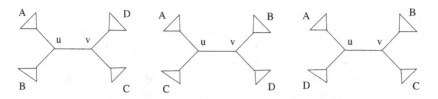

Fig. 14.18 Nearest-neighbor interchange of four subtrees labeled A, B, C, and D

a number for them to each process p_i and have each p_i work out the labeling of intermediate nodes using Fitch's or Sankoff's algorithm. However, this method will not suffice even for small input sizes as the number of trees grow exponentially. We will therefore search ways of parallelizing the schemes we have outlined for large parsimony, namely, the 2-approximation algorithm, and B&B and the NNI methods.

Parallelizing the 2-Approximation Algorithm

Let us briefly review the steps of the 2-Approximation algorithm and sketch possible ways of parallel processing using distributed memory processors in these steps. There is a central process p_0 and there are k processes $p_0, \ldots, p_{k-1}$ as before.

1. We need to compare characters of each taxa pair and this needs to be done $O(n^2)$ times. We can simply distribute n taxa to k processes where each p_i is responsible for comparing the taxa in its T_i to all other taxa. Each p_i finds the Hamming distances D_i for taxa in T_i and sends the partial distance matrix D_i to the root process p_0.
2. The root p_0 gathers all partial results and forms the upper triangle matrix D. It can then perform construction of the MST in linear time using any of the MST algorithms.
3. The root can now build the approximate parsimonious tree using the MST information.

Various optimization and load balancing methods can be implemented in step 1. Since we are trying to form an upper triangular matrix, some of the processes will have less work to do if we do not want to duplicate the executions. Therefore, a load balancing strategy can be employed. Construction of the MST can also be parallelized by implementing the parallel Boruvka algorithm outlined in Sect. 11.3.

Parallel Branch and Bound

A parallel implementation of B&B method for phylogenetic tree construction was proposed by Zomaya in cache-coherent uniform memory access SMPs [41]. The single program, multiple data (SPMD) asynchronous model was used by having each process working on different graph partitions. Since the B&B search may take irregular paths, static load balancing in this method is difficult to achieve and this was the reason that the authors in this study used shared memory SMP model. The shared memory was protected by locks and a relaxed DFS method for searching the tree was implemented. The method proposed was tested using data sets consisting of 24 sequences, each with a length of 500 base pairs and was shown to be 1.2–7 times faster than the phylogenetic analysis using parsimony (PAUP) tool [16] which is used for inferring and interpreting phylogenetic trees. The NNI algorithm also has potential to be executed in parallel. After exchanging subtrees, we would have a number of trees which can be evaluated for parsimony in parallel.

14.3.3 Maximum Likelihood

Maximum likelihood (ML) is a probabilistic method and was proposed by Felsenstein in 1981 [10] to construct phylogenetic trees. The general idea of this method is to find the tree that has the highest probability of generating the observed data. We have a data set D, a model M and search the tree with the maximum value $P(D|T, M)$. Assuming characters are pairwise independent, the following can be stated:

$$P(D|T, M) = \prod_i P(D_i|T, M) \tag{14.7}$$

In order to select the tree with the highest probability to explain data, we need to search all tree topologies and determine their branch lengths. Felsenstein provided a dynamic algorithm to find the ML score of a tree [10]. The likelihood function L under a fixed model M for given data D returns the likelihood that D was produced under this model which can be written as $L(D) = P(D|M)$. The ML method has very high computation cost and heuristics are frequently used. The quality of the tree depends on the model adopted.

14.3.3.1 Distributed ML Algorithms

The ML method for phylogeny consists of two main steps: Tree generation and evaluating the ML scores for these trees. We need to generate all possible trees and since the number of trees grows exponentially, heuristics are widely used. From parallel processing point of view, we can simply have one of the processes designated as the root (p_0) form all possible trees and distribute them evenly to a number of processes. Each process p_i can then evaluate the ML score of the trees it is responsible and send the one with the best score to the root at the end of the round. The root p_0 can then choose the tree with the best score among the ones received and output it. However, this scheme does not eliminate the overhead associated with the generation of trees and we also need ways of parallelization in this first step.

In *simple stepwise addition* (SSA) method, a tree with three leaves is first formed and taxa are added to this tree incrementally similar to the B&B method used to solve the large parsimony problem. However, different than B&B which stops search of subtrees below an edge that results in a higher score, a number of *good* trees that have better ML scores than others are selected and addition of new taxa is continued with the good trees only. A parallel formation of this approach was proposed by Zhou et al. [41]. There is a root process p_0 which starts tree generation with tree nodes and evaluates the goodness of the trees as new edges representing new taxa are added. It then distributes evenly, only the good trees to the worker processes. These processes then add new taxa to the received trees and evaluate the ML scores for the newly added trees and send the root to only the good trees. The root, having received good trees from all workers, determines a number of good trees among them and sends these to the processes to be evaluated for the next round. The parallel algorithm proposed in [5] follows a similar strategy in which a reference tree is broadcasted to the workers by the root process which generate a number of trees, evaluate their ML scores, and return these to the root to be processed further to find the best tree.

For parallel ML analysis, a number of tools have been developed. A distributed ML algorithm is presented in [24]. TREE-PUZZLE is based on MPI [37] and uses quartet puzzling method, and RAxML [40] is another parallel tool based on ML method. MultiPhyl [25] and DPRml [24] are also based on ML and can be implemented on heterogeneous platforms. Distributed and parallel algorithms for inference of huge phylogenetic trees based on the ML method are described in [36].

14.4 Phylogenetic Networks

Phylogenetic trees are widely used to infer evolutionary relationships between organisms as we have seen and they represent speciation events and descent with modification events. However, they do not represent reticulate events such as gene transfer, hybridization, recombination or reassortment [21]. A phylogenetic network can be used to discover evolutionary relatedness of organisms when reticulate events in these organisms are significant [8,15,29]. Formally, a phylogenetic network is a directed acyclic graph (DAG) $G(V, E)$ with $V = V_L \cup V_T \cup V_N$ where V_L, V_T, and V_N are defined as follows:

- V_L: Set of leaves of G
- V_T: The *tree nodes* where each node has in-degree of 1 and out-degree of 2
- V_N: The *network nodes* where each node has in-degree of 2 and out-degree of 2

Tree nodes represent mutations and the network nodes describe reticulation events and they are sometimes called as *reticulate nodes* and any input edge to these nodes are the *reticulate edges*. The phylogenetic networks can also be rooted and unrooted as the phylogenetic trees. In a rooted phylogenetic network, a special node with an in-degree of 0 should be present. Figure 14.19 displays a rooted and an unrooted phylogenetic network.

An unrooted phylogenetic network is basically an unrooted graph with leaves labeled by the taxa [22]. The two main types of such networks are the *split networks* and the *quasi-median networks*. A *split* is a partition of taxa into two nonempty sets and a split network is a connected graph where edges are splits.

(a)   **(b)**

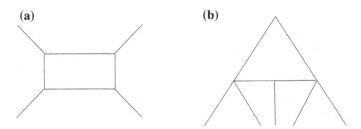

Fig. 14.19 a An unrooted phylogenetic network, **b** A rooted phylogenetic network. All of the intermediate nodes in both networks are network nodes

Phylogenetic networks conceptually can be *data display networks* or *evolutionary/explicit networks* and in some cases both. In data display networks, there is no explicit model of evolution and data is represented graphically whereas evolutionary networks attempt to model the reticulation events. Split networks are one kind of such networks. These events are mainly hybridization, horizontal gene transfer, and recombination. The evolutionary phylogenetic networks are usually rooted, directed acyclic graphs.

MP and ML analysis of phylogenetic networks can be done by extending these concepts in trees to networks. The general idea in the network application is to find the parsimony or the likelihood score of a phylogenetic network as a function of the set of trees it contains [23]. Multiple trees can be combined into a single species network and our aim in this case will be to find the phylogenetic network that has the minimum number of reticulation events.

14.4.1 Reconstruction Methods

Phylogenetic network reconstruction can be realized in four different ways as follows [28]:

1. The genes received by horizontal gene transfers (HGTs) can be identified and removed. The tree can then be constructed with the remaining genes.
2. A tree can be constructed first and a network from this tree can be obtained by iteratively adding non-tree edges to it by optimizing some cost criterion [1,17].
3. Using the distance matrix, incompabilities in the data set can be discovered and solutions can be investigated using reticulation. Split-based methods employ this approach [19,20].
4. A set of trees can be constructed first by possibly using different data sets and these can be reconciled. Problems with reconciliation may indicate a reticulation event. *Median networks* are constructed using this method [2–4].

We will now take a closer look at median networks as they are frequently used to find evolutionary relationships of species. Median networks were developed to help visualize the mtDNA sequences [2]. A node in a median network represents a sequence of a multiple sequence alignment and two nodes are connected by an edge if they differ by only one mutation [22]. The median-joining method proposed by Bandelt et al. constructs phylogenetic networks conveniently [3]. In this method, minimum spanning trees that contain only observed taxa as nodes are first constructed. In other words, these trees do not have hypothetical nodes as in an ordinary phylogenetic tree or a network. The distances between the taxa are the Hamming distances between the sequences. These MSTs are then combined into a minimum spanning network.

Basically, the median-joining method constructs a subset of full quasi-median network using the minimum spanning network. The network constructed is not as complex as a quasi-median network but it is not as simple as a minimum spanning network either. The nodes in the median-joining network are the taxa and the hypothetical nodes. Relationships are evaluated only for taxa that are close to each other in the minimum spanning network. Median-joining networks are commonly used to study the relationships between haplotypes.

14.5 Chapter Notes

Phylogeny is the study of the evolutionary relationships between organisms. The aim is to infer the evolutionary relatedness using data of living organisms or sometimes data obtained from nonliving species. The input data is either DNA/RNA nucleotide sequences or protein amino acid sequences, and sometimes both. We need to find the similarities of organisms first by comparing their sequence data which is frequently accomplished by multiple sequence alignment methods reviewed in Chap. 6.

We have analyzed two main algorithms for distance-based tree construction as the UPGMA and the NJ algorithms. In both methods, the most similar taxa are grouped together and the distances of the remaining clusters are computed to this new cluster. UPGMA assumes ultrametric tree property where two taxa are equidistant to their any common ancestor which assumes the existence of a molecular clock with constant evolution rate. The NJ algorithm does not assume the ultrametric property and uses the additive tree concept when computing the distances. The NJ algorithm is more frequently used in constructing phylogenetic trees than the UPGMA, however, neither of these algorithms depend on the site of the mutation, resulting in decreased accuracy.

Character-based tree construction methods in general have a higher time complexity than distance-based methods. Parsimony-based methods search for a tree that has the minimum number of mutations to explain the input taxa. Maximum likelihood and Bayesian trees are the probability-based methods to construct evolutionary trees. The likelihood of each tree is computed in the first method and the tree with the maximum likelihood is selected. PHYLIP [11] and MEGA [26] are tools that implement both distance-based and character-based tree construction algorithms on sequence data. MPI-PHYLIP is an MPI-based tool which supports MP, ML, and distance-based methods in a distributed environment [32]. Its limitation is the usage for protein sequences only. The branch and bound method can also be used to obtain tree topologies for taxa

All of these methods have merits and demerits. Given a set of taxa sequences, we would need to implement a multiple sequence alignment first to determine their similarities. If they are highly similar, maximum parsimony is probably the best method to implement. If they have some similarity, distance methods would be adequate. If there is very low similarity, maximum likelihood may give a more accurate result at the expense of increased computation time.

We then described briefly the phylogenetic networks which are used to model evolutionary structures of organisms with reticulate events. These networks can be unrooted or rooted. Split networks and median networks are the two types of unrooted phylogenetic networks. We also briefly described the widely used median-joining algorithm to construct median-joining networks in this section.

Phylogenetic trees have been the focus of intensive research studies and it is not possible to cover all aspects of this topic in our context. The reader is referred to [27] for a general introduction to phylogenetic trees. More recently, phylogenetic networks have been the focus of a number of research efforts, and a detailed analysis of these networks is provided by Huson et al. [22]. Although phylogenetic tree construction methods can be considered to be mature, we cannot say the same for phylogenetic networks. Phylogenetic networks cover a wide spectrum of evolutionary events, many of which are not understood completely. The problems encountered are NP-hard most of the time and there is a great need to develop standard algorithms and tools for phylogenetic network analysis.

Our general conclusion is that although phylogenetic tree construction methods have been in existence for a long while, there are only few distributed algorithms as the ones we have analyzed. In constructing phylogenetic networks such as median-joining networks, there is hardly any distributed algorithm and this area is a potential area of research for distributed processing with virtually no studies reported.

Exercises

1. Given the following distance matrix for five taxa, work out the phylogenetic tree using the UPGMA. Check the consistency of the constructed tree with the distance matrix.

	A	B	C	D	E
A	0	4	8	8	6
B		0	8	8	5
C			0	2	5
D				0	3
E					0

Fig. 14.20 The phylogenetic
tree for Exercise 4

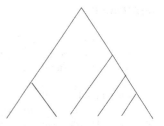

2. The following distance matrix for four taxa is given. Construct the phylogenetic
   tree using the NJ algorithm by showing the partial trees constructed at each
   iteration. Check the taxa additive tree property first. Describe also the distributed
   implementation of NJ algorithm for this set of taxa using two processes.

	A	B	C	D
A	0	4	8	8
B		0	8	8
C			0	2
D				0

3. For the NJ algorithm example which resulted in the tree of Fig. 14.14, show that
   choosing any other cluster pair in the last iteration of the algorithm will result in
   the same phylogenetic tree output.
4. The six input taxa are A, G, G, T, and A. Use Fitch's algorithm to construct a
   maximum parsimony tree for these taxa using the tree of Fig. 14.20. Workout the
   total number of mutations for each labeling of the tree.
5. Find the labeling of the vertices of the phylogenetic tree of Fig. 14.21 using
   Sankoff's algorithm. Show the vectors at each node including taxa. The weights
   for mutations are as shown in the table.
6. Discuss briefly the relation between the similarity of input taxa and the choice of
   tree construction method.

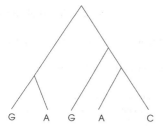

	A	C	G	T
A	0	3	1	2
C		0	2	1
G			0	3
T				0

Fig. 14.21 The phylogenetic tree for Exercise 5

References

1. Addario-Berry L, Hallett MT, Lagergren J (2003) Towards identifying lateral gene transfer events. In: Proceedings of 8th pacific symposium on biocomputing (PSB03), pp 279–290
2. Bandelt HJ, Forster P, Sykes BC, Richards MB (1995) Mitochondrial portraits of human populations using median networks. Genetics 141:743–753
3. Bandelt HJ, Forster P, Rohl A (1999) Median-joining networks for inferring intraspecific phylogenies. Mol Biol Evol 16(1):37–48
4. Bandelt HJ, Macaulay V, Richards M (2000) Median networks: speedy construction and greedy reduction, one simulation, and two case studies from human mtDNA. Mol Phyl Evol 16:8–28
5. Blouin C, Butt D, Hickey G, Rau-Chaplin A (2005) Fast parallel maximum likelihood-based protein phylogeny. In: Proceedings of 18th international conference on parallel and distributed computing systems, ISCA, pp 281–287
6. Colijn C, Gardy J (2014) Phylogenetic tree shapes resolve disease transmission patterns. Evol Med Public Health 2014:96–108
7. DasGupta B, He X, Jiang T, Li M, Tromp J, Zhang L (2000) On computing the nearest neighbor interchange distance. Proceedings of DIMACS workshop on discrete problems with medical applications 55:125–143
8. Doolittle WF (1999) Phylogenetic classification and the Universal Tree. Science 284:2124–2128
9. Dunn JC (1974) Well separated clusters and optimal fuzzy partitions. J Cybern 4:95–104
10. Felsenstein J (1981) Evolutionary trees from DNA sequences: a maximum likelihood approach. J Mol Evol 17(6):368–376
11. Felsenstein J (1991) PHYLIP: phylogenetic inference package. University of Washington, Seattle
12. Felsenstein J (2004) Inferring Phylogenies. 2nd edn. Sinauer Associates Inc., Chapter 2
13. Fitch WM (1971) Toward defining course of evolution: minimum change for a specified tree topology. Syst Zool 20:406–416
14. Gast M, Hauptmann M (2012) Efficient parallel computation of nearest neighbor interchange distances. CoRR abs/1205.3402
15. Griffiths RC, Marjoram P (1997) An ancestral recombination graph. In: Donnelly P, Tavare S (eds) Progress in population genetics and human evolution, volume 87 of IMA volumes of mathematics and its applications. Springer, Berlin (Germany), pp 257–270
16. http://paup.csit.fsu.edu/
17. Hallett MT, Lagergren J (2001) Efficient algorithms for lateral gene transfer problems. Proceedings 5th annual international conference on computational molecular biology (RECOMB01). ACM Press, New York, pp 149–156
18. Hendy MD, Penny D (1982) Branch and bound algorithms to determine minimal evolutionary trees. Math Biosci 60:133–142
19. Huber KT, Watson EE, Hendy MD (2001) An algorithm for constructing local regions in a phylogenetic network. Mol Phyl Evol 19(1):1–8
20. Huson DH (1998) SplitsTree: a program for analyzing and visualizing evolutionary data. Bioinformatics 14(1):68–73
21. Huson DH, Scornavacca C (2011) A survey of combinatorial methods for phylogenetic networks. Genome Biol Evol 3:23–35
22. Huson DH, Rupp R, Scornavacca C (2010) Phylogenetic networks. Cambridge University Press
23. Jin G, Nakhleh L, Snir S, Tuller T (2007) Inferring phylogenetic networks by the maximum parsimony criterion: a case study. Mol Biol Evol 24(1):324–337
24. Keane TM, Naughton TJ, Travers SA, McInerney JO, McCormack GP (2005) DPRml: distributed phylogeny reconstruction by maximum likelihood. Bioinformatics 21(7):969–974
25. Keane TM, Naughton TJ, McInerney JO (2007) MultiPhyl: a high-throughput phylogenomics webserver using distributed computting. Nucleic Acids Res 35(2):3337

26. Kumar S, Tamura K, Nei M (1993) MEGA: molecular evolutionary genetics analysis, ver. 1.01. The Pennsylvania State University, University Park, PA

27. Lemey P, Salemi M, Vandamme A-M (eds) (2009) The phylogenetic handbook: a practical approach to phylogenetic analysis and hypothesis testing, 2nd edn. Cambridge University Press. ISBN-10: 0521730716. ISBN-13: 978-0521730716

28. Linder CR, Moret BME, Nakhleh L, Warnow T (2004) Network (Reticulate) Evolution: biology, models, and algorithms. School of Biological Sciences. In, The ninth pacific symposium on biocomputing

29. Nakhleh L (2010) Evolutionary phylogenetic networks: models and issues. In: Heath L, Ramakrishnan, N (eds) The problem solving handbook for computational biology and bioinformatics. Springer, pp 125–158

30. Nasibov EN, Ulutagay G (2008) FN-DBSCAN: a novel density-based clustering method with fuzzy neighborhood relations. In: Proceedings of 8th international conference application of fuzzy systems and soft computing (ICAFS-2008), pp 101–110

31. Robinson DF (1971) Comparison of labeled trees with valency three. J Comb Theory Ser B 11(2):105–119

32. Ropelewski AJ, Nicholas HB, Mendez RR (2010) MPI-PHYLIP: parallelizing computationally intensive phylogenetic analysis routines for the analysis of large protein families. PLoS ONE 5(11):e13999. doi:10.1371/journal.pone.0013999

33. Ruzgar R, Erciyes K (2012) Clustering based distributed phylogenetic tree construction. Expert Syst Appl 39(1):89–98

34. Saitou N, Nei M (1987) The neighbor-joining method: a new method for reconstructing phylogenetic trees. Mol Biol Evol 4(4):406–425

35. Sankoff D (1975) Minimal mutation trees of sequences. SIAM J Appl Math 28:35–42

36. Stamatakis A (2004) Distributed and parallel algorithms and systems for inference of huge phylogenetic trees based on the maximum likelihood method. Ph.D. thesis, Technische Universitat, Munchen, Germany

37. Schmidt HA, Strimmer K, Vingron M, Haeseler A (2002) Tree-puzzle: maximum likelihood phylogenetic analysis using quartets and parallel computing. Bioinformatics 18(3):502–504

38. Studier J, Keppler K (1988) A note on the neighbor-joining algorithm of Saitou and Nei. Mol Biol Evol 5(6):729–731

39. Sung W-K (2009) Algorithms in bioinformatics: a practical introduction. CRC Press (Taylor and Francis Group), Chap 8

40. Yang Z (2007) PAML 4: phylogenetic analysis by maximum likelihood. Mol Biol Evol 24(8):1586–1591

41. Zhou BB, Till M, Zomaya A (2004) Parallel implementation of maximum likelihood methods for phylogenetic analysis. In: Proceedings of 18th international symposium parallel and distributed processing (IPDPS 2004)

Epilogue

15

15.1 Introduction

We have described, analyzed, and discussed biological sequence and network problems from an algorithmic perspective until now. Our aim in this epilogue chapter is to first observe the bioinformatics problems from a slightly far and more philosophical level and then review the specific challenges in more detail. We start by coarsely classifying the current bioinformatics challenges in our view. Management and analysis of big data is one such important task and is confronted in all areas of bioinformatics. Understanding diseases in search of cures is probably the grand challenge of all times as we discuss. Finally, bioinformatics education, its current status, needs, and prospects are stated as other current challenges as there seem to be different views on this topic.

We then provide specific challenges in technical terms which is in fact a dense overview of the book showing possible research areas to the potential researcher in the field; namely, the distributed algorithm designer and implementer in bioinformatics. We conclude by describing the possible future directions we envisage. The big data will get bigger and efficient methods for handling of big data will be needed more than before; and understanding mechanisms and searching cures for diseases will always be a central theme for bioinformatics researchers in collaboration with other disciplines. Personal medication is the tailoring of care to meet individual needs and will probably receive more attention than now in the near future as it is needed for better treatment and also for economical reasons. Population genetics is already a popular topic of interest for individuals and understanding population genetical traits will help design better therapies in future. Finally, understanding of the mechanics and operation of the cell from a system point of view as a whole rather than its individual parts such as DNA and proteins will probably help to solve many bioinformatics problems including disease analysis.

© Springer International Publishing Switzerland 2015   351
K. Erciyes, *Distributed and Sequential Algorithms for Bioinformatics*,
Computational Biology 23, DOI 10.1007/978-3-319-24966-7_15

15.2   Current Challenges

We describe and classify coarsely the current challenges we envisage in bioinformatics from an algorithmic perspective, excluding biotechnological tasks that need to be overcome. We can see three main areas of interest; analyzing big data, disease analysis, and bioinformatics education as described below.

15.2.1   Big Data Analysis

The *big data* is a collection of large and complex data which cannot be efficiently processed using conventional database techniques. High-throughput techniques such as next-generation sequencing provided big data in bioinformatics in the last decade. This data comes in basically as sequence, network, and image data. The basic requirement in bioinformatics is to produce knowledge from the raw data. Unfortunately, our ability to process this data does not grow in proportion with the production of data. We can classify the management of big data as follows [17]:

- *Access and management*: The basic requirements for access and management of big data are its reliable storage, a file system, and efficient and reliable network access. Client/server systems are typically employed to access data which is distributed over a network. The file systems can be distributed, clustered, or parallel [21].
- *The Middleware*: The middleware resides between the application and the operating system which manages the hardware, the file system, and the processes. It is impossible to generate middleware that is suitable for all applications but rather, some common functionality required by the applications can be determined and the middleware can then be designed to meet these common requirements. Message Passing Interface (MPI) can be considered as one such middleware for applications that require parallel/distributed processing.
- *Data mining*: Data mining is the process of analyzing large datasets with the aim of discovering relationships among data elements and present the user with a method to analyze data further conveniently. In practical terms, a data mining method finds patterns or models of data such as clusters and tree structures. Clusters provide valuable information about raw data as we have analyzed in Chaps. 7 and 11. A data mining method should specify the evaluation method and the algorithmic process, for example, the quality of clustering in a clustering method.
- *Parallel and distributed computing*: Given the huge size of data, the parallel/distributed computing is increasingly more required. At hardware level, one can employ clusters of tightly coupled processors, graphical processing units (GPUs), or distributed memory processor connected by an interconnection network. Our approach is using the latter as it is most versatile and can be implemented by many users more conveniently. We need distributed algorithms to exploit parallel operations on this huge data and this area has not received significant attention from the researchers of bioinformatics as we have tried to emphasize throughout this book.

A *cloud* is a computation infrastructure consisting of hardware and software resources and providing computational services to the users over the Internet. Users are charged only for the services they use and the maintenance and security of the system is granted by the provider of the cloud. Clouds can be private in which the facilities are offered to an organization only, or public where the infrastructure is made available to general public for a fee. This *pay as you use* approach is economical as the user does not have to invest on computation. Some of the public cloud computing systems in operation are Dropbox [10], Apple iCloud [9], Google Drive [12], Microsoft Azure [7], and Amazon Cloud Drive [8].

Cloud computing is merging as a commonly used platform to handle biological data. The traditional way of managing biological data is to download it to the local computer system and use algorithms/tools in the local system to analyze data which is not efficient with big data sizes. The general idea of using cloud computing for bioinformatics applications is to store big data in the cloud and move the computation to the cloud and perform processing there. Additionally, it is of interest for researchers of bioinformatics to access data of other researchers in a suitable format and easily which can be provided by a cloud. Cloud-based services for bioinformatics applications can be classified as Data as a Service (DaaS), Software as a Service (SaaS), Platform as a Service (PaaS), and Infrastructure as a Service (IaaS).

There are, however, some issues in employing cloud computing for bioinformatics applications. First, data being large necessitates the provision of efficient uploading, in some cases using hard drives should be possible. Security has to be handled by the provider and an important reason for using cloud is the scalability which requires parallel/distributed processing facilities to be provided by the cloud. The latter is possibly one of the current challenges in using cloud for bioinformatics problems as there are very few such environments such as MapReduce [3] which provides a parallel computing environment in a cloud. Hadoop [1] uses MapReduce for parallel processing of big data. An early effort of searching for SNPs using cloud computing was presented in [15] using the Crossbow system.

15.2.2 Disease Analysis

Another grand challenge in algorithmic bioinformatics is the efficient analysis of the disease states of organisms to be able to design cures and drug therapies. In order to accomplish this formidable task, especially for complex diseases like cancer, we need to understand the causes, mechanisms, and progression of diseases at molecular level. We have the basic biochemical reactions as DNA $\rightarrow$ RNA $\rightarrow$ mRNA $\rightarrow$ protein peptides called gene expression which is the central dogma in molecular biology. Considering the disease onset follows a similar pattern but with disease genes to disease proteins, we need to look at this process closely using algorithmic techniques.

The classical view of disease genes causing diseases has changed significantly to considering that a group of genes getting involved in producing a disease subnetwork. Network-based approaches are therefore increasingly being used to study complex diseases. A biological network can be represented by a graph and the rich theory and algorithmic techniques for graphs can be used to analyze these networks. Two important networks in the cell that are affected by the disease states of an organism are the protein–protein interaction networks and the gene regulation networks.

Proteins are the fundamental molecules in the cell carrying out vital functions needed for life. They interact with each other forming protein–protein interaction (PPI) networks as we have reviewed in Part II and they also interact with DNA and RNA to perform the necessary cell processes. In a graph representing a PPI network, nodes represent the proteins and the undirected edges between nodes show the interaction between the nodes. Protein interactions play a key role to sustain the healthy states of an organism from which we can deduce their dysfunction may be one of the sources of disease states.

A mutation in a gene may cause unwanted new interactions in PPI networks such as with pathogens or protein misfolds to result in diseases. For example, protein misfolding due to mutations may result in lost interactions in a PPI network causing diseases [6]. Unwanted newly formed protein interactions due to mutations are considered as the main causes of certain diseases such as Huntington's disease, Alzheimer's disease, and cystic fibrosis. Moreover, some bacterial and viral infections such as *Human papillomavirus* may interact with the proteins of the host organism. In order to understand the mechanisms of disease in the cell, PPI networks can be used to discover pathways which are sequential biochemical reactions. Finding a subnetwork of a PPI network to find the corresponding pathway can aid to understand disease progression [6]. Also, by discovering disease proteins, we can predict the disease genes.

A functional module in the cell has various interacting components and can be viewed as an entity with a specific function. Gene expression is controlled by proteins via regulatory interactions, for example, transcription factor proteins bind to sites near genes in DNA to regulate them. Such formed networks which have genes, proteins, other molecules, and their interactions are called gene regulation networks (GRNs) which can be represented by directed graphs. An edge in a GRN has an orientation, for example, from the transcription factor to the gene regulated. A GRN is basically a functional network as opposed to the PPI network which is physical. An important challenge in a functional network is the identification of the subnetwork associated with the disease. An effective approach to discover functional modules in the cell is the clustering process we have seen.

Investigation of the physical and functional networks is needed to identify the disease-associated subnetworks. For example, discovery of a disease-affected pathway can be performed by first identifying the mutated genes, then finding the PPI subnetwork associated with the mutated genes and finally searching for modules associated with the disease PPI subnetwork to discover the dysfunctional pathways [2]. In conclusion, the study of PPI networks, functional networks such as GRNs in the cell using graph-theoretical analysis is needed to understand the disease states of organisms better.

15.2.3 Bioinformatics Education

Bioinformatics started as a multidisciplinary subject, it has roots in molecular biology, biochemistry, computer science, statistics, genetics, and mathematics. This question has been asked many times: what is the difference between *computational biology* and *bioinformatics*? Although there does not seem to be a consensus on the comparison of these two terms, computational biology focuses more on the application of *existing* tools to investigate biological problems whereas bioinformatics is more concerned with the *design and implementation* of computational tools for biological analysis, as more commonly perceived. For example, design of algorithms to find clusters in biological networks is a task in bioinformatics but use of these algorithms to detect dense regions in PPI and functional networks, unify the results found and use this information for disease analysis would probably be a task in computational biology. In other words, bioinformatician would like to design and implement computational methods based on some abstract model such as DNA as a string of symbols; a PPI network as an undirected graph etc., whereas computational biology researcher would investigate the problem at a more system level making use of any available computational algorithm and tool. Our view is that the bioinformatics is a subset of computational biology discipline as it covers all aspects of bioinformatics and needs the methods and techniques of bioinformatics.

Teaching of bioinformatics is a challenge from various aspects [20]. Determining the appropriate undergraduate or postgraduate curriculum is one such difficulty. Based on the above description of bioinformatics, we can state that an undergraduate curriculum in bioinformatics should borrow considerable topics from computer science curriculum. A typical current curriculum in computer science may consist of the following areas: hardware track; systems and networks track; theory track; and software engineering track. The hardware track has the basic electronics, logic design, microprocessors, embedded systems, and advanced digital design as courses in sequence. Basic hardware knowledge is essential for bioinformatics and probably the first two courses will be adequate. System track typically has operating systems, distributed systems, and computer networks as core courses and we think all of them are necessary for the bioinformatician. The theory track includes courses such as theory of computation, algorithms, parallel and distributed algorithms, and complexity. Algorithms and complexity in this track are fundamental for the bioinformatics curriculum. Software engineering track includes programming languages, database and software engineering related courses, and knowledge of programming languages like C, Perl, and Java. Knowledge and hands-on experience of using databases together with programming languages is fundamental for the bioinformatics undergraduate curriculum. Other topics outside the computer science area includes biochemistry, genetics, statistics, biotechnology, and molecular biology and the curriculum should include basic courses in these areas.

Given the low number of undergraduate degrees offered in most of the countries worldwide, the students enrolled in a master of science and sometimes Ph.D. program come from diverse backgrounds. The lecturer in such a program is confronted with the challenge of teaching the course topics to students who have varying levels of exposure to the preliminary topics described above. A possible remedy for this situation is to have non-computer science and non-bioinformatics students enroll in a preparatory year to acquire basic computation and molecular biology knowledge and skills. Dissertation topics vary but basically we can see a good proportion is about the design and implementation of algorithms for bioinformatics problems. There is an increasing interest in analysis of biological networks, occasionally this research is termed as *complex networks* with biological networks given as a case of such networks. Complex networks have large sizes and they commonly have the small-world property and power-law distribution. Biological networks are a class of complex networks and the findings in complex networks research is immediately applicable to networks in the cell in many cases.

The appropriate formation of the curriculums at undergraduate and postgraduate levels, especially from the algorithmic/tool point of view needs to be considered carefully. We will not attempt to form a curriculum here as it has already been done by the International Society of Computational Biology (ISCB). The education committee of ISCB formed a Curriculum Task Force in 2011 which provided two reports on the possible curriculum structure [22,23].

15.3 Specific Challenges

We have reviewed the basic concepts in bioinformatics until now from two perspectives: the sequence and network domains. We will review the fundamental problems here to emphasize the main points, and also to guide the beginning researcher in the field, namely distributed algorithms for bioinformatics, on the potential research topics.

15.3.1 Sequence Analysis

Sequence Alignment: Sequence alignment is the most commonly used method to compare two or more sequences. It is frequently employed as the first step for other bioinformatics problems such as clustering and phylogenetic tree construction. Dynamic algorithms provide solutions for pairwise global or local alignment in polynomial time but heuristic algorithms are preferred for the time consuming task of multiple alignment of more than two sequences. We reviewed two commonly used tools; BLAST for local alignment between two sequences and CLUSTALW for multiple alignment. The parallel and distributed versions of these two tools which provide significant speedups are available. This topic is mature enough in the sequential case and also in distributed.

Sequence Clustering: Clustering of biological sequences involves grouping the closely related ones to find their relationships. We can then analyze them efficiently and check similar functionality among the clustered sequences. BLAST and CLUSTALW can be used to find the distances between the input sequences and once these are determined, we can employ any of the well-known data clustering algorithms such as hierarchical clustering or k-means algorithm. We saw that there are only few distributed algorithms for clustering of sequences including the parallel k-means algorithm and we proposed two algorithms for this purpose. Distributed sequence clustering is a potential research area as there are only few reported studies.

Sequence Repeats: DNA/RNA and proteins contain repeats of subsequences. These sequences maybe tandem or distributed, and tandem repeats are frequently found near genes. These repeating patterns may also indicate certain diseases and therefore finding them is an important area of research in bioinformatics. The methods for this task can be broadly classified as probabilistic and combinatorial. Probabilistic approaches for distributed repeat (sequence motifs) include MEME and Gibbs sampling method. Graph-based algorithms are commonly used for combinatorial sequence motif discovery. Parallel/distributed algorithms are very scarce for the discovery/prediction of the sequence repeats in DNA/RNA or proteins and this is again a potential area of research. We proposed two distributed algorithms for this purpose.

Gene Finding: Gene finding or gene prediction is the process of locating genes in prokaryotes and eukaryotes. Finding genes in eukaryotes is more difficult due to the existence of intron regions between the exons in the genes. Hidden Markov models, artifical neural networks, and genetic algorithms are often used as statistical approaches for gene finding. We have not been able to find any parallel/distributed gene finding method in the literature. At first look, partitioning the genome into a number of processors for gene search can be implemented with ease to speedup processing.

Genome Rearrangement: The order of subsequences of genomes of organisms change during evolution. This modification in the genome structure may start new organisms and may be the cause of some complex diseases. Therefore, it is of interest to analyze these rearrangements. A common form of these mutations at large scale is the reversals. We reviewed fundamental algorithms for reversals. Again, we have not been able to find any parallel/distributed implementation of genome rearrangement algorithms reported in the literature necessitating further research in this area.

Haplotype Inference: Haplotype inference or genotype phasing refers to the process of determining data about a single chromosome from the genotype data of two chromosomes. This is needed to analyze diseases and also to find the evolutionary distances between organisms. Three main methods for this task are Clark's algorithm, expectation maximization, and perfect phylogeny haplotyping. We have not been able to find a distributed haplotype inference algorithm in the literature and we proposed two distributed algorithms.

15.3.2 Network Analysis

Network Clustering: A biological network can be modeled by a graph with nodes representing the entities such as proteins or metabolites, and edges show the relations between them. One such network of interest is the protein–protein interaction (PPI) network. Clustering, that is, finding dense regions of such networks provides insight to their operation as these regions may indicate high activity and sometimes disease areas. We looked at various general purpose graph clustering algorithms and also analyzed algorithms targeting the biological networks which have very large sizes. We proposed two distributed algorithms: a modularity-based algorithm and another one using Markov chain for clustering of biological networks.

Network Motifs: A network motif in a biological network is a statistically over-represented subgraph of such a network. These subgraph patterns often have some associated functionality. Finding network motifs helps to discover their functionality and also the conserved regions in a group of organisms. Since the network size is very large, sampling algorithms which search for a subgraph pattern in a small representative part of the target graph are frequently used. A motif m is searched in a set R of randomly generated graphs with similar properties to the target graph G, and the occurrence of a motif in G should be statistically much higher than in the elements of R. Several sequential algorithms for this problem are proposed in various studies as we have outlined but the distributed algorithms are only few. Therefore, this is another potential area of research in distributed bioinformatics.

Network Alignment: Alignment of networks has a similar purpose to the alignment of sequences, we aim to discover the similarities between graphs or subgraphs of two or more given biological networks. We can then infer phylogenetic relationships among them and also, conserved regions can be identified to search their functionality. In this case, we are trying to discover similar subnetworks rather than over-represented regions in contrast to network motif discovery. It can be performed as locally or globally; and pairwise or multiple as in the sequence alignment problem. Network alignment is closely related to bipartite graph matching problem which can be solved in polynomial time. Auction algorithms are used for bipartite matching and distributed versions of these algorithms are reported as we have reviewed. These are the only few algorithms for this purpose and we also proposed a distributed approximation algorithm for network alignment.

Phylogeny: Phlogeny is the study of evolutionary relationships among organisms. Using phylogeny, we can find the disease spreading patterns, discover the origins of diseases to help design therapies. Phylogenetic trees depict visually the evolutionary relationships between organisms. Construction of these trees from a given input set of sequences is a fundamental problem and sequence alignment is frequently employed to find the similarities of sequences as the first step of building a phylogenetic tree. We reviewed fundamental sequential methods of phylogenetic tree construction and outlined few methods of distributed construction. Phylogenetic networks are relatively more recent structures to explain the evolutionary process in a more general sense by considering events such as gene transfer and recombination. We have not been able to find any reported method for distributed phylogenetic network construction and therefore conclude this is a potential research area.

Figure 15.1 depicts the hierarchical relationship between these areas of study in bioinformatics. For example, sequence comparison mainly by alignment is the front processing of sequential data analysis along with genome rearrangements; and repeat finding can be used for clustering. Phylogeny methods may or may not use

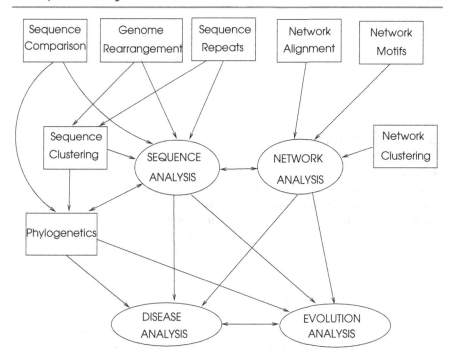

Fig. 15.1 The relationships between the algorithmic challenges in bioinformatics

clustering and disease analysis may employ phylogenetic relationships. We envisage the two interrelated big challenges are the analysis of diseases and the analysis of evolutionary relationships between organisms. In the network world, extraction of network connectivity information is the first step after which network motif search or alignment can be performed. Clustering of networks together with sequence analysis can be used for general disease or study of evolution as shown.

In summary, most of the problems encountered in the sequence and network worlds are suitable for parallel processing over a number of distributed computers connected by a communication network. In sequence-related problems, we can simply partition the DNA/RNA or protein amino acid sequences and distribute the segments to processors for parallel processing. In the case of multiple sequence analysis, a number of sequences can be distributed to processes. In the network domain, distributed processing is not so straightforward in many cases and more sophisticated algorithms are required in general.

As a final note, the potential researcher in this area, namely the distributed algorithms for bioinformatics, should have a solid background in algorithms, complexity, and parallel/distributed processing. She should then be familiar with the problems outlined in the book. The sequence algorithms have a relatively longer period of existence than the network algorithms and therefore, the network domain is a more promising area of research although many topics in the sequence domain still remain to be investigated. The long-term goal of research may address the problems described under the future directions in this chapter with keeping an eye on distributed solutions as these are needed for the problems encountered that always need high computational power.

15.4 Future Directions

Looking at the current challenges, we can anticipate the future challenges and research in bioinformatics will continue to be in these topics but with possible added orientations to new subareas. Our expectation is that the bioinformatics education will be more stabilized than the current status but the main research activities in algorithmic studies in bioinformatics will probably be on the management of big data which will grow bigger, and the analysis of diseases and evolution.

15.4.1 Big Data Gets Bigger

Big data will undoubtedly get bigger which means high computational power, parallel/distributed algorithms, tools and integrated software support will be needed more than before [16]. We envisage that there will be significant research oriented towards parallel/distributed algorithms to solve the fundamental bioinformatics problems along with analysis of disease. Data mining techniques will be imperative in the analysis and extracting meaningful knowledge out of big data, with clustering method which is one of the most investigated areas of research in various disciplines such as Computer Science, Statistics, and Bioinformatics. Therefore, we may expect to see advanced data mining methods such as intelligent clustering tailored for bioinformatics data. Cloud computing will probably be employed more for the storage, analysis, and modeling of big data with more advanced user interfaces and more integrated development tools.

15.4.2 New Paradigms on Disease Analysis

Proteins, genes, and their interactions are so far considered as the major actors of disease states of an organism. We envisage that the study of biological networks such as PPI networks and GRNs using algorithmic techniques to understand disease mechanisms will continue in the foreseeable future. However, it has been observed that small molecules such as amino acids, sugars, and lipids may also have important effects on the disease states of an organism than expected before [24]. This relatively new area is called *chemical bioinformatics* which deals with the chemical and biological processes in the cell. There are a number of databases available related to small molecules in the cell; the metabolic pathway databases include Kyoto Encyclopedia of Genes and Genomes (KEGG) [14], the Reactome database [13], and the Small Molecule Pathway Databases (SMPDB) [5]. It may be foreseen that the analysis of the role of these small molecules in disease will be needed in future along with the biological network analysis.

15.4.3 Personalized Medicine

Personalized medicine can be described as modifying and tailoring of medical treatment to suit the individual characteristics, needs, and preferences of the patient during all stages of care such as prevention, diagnosis, and treatment [11]. It uses the genome information to help medical decision-making [18]. Personalized medication may be considered as the process of finding the relation between the genotype of an individual and designing, tailoring drugs for the individual needs [4, 19]. The latter phase is called *pharmacogenomics*. We need genetic information of individuals to know how these effect the onset, diagnosis, and treatment of disease. The whole genome sequencing is still expensive and differences in the genetic structures of individuals currently can be evaluated using SNP analysis. There is a limited use of personal medication such as treatment of certain cancers using specific chemotherapy drugs. Personal medication is also a hope for patients where traditional drug response is low.

There are several challenges to be addressed for common and practical implementation of personal medication. The big data challenge, as in all other areas of bioinformatics, needs to be handled. The relationship between the phenotype and the genotype should also be established. The big challenge is then to apply the results obtained for better care of the patients in the clinic. The disease analysis using SNPs, proteins, and networks within the cell should be performed to find the connectivity between the genotype and the phenotype.

Population genetics is the study of genetic variations within populations. Populations may have genetic traits and this information together with the genotype of individual can be used to design personal drugs. We think personal medication will be increasingly used more in future due to the inadequateness and cost of using *one-drug-for-all* concept of most of the current clinical health care practice.

Disease analysis shows that neither mutated genes nor dysfunctional PPI networks or functional networks or small molecules described are alone responsible for the disease state of the cell. The disease is a result of all of these making its analysis more difficult. We see complex processes as a result of various subprocesses at almost every level of cell functioning. We think the cell will be analyzed more and more at system level in future to be able to understand all aspects of complicated processes such as disease.

References

1. Apache Hadoop: http://hadoop.apache.org/
2. Cho D-Y, Kim Y-A, Przytycka TM (2012) PLOS computational biology, translational bioinformatics collection volume 8, issue 12, chapter 5: network biology approach to complex diseases
3. Dean J, Ghemawat S (2004) MapReduce: simplified data processing on large clusters. Proceedings of the 6th symposium on operating systems design and implementation: 6–8 Dec 2004; San Francisco, California, USA, vol 6. ACM, New York, USA, pp 137–150
4. Fernald GH, Capriotti E, Daneshjou R, Karczewski KJ, Altman RB (2011) Bioinformatics challenges for personalized medicine. Bioinformatics 27(13):1741–1748
5. Frolkis A, Knox C, Lim E, Jewison T, Law V et al (2010) SMPDB: the small molecule pathway database. Nucleic Acids Res 38:D480–487
6. Gonzalez MW, Kann MG (2012) PLOS computational biology, translational bioinformatics collection, volume 8, issue 12, chapter 4: protein interactions and disease
7. https://azure.microsoft.com
8. https://www.amazon.com/clouddrive
9. https://www.icloud.com/
10. https://www.dropbox.com/
11. http://www.fda.gov/scienceresearch/specialtopics/personalizedmedicine/
12. https://drive.google.com/drive
13. Joshi-Tope G, Gillespie M, Vastrik I, DEustachio P, Schmidt E, et al (2005) Reactome: a knowledgebase of biological pathways. Nucleic Acids Res 33:D428–432
14. Kanehisa M, Goto S, Hattori M, Aoki-Kinoshita KF, Itoh M et al (2006) From genomics to chemical genomics: new developments in KEGG. Nucleic Acids Res 34:D354–357
15. Langmead B, Schatz MC, Lin J, Pop M, Salzberg SL (2009) Searching for SNPs with cloud computing. Genome Biol 10(11):R134
16. Marx V (2013) Biology: the big challenges of big data. Nature 498:255–260
17. Merelli I, Perez-Sanchez H, Gesing S, DAgostino D (2014) Managing, analysing, and integrating big data in medical bioinformatics: open problems and future perspectives. BioMed Res Int 2014, Article ID 134023
18. Olson S, Beachy SH, Giammaria CF, Berger AC (2012) Integrating large-scale genomic information into clinical practice. The National Academies Press, Washington
19. Overby CL, Tarczy-Hornoch P (2013) Personalized medicine: challenges and opportunities for translational bioinformatics. Per Med 10(5):453–462
20. Ranganathan S (2005) Bioinformatics education-perspectives and challenges. PLoS Comput Biol 1:e52
21. Thanh TD, Mohan S, Choi E, SangBum K, Kim P (2008) A taxonomy and survey on distributed file systems. In: Proceedings of the 4th international conference on networked computing and advanced information management (NCM'08), vol 1, pp 144–149
22. Welch LR, Schwartz R, Lewitter F (2012) A report of the curriculum task force of the ISCB Education Committee. PLoS Comput Biol 8(6):e1002570
23. Welch L, Lewitter F, Schwartz R et al (2014) Bioinformatics curriculum guidelines: toward a definition of core competencies. PLoS Comput Biol 10(3):e1003496
24. Wishart DS (2012) PLOS computational biology translational bioinformatics collection, volume 8, issue 12, chapter 3: small molecules and disease

Index

© Springer International Publishing Switzerland 2015
K. Erciyes, *Distributed and Sequential Algorithms for Bioinformatics*,
Computational Biology 23, DOI 10.1007/978-3-319-24966-7

Printed in the United States
By Bookmasters